Statistical Methods for Six Sigma

Statistical Methods for Six Sigma

In R&D and Manufacturing

Anand M. Joglekar

WILEY-INTERSCIENCE

A JOHN WILEY & SONS, INC., PUBLICATION

Published by John Wiley & Sons, Inc., Hoboken, New Jersey.
Published simultaneously in Canada.

Library of Congress Cataloging-in-Publication Data:

Joglekar, Anand M.
Statistical methods for six sigma / Anand M. Joglekar.
p. cm.
ISBN 0-471-20342-4 (Cloth)
1. Quality control—Statistical methods. 2. Process control—Statistical methods. I. Title.
TS156.J64 2003
658.5′62—dc21 2003003583

Printed in the United States of America.

10 9 8 7 6 5 4 3

To the memory of my parents
and to Jaya, Nikhil, and Neha

Contents

6

Other Useful Charts 153

7

Variance Components Analysis 177

8

Quality Planning with Variance Components 201

Preface

Over the past several years, in my attempts to implement statistical methods in industry, I have had the pleasure of working with a variety of people. Some were curious and asked penetrating questions about all the assumptions behind each statistical method. Others did not need the details and were satisfied with the recipe to solve specific design or analysis problems. Still others, at first blush, seemed not to care, but once convinced, became some of the best users of statistical methods. This book owes a lot to all these interactions.

When I first came to America, from IIT, in Mumbai, India, I was fortunate to be an engineering graduate student at the University of Wisconsin, Madison. There was a strong collaboration between the engineering and the statistics departments at Madison, which allowed us to learn how to apply statistical methods to engineering and scientific problems. Later, as an internal and external consultant, I collaborated with a large number of engineers, scientists, managers, and other technical people from a wide variety of industries. I started teaching statistical methods to industry participants. For a number of years, I have used earlier versions of this book to teach statistical methods to thousands of industry participants. The practical problems the participants raised and the questions they asked have helped shape the selection of topics, examples, and the depth and focus of the explanations in this book. I wish to thank the many collaborators and seminar participants.

This book on the applications of statistical methods is specifically written for technical professionals in industry. This community of bright and busy individuals is typically short on time but generally willing to learn useful methods in order to quickly and effectively solve practical problems. Many have taken courses is statistics before. Unfortunately, what they learned was often unused and long forgotten, primarily because the subject was taught as an exercise in mathematics, and a connection between statistics and real world problems was never established. Consequently, this book does not presume any prior knowledge of statistics. It includes many practical applications. It covers a large number of useful statistical methods compactly, in a language and at a depth necessary to make successful applications using statistical software packages when necessary. This book can also be used as a text for an engineering-statistics course or a quality-engineering course for seniors and first year graduate students. It should help provide university students with a much-needed connection between statistical methods and industrial problem solving.

This book contains nine chapters. After some introductory material in Chapter 1, Chapter 2 explains basic statistical concepts useful in everyday work and necessary to understand the rest of the book. It describes how to make decisions by meaningfully summarizing information contained in the data. Chapter 3 presents two statistical methods: comparative experiments to compare alternatives and regression analysis to build equations relating inputs and outputs. Chapters 4, 5, and 6 deal with the subject of control charts and process capability aimed at determining whether the process is stable and capable of meeting specifications and, if not, how to improve it. Chapters 7 and 8 explain the underutilized subject of variance components, namely, identifying the contribution of each principal cause of variability to the total variance so that the variance reduction efforts can be properly targeted. Chapter 9 presents measurement systems analysis to evaluate whether the measurement system is satisfactory and, if not, how to improve it. An important topic of design of experiments is not considered in this book and the reader is referred to several available books on the subject. Chapter 10 includes a test and answers to the test. People familiar with the subject may take the test prior to reading the book and then decide where to focus their attention, whereas others may read the book and then take the test to prove that they have understood the ideas.

I wish to particularly thank three individuals: Mr. Alfred T. May, Ms. Paula Rosenfeld, and Ms. Karen King. Al and Paula are colleagues from the days when we worked together at The Pillsbury Company. Al and I first collaborated to write a book on design of experiments for use within the company and have spent many hours discussing applications of statistics. Paula is now a statistics consultant at General Mills and made many constructive comments on an earlier draft of the book.

Karen and I have worked together for a number of years and Karen diligently typed the manuscript.

I hope you, the reader, will find this book useful. If you have suggestions or comments, you may reach me at my website, *www.joglekarassociates.com.*

ANAND M. JOGLEKAR

Plymouth, Minnesota

1 Introduction

Much work in manufacturing and in research and development (R&D) is directed toward improving the performance of products, processes, and measurement systems. The words six-sigma embody this drive toward business improvements. Six-sigma is a structured approach that combines understanding of customer needs, a disciplined use of statistical methods of data collection and analysis, and careful attention to process management in order to improve business performance. The performance requirements are often stated in terms of preliminary or final targets and specifications with consistent, on-target performance being a measure of performance excellence. Achieving and sustaining product, process, and measurement system performance excellence, and doing so profitably, requires a clear definition of the specific performance measures to be improved followed by answers to six questions that form the steps of the improvement process.

Depending upon the stage of the project, the performance measures to be improved may be performance characteristics, product characteristics, or process characteristics. Performance characteristics are those measurable characteristics that define the expectations of all key customers. They may be grouped into quality, cycle time, and cost characteristics. Examples of performance characteristics include the degree of liking for a food product, the clarity of picture on a TV screen, the time taken by a car to accelerate from zero to 60 mph, and the length of time to place a purchase order. Product characteristics are features of the

Statistical Methods for Six Sigma by Anand M. Joglekar
ISBN 0-471-20342-4 © 2003 John Wiley & Sons, Inc.

product that control performance characteristics. Weight, composition and particle size are examples of product characteristics for a drug tablet. The product characteristics are in turn controlled by manufacturing process and purchased material characteristics such as compressive force, mix speed, mix time, and so on. For early-stage projects in R&D the selected performance measures may be performance characteristics. For improvement projects in manufacturing, the selected performance measures are likely to be product or process characteristics.

To improve the selected performance measures, the following six questions need to be answered. Answers to the first three questions define the strategic improvement plan. The remaining questions deal with the execution of the plan, verification that the desired improvements have occurred, and control to ensure that performance deterioration does not occur.

1. Where are we now? This is the assessment of the current performance of the product, process, or measurement system regarding the selected performance measures. Methods to quantify each performance measure need to be established. For example, if the degree of liking for a food product is the selected performance measure, then it may be evaluated on a specially developed hedonic scale. Each participant in a consumer test could be asked to rate the food product in terms of degree of liking on this hedonic scale. The resultant data can be analyzed to provide an assessment of the current performance. The output of this step is a list of key performance measures and a data-based assessment of current performance regarding each of these measures.

2. Where do we want to be? This is the setting of targets and specifications for each selected performance measure. It requires an understanding of customer expectations, competitive performance, and benchmark performance. A comparison of current performance with targets and specifications identifies gaps that must be filled to achieve business success.

   For example, a direct comparison of the food product with competitive products helps set a target for minimum hedonic score. The relationship between product characteristics and hedonic score helps set the targets and specifications for product characteristics. Benchmarking the purchase order system could lead to maximum cycle-time targets. For a pharmaceutical product, a clinical trial can provide information on both the target amount of drug required by the patient along with a specification around the target that must be met for desired performance. For products currently in production, targets and specifications for performance measures may already exist. The output of this step is the definition of targets and specifications for each performance measure, the identification of gaps between the current performance and the desired performance, and the establishment of improvement priorities. Usually, the gaps have to do with meeting targets or meeting variability around the target.

3. How do we get there? This is the development of a strategic improvement plan. For early-stage projects in R&D, the plan may consist of ways to identify key product characteristics, development of methods to measure these characteristics and the selection of an approach to establish targets and specifications for key product characteristics. For projects in manufacturing, the plan may consist of a list of improvement strategies based upon data. These strategies will often deal with whether the specific product characteristic is properly targeted, whether the process is stable, whether the process exhibits excessive short-term variation and, in general, the identification of likely sources of variability. The output of this step is a strategic improvement plan that defines how the performance gaps will be bridged.

4. Are we making progress? This is the development and execution of action plans to implement the strategic improvement plan. In early-stage projects, this involves the generation of design alternatives and the approach to product optimization. Projects in manufacturing often begin with the identification of key process characteristics that control the product characteristics being improved, along with methods to measure the process characteristics. The next step is to develop action plans to achieve the desired improvements. If the strategy is to make the process stable, the action plan may involve implementing an appropriate control chart, identifying causes of process shifts, and taking corrective actions to prevent recurrence of such shifts. If the strategy is to reduce variability, the action plan will require the determination of the causes of variability and the manner in which the impact of these causes is to be reduced or eliminated. For example, to reduce variability, is the best course of action to tighten tolerances, procure more homogeneous raw material, control the manufacturing process because it is unstable, change the product formulation to make it more robust, or improve the measurement system to reduce measurement error? Development of action plans requires idea generation based upon knowledge of the product, process, and measurement system, coupled

with statistical data collection and analysis. The output of this step is a specific cost-effective action plan to make improvements happen and a timely execution of that action plan.

5. Are we there yet? This consists of validating the improvements and is similar to the first question. It provides a data-based assessment of the improved product, process, or measurement system to determine if the improvement objectives have been met.
6. Are we staying there? This consists of ensuring that the benefits of improvement continue to be realized and continuous improvements occur by controlling the process and by conducting troubleshooting.

1.1 ROLE OF STATISTICAL METHODS

What is the role of statistical methods in achieving and sustaining performance excellence? How do statistical methods help in implementing the six-sigma approach? The above six steps of the improvement process depend greatly on the scientific, engineering, and business understanding of the product, process, and measurement system. Each step of the improvement process requires generation of new knowledge through iterative cycles of data collection, data analysis, and decision making. This task is made difficult because all processes have variability, and variability causes uncertainty regarding the information necessary to answer the six questions unambiguously. For example, due to variability, it is not possible to estimate the process mean precisely. We do not know if the mean is perfectly centered on the target or not. If we decide to adjust the mean to bring it on target, the decision becomes risky because of the uncertainty associated with the estimate of the process mean. Statistics provides the scientific discipline to make reasoned decisions under uncertainty. In improving products, processes, and measurement systems, statistical methods play an important role by providing a way of thinking about problems, by identifying conditions under which data should be collected, by estimating how much data should be collected and how the data should be analyzed to make decisions with acceptable risks of erroneous conclusions.

Table 1.1 shows 12 statistical methods considered in this book. Some of these methods are simple, others are complex. Each plays an important role in answering the six questions posed earlier. In Table 1.1, the six questions are restated as steps of the improvement process. Based upon the specific problem at hand, appropriate statistical methods need to be used at each step. The dots in Table 1.1 indicate the statistical

Table 1.1 Role of Statistical Methods in Performance Improvement

Statistical Methods	Assess Performance	Set Goals	Define Strategy	Improve	Validate	Control and Troubleshoot
1. Descriptive statistics	•	•				
2. Confidence intervals				•		
3. Sample size	•			•	•	
4. Tolerance intervals					•	
5. Comparative experiments				•		
6. Corrrelation and regression	•			•		
7. Control charts	•		•		•	•
8. Capability analysis	•	•			•	
9. Economic loss function		•				
10. Variance components analysis			•			
11. Variance transmission analysis				•		
12. Measurement systems analysis			•			

Note: The dots indicate the specific statistical methods used to improve the pizza manufacturing process.

methods that could be used to improve a pizza manufacturing process as discussed below.

Consider the case of frozen pepperoni pizza currently in production. Through consumer testing, it has been established that each pizza should have a minimum of 30 g of pepperoni to achieve desired consumer preference. Such specifications can be set by conducting consumer tests with varying amounts of pepperoni and measuring consumer preference. Regression analysis helps in this regard. To assess current process performance, a large number of pizzas could be evaluated for pepperoni weight using control charts. Sample-size formulae suggest the number of pizzas to be evaluated. The capability of the production process could be assessed by comparing the statistical distribution of weight to 30 g and by computing capability indices. If a large proportion of pizzas have pepperoni weights less than 30 g, then the variability will have to be reduced and/or the process mean will have to be increased to avoid this situation.

The current situation may be that due to large variability in pepperoni weight; the mean pepperoni weight has to be targeted at 50 g to ensure that a very small proportion of pizzas will have less than 30 g of pepperoni. However, this causes an average of 20 g of pepperoni per pizza to be given away for free and the economic consequence can be assessed using economic loss functions. Because the losses are large, specific improvement goals can be set to reduce the variability of pepperoni weight and retarget the mean closer to 30 g.

In order to better define this weight control strategy, we need to determine the causes of weight variability. Using control charts, variance components analysis, and measurement systems analysis of existing data, it may turn out that this multilane manufacturing process exhibits large lane-to-lane variability and that this lane-to-lane variability changes as a function of time of manufacture. The measurement system variability and the variability due to procured batches of pepperoni turn out to be small. The improvement strategy becomes one of reducing the lane-to-lane variability.

Accomplishing the desired improvement requires an engineering understanding of the pepperoni deposition process. Based upon this understanding, experiments are performed to identify and optimize key process factors. Comparative experiments (also designed experiments), confidence intervals, regression analysis, and variance transmission analysis help in this regard.

After the engineering improvements are implemented, a validation experiment is done to show that the desired reduction in variability has occurred and the mean weight could be retargeted much closer to 30 g. Control charts, tolerance intervals, and capability analysis are particularly helpful here.

After the new process is successfully validated, real-time control charts are implemented to ensure that the process mean and variability do not degrade and the profit improvement continues to be realized and improved upon.

1.2 IMPLEMENTING STATISTICAL METHODS

There has always been a need for scientists, engineers, managers, and other technical individuals to use statistical methods to effectively research, develop, and manufacture products. The six-sigma approach recognizes this need. Despite many efforts, there continues to be a wide gap between the need for statistical applications and actual applications. What should be done to bridge this gap? How can statistical methods be implemented for bottom line impact?

It is the author's experience that there are four key ingredients of success: management leadership, statistics education, software, and consulting support. This combination has resulted in a large number of success stories and an organizational change toward statistical thinking.

Management Leadership

Managers need to understand the benefits of statistical methods and lead the implementation effort through knowledge and conviction. This requires education on the part of managers. Additionally, they need to communicate a strong rationale and vision, select major projects, provide necessary resources, ask the right questions during review meetings, make effective use of metrics, and do so with a constancy of purpose. Otherwise, it is unlikely that statistical methods will become a part of the culture and result in business successes on a daily basis. The six-sigma approach formalizes this leadership role through the establishment of a leadership council, implementation leader, master black belt, black belts, team leaders, and green belts (Pande, Neuman, Cavanaugh), each with specific roles and responsibilities.

Education

Some companies have internal statisticians who help with the planning of experiments and the analysis of data. There are not enough applied statisticians in industry to get the job done in this manner. Also, successful implementation requires an interplay between technical understanding of the subject matter and an understanding of statistical methods. The solution is to teach necessary statistical methods to the technical personnel. This job has become considerably easier due to the availability of user-friendly statistics software packages. With these software packages, technical personnel can plan experiments and analyze data themselves; however, it requires them to take the responsibility to use appropriate statistical methods and to correctly interpret the results of statistical analyses.

These trends dramatically increase the need for statistics education in industry. Good educational seminars emphasize clear and concise communication of practically useful statistical methods, discussion of the assumptions behind these methods, industry applications, use of software, and a focus on interpretation and decision making. They have resulted in major increases in the implementation of statistical methods for performance improvement.

Software

Considerable PC software is now available to design data collection schemes, to collect data, and to analyze data. With appropriate education and some continuing support, technical personnel have effectively used statistical software to accelerate product and process development; to improve products, processes, and measurement systems; to reduce costs; to control and troubleshoot manufacturing; and to make good decisions based upon data.

Consulting Support

Even with education and software, some consulting support has often been necessary to jump-start applications, to help structure problems correctly, to deal with particularly special or difficult issues, and to provide course corrections from time to time. It is critical that the individuals providing education and consulting support be well versed in statistics, have an engineering or scientific background, be oriented to solving real problems, and possess excellent communication skills.

If some of the above success factors are missing, failures result. As an example, suppose that data analysis suggests that factor A influences the output but factor B does not. What are the possible explanations? Perhaps what meets the eye is correct. On the other hand, the effect of factor B may not be visible because of its narrow range, small sample size, curvilinearity, settings of other factors, and so on. It is also possible that the effect of factor A may merely be correlation without causation. What precautions should be taken to find the correct answers? This requires the technical individual to either have good statistics education or have access to statistical consulting support. As another example, if the first plotted point on an $\overline{X}$ chart falls outside the control limits, the software will identify this as a special cause that has shifted the average. However, there could be many other reasons such as bad data, increased variance, wrong control limits, need to evaluate subgrouping scheme, etc. These require the software user to also have appropriate statistics education. Managers need to be able to review the work and ask the right questions, and so on.

All four success factors are necessary. The combination of statistics education, software, consulting support, and management leadership empowers technical people to use statistical methods to make business successes happen.

1.3 ORGANIZATION OF THE BOOK

The purpose of this book is to help provide widely useful statistics education in a clear and concise manner. The book is divided into five parts and also includes a test (Chapter 10). The first part (Chapter 2) deals with basic statistical concepts. No prior knowledge of statistics is assumed. Chapter 2 describes how to meaningfully summarize information contained in the sample data from a single population and how to translate this information into an understanding of the population in order to make decisions. It begins with ways to estimate population mean, variability, and distribution. The commonly encountered statistical distributions, estimation of confidence intervals for the parameters of these distributions, and the sample sizes necessary to estimate the parameters with a prespecified degree of precision are described. Tolerance intervals that contain a certain percent of the population are considered next. Finally, the three key assumptions made in most statistical analyses, namely the assumptions of normality, independence, and constancy of error variance are explained along with ways to check these assumptions and remedies if the assumptions are not satisfied. Knowledge of this chapter is a prerequisite to understand the other chapters. Also, the methods contained in this chapter are applicable to a very wide cross section of employees in a company.

The second part of the book (Chapter 3) presents two statistical methods: comparative experiments and regression analysis. The chapter begins with a discussion of the hypothesis-testing framework used to design and analyze comparative experiments. Statistical tests, including the t-test, F-test, and ANOVA, are presented to compare one, two, and multiple populations. The use of hypothesis-test-based sample size formulae to design the experiment and the confidence interval approach to analyze the experiment are recommended in practice. The chapter also considers the topic of correlation and regression analysis, whose main purpose is to obtain an equation relating independent and dependent factors. Applications of such equations to setting specifications and designing accelerated stability (shelf life) tests are considered.

The third part of the book (Chapters 4, 5, 6) deals with the subject of statistical process control; in particular, control charts and process capability. Chapter 4 starts by defining the role of control charts and the basic principles behind determining control limits. It presents formulae to design the commonly used variable and attribute control charts with examples. The out-of-control rules to detect special causes and their rationale are explained. Finally, key success factors to implement effective charts are discussed. Chapter 5 deals with the quantification of process capability, for both stable and unstable processes, in terms of capability and performance indices. Methods to estimate these indices and their associated confidence intervals are described. The connection between a capability index and tolerance interval is established. The use of capability and performance indices to assess the current process, to set goals, and to identify improvement actions is explained. The rationale behind the six-sigma goal is explained. Chapter 6 describes five additional control charts: risk-based charts, modified limit charts, moving average charts, short-run charts, and charts for nonnormal distributions. These charts are useful in many practical applications. Risk-based charts explicitly manage the two risks of making wrong decisions. Modified limit charts are useful when it becomes uneconomical to adjust the process every time it goes out of control. Such is the case when the process capability is high and the cost of adjustment is also high. Moving average charts are useful when it is important to rapidly detect small but sustained shifts, as would be the case when process capability is low. Short-run charts deal with the situation in which the same process is used to produce multiple products, each for a short period of time. In this case, keeping a separate chart for each product becomes ineffective and inefficient. Charts for nonnormal distributions apply when the distribution of the plotted point departs significantly from a normal distribution.

The fourth part of the book (Chapters 7 and 8) deals with the underutilized subject of variance components. Chapter 7 begins by introducing the idea of variance components, first for an $\bar{X}$ chart and then for a one-way classification with a fixed factor. These two simple applications of variance components prove very useful in practice. The topic of structured data collection or structured capability studies is considered next. The differences between fixed and random factors, nested and crossed classifications, along with the mathematics of variance components analysis, are explained. By matching data collection to the likely major causes of variability, structured studies permit data to be collected and analyzed in a manner suitable for understanding not only the total variability but also the variability due to each major cause of variation. Chapter 8 illustrates the many applications of variance components analysis, including applications to single-lane and multi-lane processes, factorial designs, and the question of allocating specifications. It introduces the classical and the quadratic loss functions to translate variability into its economic consequence and shows how the ideas of variance components, process capability, and economic loss can be coupled through "what if" analysis to plan improvement efforts and to make improvement decisions on an economic basis. The chapter also introduces two additional tools: the multivari chart, which is a graphical approach to viewing variance components; and variance transmission analysis, which is an equation-based approach to computing variance components.

The fifth part of the book (Chapter 9) deals with measurement systems analysis. This chapter begins by defining the statistical properties of measurement systems. These properties are called stability, adequacy of measurement units, bias, repeatability, reproducibility, and linearity. The detrimental effects of measurement variation are then considered and lead to acceptance criteria to judge the adequacy of a measurement system. The design and analysis of calibration studies, stability and bias studies, repeatability and reproducibility studies, intermediate precision and robustness studies, linearity studies, and method transfer studies are presented with examples to illustrate the assessment and improvement of measurement systems. Finally, an approach to compute the number of significant figures is presented.

1.4 HOW TO USE THIS BOOK

This book could be used in many ways. The author has used prior, informal versions of the book to teach thousands of seminar participants from numerous industries, and continues to do so today. This book could form the lecture notes in future educational seminars. The book could be used as a reference book, to look up a formula, a table, or to review a specific statistical method. The book could also be used for self-study. Those who wish to study on their own should first review the table of contents, decide whether they are or are not generally familiar with the subjects covered in the book, and then take the appropriate one of the following two approaches.

For those generally not familiar with the subject

1. Start reading the book from the front to the back. If you want to get to later chapters faster, you may be able to skip Chapters 3 and 6 initially and return to them later. Go over a whole chapter, keeping track of topics that are not clear at first reading.

2. Read through the chapter again, paying greater attention to topics that were unclear. Wherever possible, try to solve the examples in the chapter manually and prove the concepts independently. Makes notes on the key points learned from the chapter.

3. If you feel that you have generally understood the chapter, go to Chapter 10, which contains test questions. These questions are arranged in the order of the chapters. Solve the questions pertaining to your chapter. Compare your answers and reasons to those given in Chapter 10. If there are mistakes, review those sections of the book again.

4. Obtain an appropriate software package, type in the data from various examples and case studies given in the book and ensure that you know how to get the answers using the software of your choice.

5. Think about how these statistical methods could be applied to your company's problems. You are bound to find applications. Either find existing data concerning these applications or collect new data. Make applications.
6. Review your applications with others who may be more knowledgable. Making immediate applications of what you have learned is the key to retaining the learning.

For those generally familiar with the subject

1. Start by taking the test in Chapter 10. Take your time. Write down your answers along with the rationale. Do not read the answers given in Chapter 10 yet, but simply compare your answers to the answer key. Circle the wrong answers. Only for the questions you answered correctly, read the answers in Chapter 10 to ensure that your rationale matches that given in Chapter 10.
2. Based upon the above assessment, identify the chapters and sections you need to study. For these chapters and sections, follow the six steps outlined above.

2

Basic Statistics

Statistics is the science of making decisions based upon data. The data we collect are usually a sample from a population, and the decisions we make are based upon our understanding of the population gained by analyzing the collected data. For example, a shoe manufacturer wishes to decide what the shoe sizes should be and what the likely market is for each size. For this purpose, the manufacturer collects data regarding foot sizes for a selected set of individuals. However, the manufacturer's decision rests on an understanding, based upon the collected data, of the foot size for the entire population of individuals. This chapter describes how to meaningfully summarize the information contained in the collected data and how to translate this information into an understanding of the population. The following statistical tools are discussed: descriptive statistics, statistical distributions, confidence intervals, sample size determination, tolerance intervals, and the key assumptions made in statistical analyses, namely, normality, independence, and constancy of variance.

Data Collection

Before collecting data, the objectives of data collection and analysis need to be clearly defined. Are we interested in determining whether the

Statistical Methods for Six Sigma by Anand M. Joglekar
ISBN 0-471-20342-4 © 2003 John Wiley & Sons, Inc.

process is on target? Do we wish to assess if the desired variance reduction has been achieved? Is the objective to determine what proportion of future production will meet specifications? Do we wish to know whether our measurement system is adequate for the task? How precisely do we need to answer these questions? Specificity in defining objectives is crucial.

Having defined the objectives, we need to decide what data to collect and the conditions under which the data will be collected. If the objective is to assess whether the products meet specifications or not, what are all the key product characteristics we need to collect data on? If the objective is to validate the process, will the data be collected using a single lot of raw material or will multiple lots be used? The data we decide to collect may be variable data or attribute data. Variable data are measurements made on a continuous scale, such as weight, temperature, dimensions, moisture, and viscosity. Attribute data are count data. Classification of a product as acceptable or not acceptable, broken or not broken, is attribute data. Number of defects per product is also attribute data. Attribute data contain far less information than variable data; consequently, a large number of observations are required to draw valid conclusions. The use of variable data is preferred.

Most of the data that we collect is on a sample from a population. A population is all elements of a group that we wish to study or for which we plan to make a decision. A population may be finite, such as all the employees of a company, or it may be infinite, such as all products that could be produced by a particular process. The number of frozen pizzas to be produced under current manufacturing conditions may be very large and the population may be considered to be infinite. Testing every unit of the population is costly and impractical; therefore, sampling is necessary. The sample needs to be a representative portion of the population, usually a very small portion. The results obtained from analyzing the sample will approximate the results we would get by examining the entire population. One way to obtain a representative sample is to take a random sample in which every unit in the population has an equal chance of being selected. As one example of population and sample, if we wish to determine the percent of adults in the United States who prefer diet soft drinks, the population is all adults in the United States. The sample is the randomly chosen adults in the study. How big should the sample size be? How many people should be selected to participate in the diet soft drink study? The answer depends upon the objectives of data collection, the precision needed to make the decision, the kind of data, and the associated variability, as we will see in later sections.

2.1 DESCRIPTIVE STATISTICS

Once the data are collected, the next step is to analyze the data and draw correct conclusions. We begin by considering some simple ways of numerically and graphically summarizing the information in the data. The numerical summaries are the measures of central tendency and the measures of variability. The graphical summary is a histogram. Together, they are called descriptive statistics.

Suppose the following data were collected on moisture in a food product. Each observation is the amount of moisture, measured in grams, for six randomly selected products from a production line:

Moisture (grams): 4.9, 5.1, 5.1, 5.2, 5.5, 5.7

In practice, a hundred such data points may have been collected. How can we summarize most of the information in this data with a very few numbers? The measures of central tendency and variability provide an initial summary. For the moisture data, these measures are shown in Figure 2.1.

2.1.1 Measures of Central Tendency

Mean, median, and mode are measures of the central tendency of data; namely, they describe where the center of the data may be located.

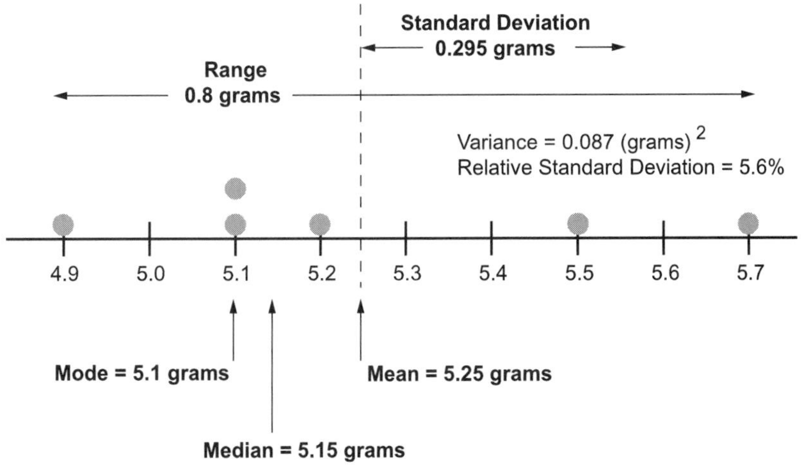

Figure 2.1 Descriptive statistics.

Mean. The sample mean is the most common measure of central tendency. The sample mean is the arithmetic average of the data, computed as the sum of all observations divided by the number of observations. The mean has the same units as the original data. It provides an estimate of the population mean, which is usually unknown. If the data are symmetrically distributed, the mean locates the central value of the data. The following nomenclature is used:

x_i = Individual observations

n = Number of observations in the sample or the sample size

$\bar{x}$ = Sample mean (estimates μ)

μ = Population mean (usually unknown)

$$\text{Sample mean} = \bar{x} = \frac{\Sigma x_i}{n} \tag{2.1}$$

For the moisture data, the sample size is six and $\bar{x} = 5.25$ (grams).

Median. This is a number such that half of the observations are below this number and half are above it. The median has the same units as the original data. For the moisture data, the median is 5.15 (grams), because there are three observations on either side of this number.

Mode. This is the observation that occurs most frequently. The mode has the same units as the original data. For the moisture data, the mode is 5.1 (grams), because this number occurs most frequently in the sample data.

2.1.2 Measures of Variability

Range, variance, standard deviation, and relative standard deviation (coefficient of variation) are measures of variability.

Range. This is the simplest measure of variability. Range is defined as the largest observation minus the smallest observation. The larger the range, the greater the variability. Range has the same units as the original data:

$$\text{Range} = R = \text{Largest observation} - \text{Smallest observation} \tag{2.2}$$

For the moisture data, $R = 5.7 - 4.9 = 0.8$ (grams). Range is easy to calculate. However, it has some drawbacks. It is greatly influenced by outliers. It does not explicitly use any observations other than the largest and the smallest. For example, the ranges for the data set (10, 5, 7, 8, 6, 4, 2) and for the data set (10, 3, 3, 3, 3, 3, 2) are equal, namely, (10 – 2) = 8. Intuitively, the variability in the two data sets is different. Therefore, another measure of variability that responds to a change in any observation is needed.

Variance. One approach to make use of all observations in computing variability is to first calculate $(x_i - \bar{x})$, the deviation of each observation from the mean. There will be n such deviations. In order to obtain a single number as a measure of variability, we could calculate the average of these n deviations. Unfortunately, this average will always turn out to be zero because some of these deviations will be positive; others will be negative and they will cancel each other. If we square the deviations, all signs will be positive and no cancellations will occur. This concept leads to the definition of variance. The following nomenclature is used:

σ^2 = Population variance (usually unknown)

s^2 = Sample variance (estimates σ^2)

The sample variance is defined as follows:

$$s^2 = \frac{\Sigma(x_i - \bar{x})^2}{n - 1} \qquad (2.3)$$

Variance does not have the same units as data. It has squared units. For the moisture data,

$$s^2 = \frac{(4.9 - 5.25)^2 + \ldots + (5.7 - 5.25)^2}{(6 - 1)} = 0.087\ (\text{grams})^2$$

In the formula for sample variance, why divide by $(n - 1)$ and not by n? The reason is as follows. The original sample consists of n observations. Each observation can take any value it wants to take, so it is said to have one degree of freedom. Hence, the sample has n degrees of freedom. To calculate variance, we must first calculate $\bar{x}$. Given this fixed value of $\bar{x}$, all observations are no longer free to be what they want to be; one of the observations becomes dependent upon the other $(n - 1)$ observations. We only have $(n - 1)$ independent observations to calculate variance. Therefore, we divide by $(n - 1)$.

In defining sample variance, squaring $(x_i - \bar{x})$ is only one way to get rid of minus signs. We could have taken the absolute value or any even power of $(x_i - \bar{x})$. Why the power of 2? A practical reason is that the bad consequences of a characteristic not being on target are likely to be proportional to the square of the distance from the target. This implies that for a process with mean on target, the loss (bad consequences) due to variability is directly proportional to variance, if variance is defined as a squared function of $(x_i - \bar{x})$. For an on-target process, if we reduce variance by half, we reduce the bad consequences due to variability by half. Therefore, variance reduction is important and is a key goal of six sigma and associated statistical methods. These ideas are illustrated with examples in Chapter 8.

Standard Deviation. Variance has squared units, which are meaningless. Therefore, we take the square root of variance, which is called the standard deviation. The larger the standard deviation, the larger the variability. The standard deviation has the same units as the original data and is easier to interpret. The following nomenclature is used:

σ = Population standard deviation (unknown)

s = Sample standard deviation (estimates σ)

The sample standard deviation is calculated as

$$s = \sqrt{s^2} = \sqrt{\frac{\Sigma(x_i - \bar{x})^2}{n-1}} \tag{2.4}$$

For the moisture data, $s = \sqrt{0.087} = 0.295$ (grams). The practical interpretation of standard deviation is explained in Section 2.2.1.

Relative Standard Deviation (*RSD*). This is also known as the coefficient of variation (*CV*). It is defined as follows:

$$RSD = \frac{\sigma}{\mu} \text{ and is estimated as } \frac{s}{\bar{x}} \tag{2.5}$$

RSD is dimensionless and is often expressed as a percentage by multiplying by 100. For the moisture data, % $RSD = (0.295/5.25)100 = 5.6\%$.

Since *RSD* is dimensionless and not subject to change due to changes in units of measurement, it is more comparable across different situations. For example, let us say that $\sigma_A = 10$ for process *A*, and $\sigma_B = 20$ for process *B*. Which process is better? One may be tempted to say that process *A* is better. However, if $\mu_A = 100$ and $\mu_B = 1000$, then our conclusion may reverse, which is correctly reflected by $RSD = 10\%$ for process *A* and $RSD = 2\%$ for process *B*. A disadvantage of *RSD* is that if it changes, by simply looking at the *RSD*, it is not possible to say whether the change is due to a change in σ, a change in μ, or a change in both.

Properties of Variance. If variance has wrong units, why does it continue to have a prominent place in statistics? The reason is that although variance has the wrong units, it has the right mathematics.

Standard deviation has the right units but the wrong (more difficult) mathematics. Therefore, all calculations are done in variance and, at the last step, we convert the answer into standard deviation so it can be understood.

Three properties of variance are useful in practice. In the following, X and Y are two independent random factors and k is a constant.

$$\begin{aligned} &1.\ \textit{Variance}(k\,X) = k^2\,\textit{Variance}(X) \\ &2.\ \textit{Variance}(X+Y) = \textit{Variance}(X) + \textit{Variance}(Y) \qquad (2.6) \\ &3.\ \textit{Variance}(X-Y) = \textit{Variance}(X) + \textit{Variance}(Y) \end{aligned}$$

The first property is useful when we want to convert X from one unit of measurement to another. For example, if X represents length and has a variance of 0.2 $(\text{cm})^2$, to compute what the variance would be if length were to be measured in millimeters, $k = 10$ and the variance will be 20.0 $(\text{mm})^2$.

The second property finds widespread use in statistics. Suppose a person eats two candy bars each day. Each bar comes from a population with $\mu = 100$ calories and $\sigma = 4$ calories. What will be the mean and standard deviation of calories due to the daily consumption of two candy bars? The average calorie intake per day will be 200 calories. It may be tempting to think that the standard deviation per day will be $4 + 4 = 8$ calories, but that would be incorrect. We should remember that standard deviations do not add, but variances do. So the variance per day will be $16 + 16 = 32$ $(\text{calories})^2$ and the standard deviation per day will be $\sqrt{32} = 5.7$ calories.

Another very important consequence of the first two properties is the following. Let $X_1, X_2 \ldots X_n$ represent first, second ... *nth* observation on a process, each with variance σ^2. Let $\bar{X}$ represent the average. Using the first two properties of variance

$$\textit{Variance}(\bar{X}) = \sigma_{\bar{X}}^2 = \textit{Variance}\left(\frac{X_1 + X_2 \ldots + X_n}{n}\right) = \frac{n\sigma^2}{n^2} = \frac{\sigma^2}{n} \qquad (2.7)$$

and

$$\sigma_{\bar{X}} = \frac{\sigma}{\sqrt{n}}$$

This important result says that as the sample size increases, the standard deviation of the average becomes smaller. It reduces by a factor of square root of sample size.

The following example illustrates the use of the third property by showing that if the quantity of interest is computed as a small difference between two large measurements, it is likely to have a large *RSD*, making the measurement potentially unsatisfactory. Let us suppose that product moisture was measured as a difference between wet weight and dry weight:

$$\text{Moisture} = \text{wet weight} - \text{dry weight}$$

1. If μ (wet weight) = 100 grams, μ (dry weight) = 95 grams and σ (wet or dry weight) = 0.1 grams, then *RSD* (wet weight) = 0.1% and *RSD* (dry weight) $\approx$ 0.1%. Both measurements appear to have reasonably low *RSD*s.

2. Our interest centers on product moisture, for which μ (moisture) = 100 − 95 = 5 grams. From the third property of variance, σ^2 (moisture) = $(0.1)^2 + (0.1)^2 = 0.02$ (grams)^2. Therefore, σ (moisture) = 0.14 (grams) and *RSD* (moisture) = (0.14/5)100 = 2.8%. The *RSD* for moisture is nearly 30 times the *RSD* for primary weight measurements.

2.1.3 Histogram

The mean and standard deviation are two powerful ways of numerically summarizing data on a single factor. However, they do not capture the overall shape of the frequency distribution of data. A histogram graphically shows the frequency distribution of data and adds to the information contained in the mean and standard deviation. The values taken by the random factor X are plotted along the horizontal axis, whereas the vertical axis shows the frequency with which these values occur.

Figure 2.2 shows the frequency distribution of the diameters of steel rods, this being the result of measuring 1000 rods. The lower specification limit (*LSL*) is 1.000 centimeters and the upper specification limit (*USL*) is 1.010 centimeters. Rods smaller than 1.000 centimeter are rejected and rods larger than 1.010 centimeters are reworked. The cost of rework is considerably smaller than the cost of rejection.

Figure 2.2 is called a histogram. The horizontal axis shows the diameter in centimeters. The axis is subdivided into several equal intervals called cells. The vertical axis shows the frequency, namely, the number of rods that fall in each cell. For example, there were 40 rods with diameters between 0.997 and 0.998 centimeters. A total of 120 rods were rejected out of the 1000 examined, for a rejection rate of 12%. No rods were rejected for being too large. The process mean appears to be around 1.003 and needs to be increased to not produce scrap. Some rework will result and the process variability will have to be reduced next.

This figure is also trying to tell us something more. It is rather peculiar that there is a gap just to the left of 1.000. Also, the cell just to the right of 1.000 is too tall. It appears as if the inspectors were passing rods that were just below the lower specification, recording them as 1.000 in-

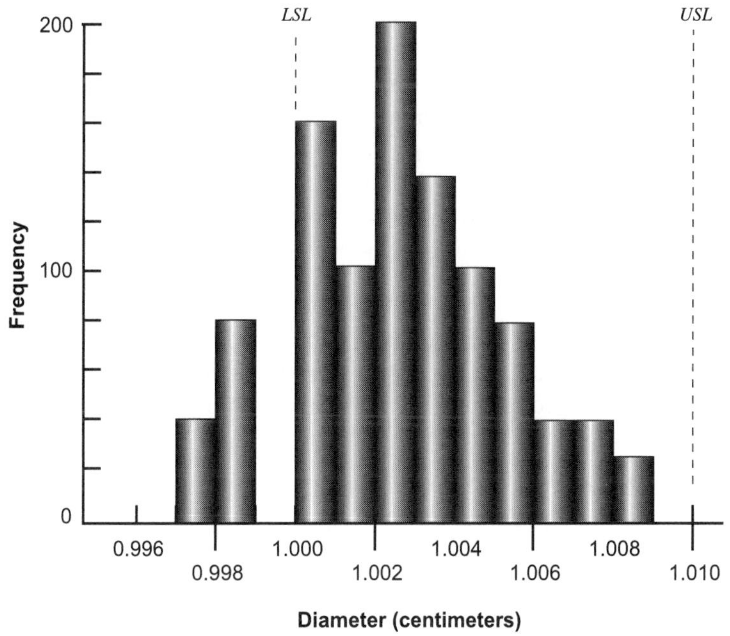

Figure 2.2 Histogram.

stead. When these rods were correctly recorded, the empty interval filled up, for a total reject rate of 17%. By simply knowing the mean and standard deviation of data, we would not have been able to predict the existence of the gap. The histogram gave us new information.

Thus, a histogram provides a visual picture of the mean, the variability, the shape of the distribution, whether there is anything unusual about the shape, and, if specifications are known, the proportion outside the specification. It provides a large amount of information in an easy to understand graphical format. Two precautions should be taken to obtain a meaningful histogram. The sample size n should preferably be 50 or more. The number of cells k should not be arbitrarily chosen but should be selected to approximately satisfy the following relationship:

$$2^k \approx n$$

For example, if $n = 50$ then $k = 5$ or 6.

Interpreting Histograms. We are looking to see if the shape of the histogram suggests something surprising, as was the case with the empty cell for the histogram of steel rod diameters. To judge whether the observed shape is surprising or not, we need to know the expected shape of the histogram for the data being collected. As an example, if a person goes to the same restaurant over and over, and plots a histogram of the length of time it took to get served once entering the restaurant, called the waiting time, what is the expected shape of this histogram? If we knew the answer, then we could decide whether the plotted histogram contains a surprise or not. If the shape of the histogram is unexpected, the reasons for this should be investigated. The commonly observed shapes are shown in Figure 2.3 and are discussed below (Ishikawa), assuming that the histogram is based upon a sufficiently large sample size and the correct number of cells.

Normal. Many measured characteristics follow a normal distribution (see Section 2.2.1). The histogram is bell-shaped. Normal distribution is so common that if the histogram is not bell-shaped, we should ask ourselves "why not?"

Bimodal (or Multimodal). These histograms have two (bimodal) or many (multimodal) peaks. Such histograms result when the data come from two or more distributions. For example, if the data

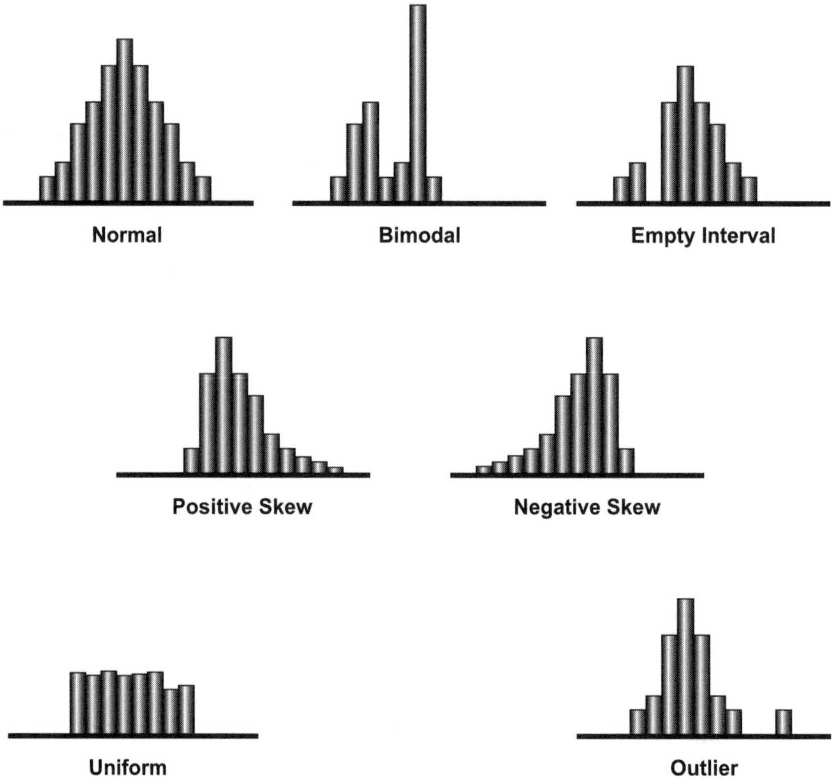

Figure 2.3 Interpreting histograms.

came from different suppliers, machines, shifts, and so on, a bimodal (or multimodal) histogram will signal large differences due to these causes.

Empty Interval. In this case, one of the intervals has zero frequency. This may result from prejudice in data collection.

Positive Skew. Positive skew means a long tail to the right. This is common when successful efforts are being made to minimize the measured value. Also, variance has a positively skewed distribution. The distributions of particle size and waiting time are positively skewed. A positive skew could result if the observed data have many values close to zero. Such a histogram may also result if sorting is taking place.

Negative Skew. Negative skew means a long tail to the left. This is common when successful efforts are being made to increase the measured value. Such a histogram may also result if sorting is taking place.

Uniform. This histogram looks more like a rectangular distribution. Such a histogram can result if the process mean is not in control, as in the case when tool wear is taking place.

Outlier. Here one or more cells are greatly separated from the main body of the histogram. Such observations are often the result of wrong measurement or other mistakes.

The mean, standard deviation, and histogram provide extremely useful summaries of the data. However, they do not contain all the information in the data. In particular, data are often collected over time and any time trends are lost in the summaries considered so far. We need to plot the data over time. We will return to this topic in Chapter 4.

2.2 STATISTICAL DISTRIBUTIONS

A histogram shows the frequency distribution of the collected data, which usually are a sample from the population. A statistical distribution is a probability distribution that describes the entire population. The histogram estimates the population distribution. As the number of observations in the sample increases, the histogram becomes a better and better approximation of the true population distribution. This section describes normal, binomial, and Poisson distributions that are of particular interest in practical applications. Normal distributions deal with continuous data, whereas binomial and Poisson distributions deal with count data.

2.2.1 Normal Distribution

The normal distribution is the most important distribution because it occurs very frequently. The reason for this frequent occurrence is explained by the central limit theorem, which states that the sum of n independent random factors is approximately normally distributed, regardless of the distributions of individual factors, so long as no one factor dominates. The central limit theorem has the following two important implications.

First, there would be no distribution without variation. Variation is the result of a large number of causes. The final variability that we observe is the sum of the variability due to each cause. If none of the causes dominate, then the central limit theorem says that regardless of the type of distribution due to each cause, the sum has a distribution that approximates a normal distribution. This is the reason why normal distribution occurs so frequently. The second implication of the central limit theorem has to do with the distribution of the average. Since the average is proportional to the sum, the distribution of the average always tends to be normal regardless of the distribution of individual values.

Normal distribution, shown in Figure 2.4, is a continuous, bell-shaped, symmetric distribution. It is characterized by two parameters: the population mean $\mu(-\infty < \mu < \infty)$ and the population standard deviation $\sigma(\sigma > 0)$. The fact that X has such a normal distribution is frequently denoted by $X \sim N(\mu, \sigma)$. The parameters μ and σ are independent of each other. Knowledge of the mean tells us nothing about the standard deviation.

For a normal distribution, σ has a simple interpretation, as shown in Figure 2.4. Mean plus and minus one standard deviation ($\mu \pm 1\sigma$) includes 68.26% of the area under the bell curve. The total area under the bell curve is 100%. This means that 68.26% of the population is expected to be in the interval $\mu \pm 1\sigma$. Similarly, 95.46% of the population is in the interval $\mu \pm 2\sigma$ and 99.73% of the population is within $\mu \pm 3\sigma$. The probability that an observation will be more than 3σ away from the mean is approximately 0.3% or 3 in 1000.

Calculating Probabilities. We may be interested in finding the probability of a product being outside specifications. Such probability computations cannot be done by a direct integration of the normal distribution because the integrals involved cannot be evaluated in a closed form. Instead, we first convert the normal distribution to a standard normal and

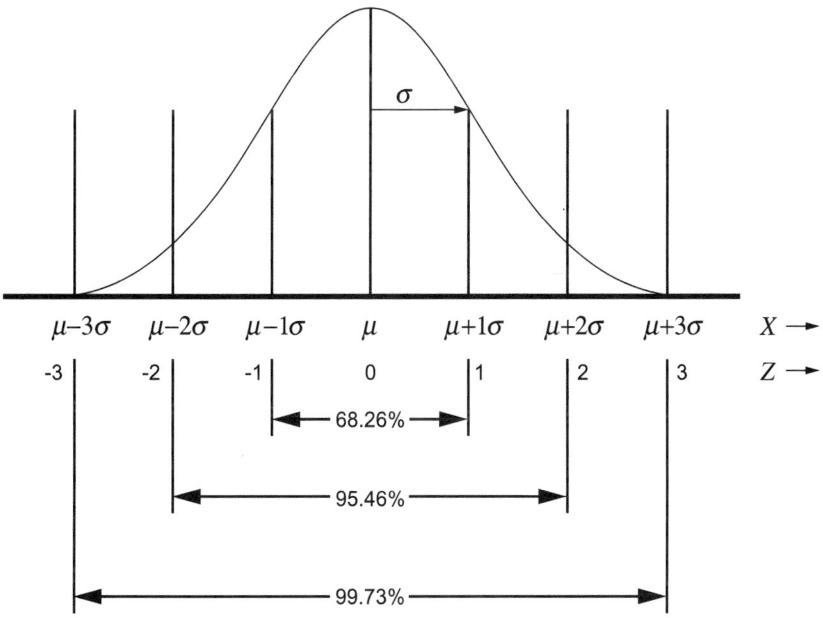

Figure 2.4 Normal distribution.

then use the normal distribution table in Appendix A. If $X \sim N(\mu, \sigma)$, then

$$Z = \frac{X - \mu}{\sigma} \sim N(0, 1) \tag{2.8}$$

i.e., Z has a standard normal distribution with zero mean and a standard deviation of one. This is easy to see using the properties of variance given by Equation 2.6. The transformed horizontal Z axis is also shown in Figure 2.4. The normal distribution table in Appendix A provides probabilities for $Z \geq z$. Probabilities for $Z \leq -z$ are equal to probabilities for $Z \geq z$ due to the symmetry of the distribution.

To illustrate, consider the following question. Percent moisture X has a normal distribution with $\mu = 10.2\%$ and $\sigma = 0.5\%$. The specification is 9% to 11%. What proportion of the product is out of specification? If P denotes probability, then we want to find

$$P(X > 11) + P(X < 9)$$

We first rewrite this in terms of Z:

$$P\left(\frac{X - \mu}{\sigma} > \frac{11 - 10.2}{0.5}\right) + P\left(\frac{X - \mu}{\sigma} < \frac{9 - 10.2}{0.5}\right) = P(Z > 1.6) + P(Z < -2.4)$$

From the normal table in Appendix A, $P(Z > 1.6) = 0.0485$ and $P(Z < -2.4) = 0.0082$. Hence, the probability of out of specification product is $0.0485 + 0.0082 = 0.0567$ or 5.67%.

2.2.2 Binomial Distribution

This is a discrete distribution. It applies when the following conditions are true:

1. Each of the n items being tested is being classified into only two categories: defective and not defective.
2. The probability p of a defective item is constant for every item being tested, regardless of whether any other item is defective or not.

If X represents the number of defectives in n items, then the probability of finding x defectives in n items is

$$P(x) = \frac{n!}{x!(n-x)!} p^x(1-p)^{n-x} \tag{2.9}$$

n (≥ 1) and p($0 \leq p \leq 1$) are the parameters of this binomial distribution. The mean, variance, and standard deviation of X are

$$\begin{aligned} \mu &= np \\ \sigma^2 &= np(1-p) \\ \sigma &= \sqrt{np(1-p)} \end{aligned} \tag{2.10}$$

If we observe x defectives in n items, then the sample fraction defective, $\bar{p} = x/n$, provides an estimate of the unknown population fraction defective p.

To observe the shape of the binomial distribution, consider the case when $p = 0.1$ and $n = 10$. The probability of obtaining x defectives can be computed from Equation 2.9 for various values of x and is plotted in Figure 2.5. The distribution is discrete and asymmetric. For this distribution $\mu = np = 1$ and $\sigma = \sqrt{np(1-p)} = 0.95$. Note that for a binomial distribution, μ and σ are related, and $\sigma = \sqrt{\mu(1 - \mu/n)}$. For a fixed sample size n, if μ is known, σ can be calculated.

2.2.3 Poisson Distribution

This is a discrete distribution. A typical application is the following. If X denotes the number of countable defects per product, then the probability of x defects in a randomly selected product is

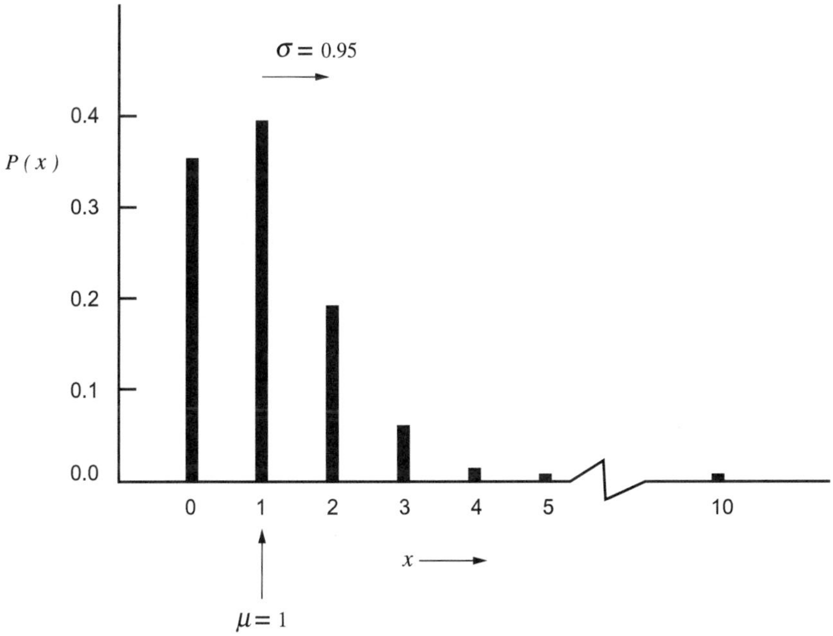

Figure 2.5
Binomial distribution (n = 10, p = 0.1).

$$P(x) = \frac{e^{-\lambda}\lambda^x}{x!} \qquad \text{for } x = 0, 1, 2, 3 \ldots \qquad (2.11)$$

$\lambda(>0)$ is the single parameter of the Poisson distribution. The mean, variance and standard deviation of X are

$$\mu = \lambda$$
$$\sigma^2 = \lambda \qquad (2.12)$$
$$\sigma = \sqrt{\lambda}$$

Obviously, μ and σ are deterministic functions of each other since $\sigma = \sqrt{\mu}$. If $x_1, x_2 \ldots x_n$ defects are observed in n products, then λ is estimated as the average number of defects per product, given by

$$\bar{c} = \Sigma x_i/n$$

The Poisson distribution may be considered to be a limiting case of the binomial distribution. The binomial distribution described by Equation (2.9) becomes the Poisson distribution given by Equation (2.11) if we let $n \to \infty$ and $p \to 0$ in a manner to keep $np = \lambda$ constant. As an example, consider the case of chocolate chip cookies containing very tiny chocolate chips. Let us suppose that the average number of chocolate chips per cookie is 20. Consider the cookie being subdivided into a large number of n parts. As $n \to \infty$, the probability p of finding a chocolate chip in the subdivision will go to zero while keeping $np = 20$. A further requirement is that the probability of finding a chocolate chip in one subdivision be not dependent on another subdivision either having or not having a chocolate chip. Under these conditions, the number of chocolate chips per cookie will have a Poisson distribution.

To observe the shape of the Poisson distribution, the probability of obtaining x defects in a product can be computed from Equation (2.11). Figure 2.6 shows the Poisson distribution for $\lambda = 2$. It has $\mu = 2$ and $\sigma = 1.4$. It is a positively skewed distribution. The Poisson distribution becomes more symmetric as λ increases.

Some Useful Approximations. As noted before, a binomial distribution can be approximated by a Poisson distribution as $n \to \infty$, $p \to 0$, and $np = \lambda$ stays constant. A Poisson distribution with $\lambda = np$ may be used to satisfactorily approximate the binomial for $p < 0.1$ and large n.

A Poisson distribution may be approximated by a normal distribution for large values of λ. If $\lambda \geq 15$, a normal distribution with $\mu = \lambda$ and $\sigma = \sqrt{\lambda}$ provides a good approximation of the Poisson distribution.

It then follows that a binomial distribution may be approximated by

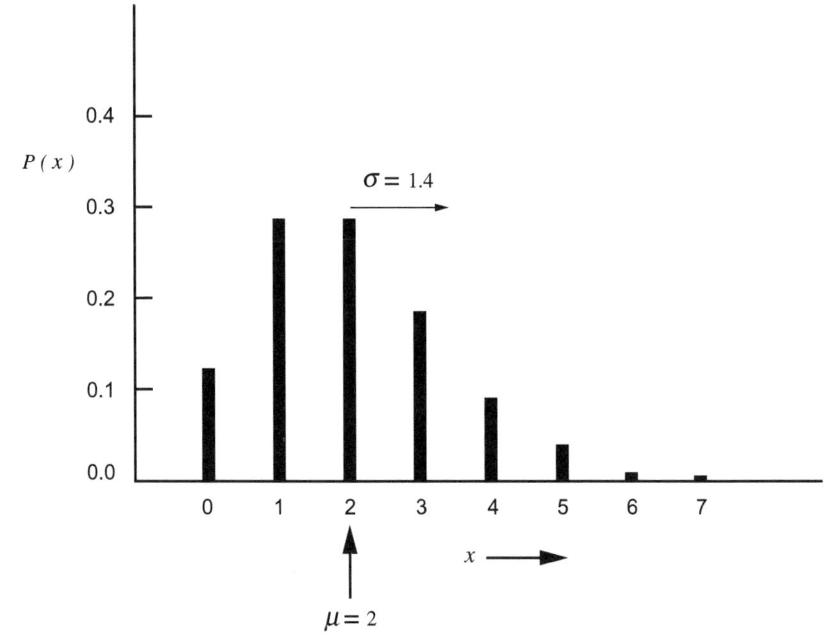

Figure 2.6 Poisson distribution ($\lambda = 2$).

a normal distribution as well. The normal approximation is satisfactory for p approximately equal to 0.5 and $n > 10$. For other values of p, larger n is required. For $p < 0.1$ and $np \geq 15$, the normal approximation with $\mu = np$ and $\sigma = \sqrt{np}$ is satisfactory.

2.3 CONFIDENCE INTERVALS

In the previous section, we considered three distributions, a normal distribution with parameters μ and σ, a binomial distribution with parameter p, and a Poisson distribution with parameter λ. All these population parameters are unknown and cannot be known exactly because that would require infinite data collection. They can, of course be estimated. For example, sample average $\bar{x}$ estimates μ and sample standard deviation s estimates σ. On the basis of 10 observations of the thickness of a candy bar, if we found $\bar{x} = 3$ mm and $s = 0.1$ mm, then our best estimate of μ is 3 mm and our best estimate of σ is 0.1 mm. However, this does not mean that μ is exactly 3 mm and σ is exactly 0.1 mm. It is easy to see that if we took another 10 observations, we would get different values of $\bar{x}$ and s. If we did observe $\bar{x} = 3$ mm, how far away from 3 mm can the real μ be? If we could say that we are 95% sure that the population mean is in the interval 2.8 mm to 3.2 mm, then this interval is called the 95% confidence interval for μ. Similar situations apply to other parameters of the distributions.

Confidence intervals are useful because decisions are made based upon our understanding of the population, not just our understanding of the sample. The width of the confidence interval is a measure of how uncertain we are regarding the population parameters. The wider the confidence interval, the greater the uncertainty. The greater the number of observations, the narrower the confidence interval. Thus, confidence intervals also provide us with a way to estimate the sample size necessary to understand the population parameters to a desired degree of precision.

2.3.1 Confidence Interval for μ

To compute the confidence interval for population mean μ, we first need to understand the distribution of $\bar{X}$, the estimator of μ.

Distribution of $\bar{X}$. If X has a normal distribution with mean μ and standard deviation σ, then $\bar{X}$ has a normal distribution with

$$\mu_{\bar{X}} = \mu$$

$$\sigma_{\overline{X}}^2 = \frac{\sigma^2}{n}$$

and

$$\sigma_{\overline{X}} = \frac{\sigma}{\sqrt{n}}$$

This means that the distribution of $\overline{X}$ is centered at μ. As n increases, the variance of $\overline{X}$ becomes smaller, being inversely proportional to the sample size n. The standard deviation of $\overline{X}$, known as the standard error, reduces inversely proportional to the square root of n. This is graphically shown in Figure 2.7. The distribution corresponding to $n = 1$ is the distribution of X. With $n = 4$, the distribution of $\overline{X}$ shrinks and continues to do so as n increases. As $n \to \infty$, the distribution of $\overline{X}$ collapses to μ.

These properties of $\overline{X}$ have been proved using the properties of variance in Section 2.1.2. The fact that $\overline{X}$ has the same mean as X should be

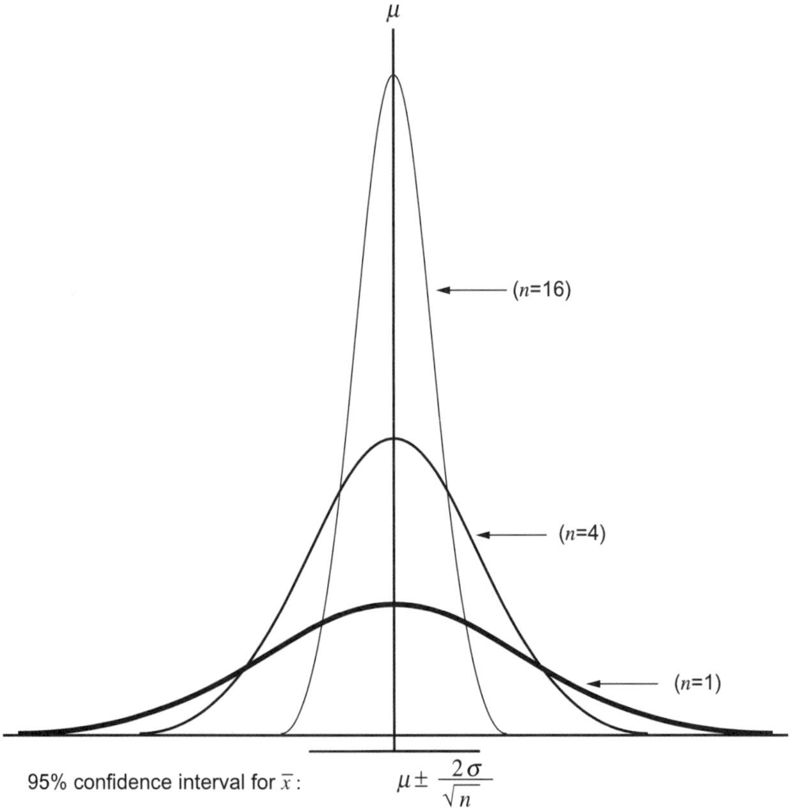

Figure 2.7
Distribution of $\overline{X}$.

intuitively obvious. That the distribution of $\bar{X}$ should shrink is also clear because the average of n observations should be closer to μ than some of the individual observations are.

Confidence Interval for μ (Sigma Known). How far away from μ can the observed $\bar{x}$ be? If we initially assume that σ is known, then we are 95% sure that $\bar{x}$ must be in the interval $\mu \pm 2\sigma/\sqrt{n}$, as shown in Figure 2.7. This means that we are 95% sure that the difference between $\bar{x}$ and μ is at most $\pm\, 2\sigma/\sqrt{n}$. Therefore, if we observe a certain value for $\bar{x}$, then with 95% confidence, μ can only be $\pm\, 2\sigma/\sqrt{n}$ away from it. Thus, the 95% two-sided confidence interval for μ is given by $\bar{x} \pm 2\sigma/\sqrt{n}$ and is shown in Figure 2.8.

Another way to state this is to realize that $(\bar{X} - \mu)/(\sigma/\sqrt{n}) = Z \sim N(0, 1)$. This is so because we are subtracting the mean from $\bar{X}$ and dividing by the standard deviation of $\bar{X}$. For this Z distribution,

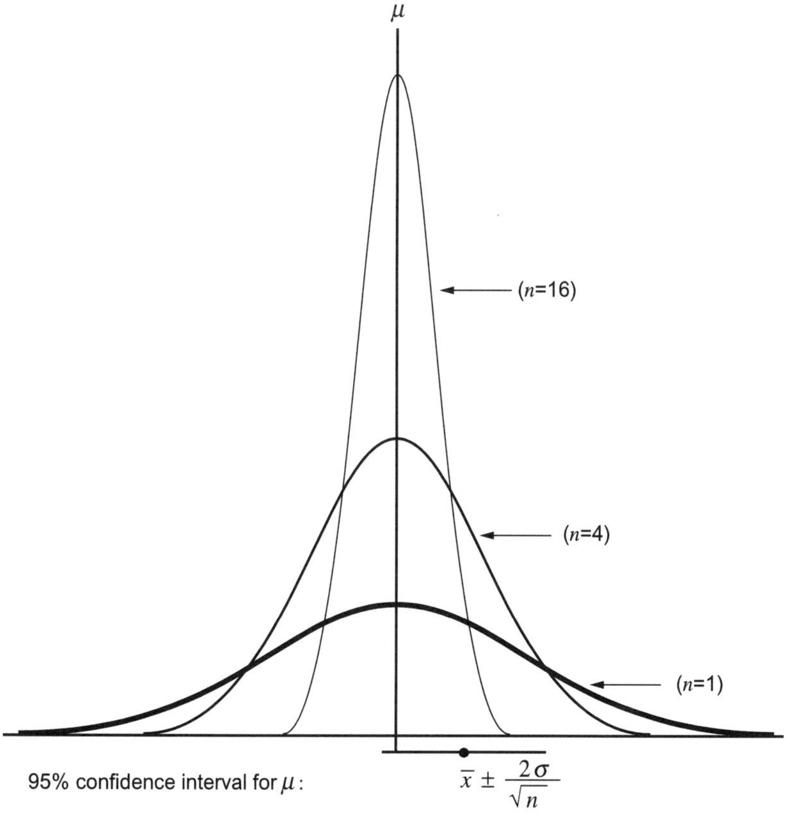

Figure 2.8 Confidence interval for μ (σ known).

$$P\left[-Z_{\alpha/2} \le \frac{\overline{X}-\mu}{\sigma/\sqrt{n}} \le Z_{\alpha/2}\right] = 1-\alpha$$

which leads to

$$100\ (1-\alpha)\%\ \text{confidence interval for}\ \mu = \bar{x} \pm Z_{\alpha/2}\frac{\sigma}{\sqrt{n}} \qquad (2.13)$$

$Z_{\alpha/2}$ denotes a value such that the probability of the Z distribution exceeding that value is $\alpha/2$. From the normal table in Appendix A, for $\alpha = 0.05$, $Z_{\alpha/2}$ is approximately 2. For $\alpha = 0.003$, $Z_{\alpha/2}$ is approximately 3.

Our statement that μ is contained in the confidence interval given by Equation 2.13 will be wrong 100 α% of the time. So α is the probability of being wrong. If $\alpha = 0.05$, then in a long-run frequency sense, 5% of the constructed confidence intervals will not contain μ. That is why, with $\alpha = 0.05$, we are only 95% sure that the confidence interval will include μ. In general, percent confidence = 100 $(1-\alpha)$%. What percent confidence should we use in practice? This is the same as asking what α risk should we assume? The answer depends upon the consequence of being wrong. The greater the consequence, the smaller should be the α risk.

To illustrate the calculation of confidence intervals, let us suppose that on the basis of $n = 4$ observations, the average candy bar thickness turns out to be $\bar{x} = 3.0$ mm. From past experience, σ is known to be 0.1 mm. What is the 95% confidence interval for population mean thickness μ? In this case, $\alpha = 0.05$ and $Z_{0.025} = 2$. From Equation 2.13, the 95% confidence interval for μ is

$$\bar{x} \pm 2\frac{\sigma}{\sqrt{n}} = 3.0 \pm 2\frac{0.1}{\sqrt{4}} = 3.0 \pm 0.1\ \text{mm}$$

This means that we are 95% sure that μ is between 2.9 mm and 3.1 mm.

Sometimes a one-sided confidence interval might be more appropriate. This is so when the consequence of being wrong is important only on one side, leading to a single-sided specification. The 100 $(1-\alpha)$% upper and lower confidence bounds for μ are

$$\mu < \bar{x} + \frac{Z_\alpha \sigma}{\sqrt{n}} \qquad \text{(upper)}$$

$$\mu > \bar{x} - \frac{Z_\alpha \sigma}{\sqrt{n}} \qquad \text{(lower)}$$

Confidence Interval for μ (Sigma Unknown). When σ is known,

$$\frac{\overline{X}-\mu}{\sigma/\sqrt{n}} = Z \sim N(0, 1)$$

When σ is unknown and is replaced by S,

$$\frac{\bar{X} - \mu}{S/\sqrt{n}} \sim t_{n-1}$$

i.e., it is a t-distribution with $(n - 1)$ degrees of freedom. The t-distribution is a symmetric distribution and looks very similar to a normal distribution but has thicker tails. As $n \to \infty$, the t distribution becomes the Z distribution. For a standard normal distribution Z, 95% of the probability is contained within ± 2. For a t distribution, since the tails are thicker, an interval wider than ± 2 is necessary to contain 95% probability. If $t_{\alpha/2,n-1}$ denotes a value such that the probability of a t_{n-1} distribution exceeding that value is $\alpha/2$, it follows that

$$P\left[-t_{\alpha/2,n-1} < \frac{\bar{X} - \mu}{S/\sqrt{n}} < t_{\alpha/2,n-1}\right] = (1 - \alpha)$$

which leads to

$$100\,(1 - \alpha)\%\text{ confidence interval for } \mu = \bar{x} \pm t_{\alpha/2,n-1}\frac{s}{\sqrt{n}} \qquad (2.14)$$

The values of $t_{\alpha/2,n-1}$ may be obtained from Appendix B for various values of α and degrees of freedom ν, in this case equal to $(n - 1)$.

Let us reconsider the candy bar thickness example with $n = 4$ and $\bar{x} = 3$ mm but let us now assume that instead of σ being known to be 0.1 mm, s was computed to be 0.1 mm based upon the four observations. For 95% confidence, $\alpha/2 = 0.025$ and from the t-table in Appendix B, $t_{0.025,3} = 3.182$. Hence,

$$95\%\text{ confidence interval for } \mu \text{ is } 3 \pm 3.182\frac{0.1}{\sqrt{4}} = 3 \pm 0.16\text{ mm}$$

As expected, this confidence interval for μ is wider than the confidence interval computed assuming σ to be known. The confidence interval has become 60% wider. A wider confidence interval means greater uncertainty regarding μ. This penalty leading to wider confidence intervals rapidly reduces as n increases and almost disappears for $n > 20$. This may be seen from the t-table as the t values rapidly approach the corresponding Z values.

The one-sided upper and lower $100\,(1 - \alpha)\%$ confidence bounds for μ are

$$\mu < \bar{x} + t_{\alpha,n-1}\frac{s}{\sqrt{n}} \qquad \text{(upper)}$$

$$\mu > \bar{x} - t_{\alpha,n-1}\frac{s}{\sqrt{n}} \qquad \text{(lower)}$$

2.3.2 Confidence Interval for σ

The quantity $(n - 1)\,S^2/\sigma^2$ has a chi-square distribution with $(n - 1)$ degrees of freedom, denoted by χ^2_{n-1}. As shown in Figure 2.9, this is a positively skewed distribution. Therefore, the confidence interval for σ^2 is asymmetric and is longer on the positive side. If $\chi^2_{\alpha/2,n-1}$ denotes a value such that the probability of a χ^2_{n-1} distribution exceeding that value is $\alpha/2$, then

$$P\left[\chi^2_{1-\alpha/2,n-1} \le \frac{(n-1)S^2}{\sigma^2} \le \chi^2_{\alpha/2,n-1}\right] = (1 - \alpha)$$

which leads to

$$100\,(1-\alpha)\%\text{ confidence interval for }\sigma^2 = \frac{(n-1)s^2}{\chi^2_{\alpha/2,n-1}} \le \sigma^2 \le \frac{(n-1)s^2}{\chi^2_{1-\alpha/2,n-1}} \tag{2.15}$$

The values of $\chi^2_{\alpha/2,n-1}$ and $\chi^2_{1-\alpha/2,n-1}$ may be obtained from the chi-square table in Appendix C. The one-sided upper and lower $100\,(1 - \alpha)\%$ confidence bounds for σ^2 are

$$\sigma^2 < \frac{(n-1)s^2}{\chi^2_{1-\alpha,n-1}} \qquad \text{(upper)}$$

$$\sigma^2 > \frac{(n-1)s^2}{\chi^2_{\alpha,n-1}} \qquad \text{(lower)}$$

The confidence intervals for σ are obtained by taking square roots.

To continue the candy bar example, if the standard deviation of thickness turned out to be 0.1 mm based upon $n = 4$, then $s = 0.1$ mm and $s^2 = 0.01\ (\text{mm})^2$. From the chi-square table in Appendix C, $\chi^2_{0.025,3} = 9.35$ and $\chi^2_{0.975,3} = 0.216$. From Equation 2.15, the 95% confidence interval for σ^2 is

$$\frac{3(0.01)}{9.35} \le \sigma^2 \le \frac{3(0.01)}{0.216}$$

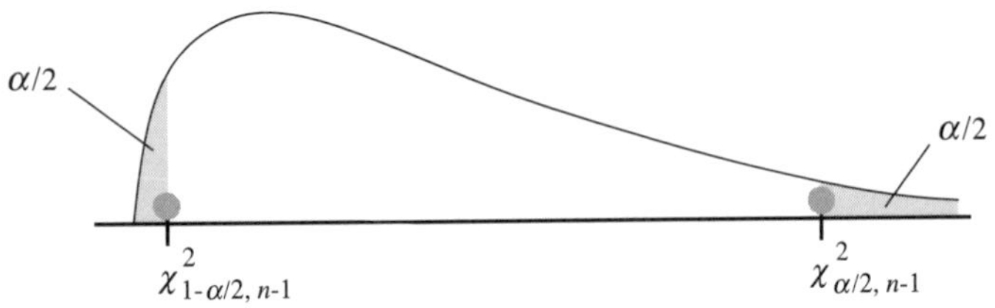

Figure 2.9 Chi-square distribution.

or

$$0.003 \le \sigma^2 \le 0.139$$

We are 95% sure that the population variance is between 0.003 and 0.139. By taking square roots, the 95% confidence interval for σ is

$$0.056 \le \sigma \le 0.373$$

The confidence intervals for σ^2 and σ are asymmetric around their estimates and are very wide. The confidence interval for σ has more than a six-fold range, indicating that we know very little about population σ based upon four observations.

2.3.3 Confidence Intervals for *p* and λ

Binomial *p*. For a binomial distribution, if x defective products are found in a random sample of n products, then the population fraction defective p is estimated by $\bar{p} = x/n$. An approximate confidence interval for p can be determined by using the normal approximation to the binomial. More exact confidence intervals are also available (Duncan; also see Box, Hunter, and Hunter).

Since the standard deviation of x is $\sqrt{np(1-p)}$, standard deviation of $\bar{p}$ is $\sqrt{p(1-p)/n}$, which is estimated by replacing the unknown p by $\bar{p}$. Hence, approximately,

$$100\,(1-\alpha)\%\text{ confidence interval for } p = \bar{p} \pm Z_{\alpha/2}\sqrt{\frac{\bar{p}(1-\bar{p})}{n}} \qquad (2.16)$$

The one-sided 100 $(1-\alpha)$% confidence bounds for p are:

$$p < \bar{p} + Z_{\alpha}\sqrt{\frac{\bar{p}(1-\bar{p})}{n}} \qquad \text{(upper)}$$

$$p > \bar{p} - Z_{\alpha}\sqrt{\frac{\bar{p}(1-\bar{p})}{n}} \qquad \text{(lower)}$$

As an example, if $\bar{p}$ is found to be 0.1 and $n = 200$, then the two-sided 95% confidence interval for p is

$$0.1 \pm 2\sqrt{0.1(0.9)/200} = 0.1 \pm 0.042 = 0.058 \text{ to } 0.142$$

This means that p is expected to be somewhere between 5.8% and 14.2% with 95% confidence. Even with 200 observations, the resultant confidence interval is very wide (a range of almost 2.5 to 1) indicating that a much larger sample size would be required to estimate p precisely.

The confidence interval formulae do not work if the number of defectives $x = 0$, because then $\bar{p} = 0$ and the confidence interval for p turns out to be zero from Equation 2.16. However, the fact that we did not observe any defective products in our random sample does not necessarily mean that p is actually zero. For the case of zero observed defectives, we need a different approach. Using the Poisson approximation to the binomial, we know that the probability of zero defectives is

$$P(x = 0) = e^{-\lambda}$$

where $\lambda = np$. If the observed $x = 0$, how large can p be? For 95% confidence, λ must be such that $e^{-\lambda}$ is at most 5%. This gives $\lambda = 3$. The upper limit for p is obtained by setting $\lambda = np_{\text{upper}} = 3$. For 90% confidence, $e^{-\lambda} = 0.1$ and $\lambda = np_{\text{upper}} = 2.3$. Hence,

$$p_{\text{upper}} = \frac{3}{n} \text{ (with 95\% confidence)}$$
$$p_{\text{upper}} = \frac{2.3}{n} \text{ (with 90\% confidence)} \qquad (2.17)$$

For example, if we observe zero defectives in 100 products, $p_{\text{upper}} = 3\%$ with 95% confidence. This means that we are 95% sure that the true fraction defective $p \leq 3\%$.

Poisson λ. For a Poisson distribution, the population mean λ, denoting mean number of defects per product, is estimated by $\bar{c} = \Sigma x_i/n$, where x_i is defects per product and n is the total number of products examined. Since the standard deviation of X is $\sqrt{\lambda}$, $\bar{c}$ comes from a distribution with a standard deviation of $\sqrt{\lambda/n}$, where the unknown λ is estimated by $\bar{c}$. Using the normal approximation of the Poisson distribution, the approximate

$$100\,(1 - \alpha)\% \text{ confidence interval for } \lambda = \bar{c} \pm Z_{\alpha/2}\sqrt{\frac{\bar{c}}{n}} \qquad (2.18)$$

The one-sided confidence bounds for λ are

$$\lambda < \bar{c} + Z_{\alpha}\sqrt{\frac{\bar{c}}{n}} \qquad \text{(upper)}$$

$$\lambda > \bar{c} - Z_{\alpha}\sqrt{\frac{\bar{c}}{n}} \qquad \text{(lower)}$$

What if after examining n products no defects are found? The probability of zero defects in a single product is $e^{-\lambda}$, hence the probability of zero defects in n products is $e^{-n\lambda}$. Therefore, the upper limit for λ is

$$\lambda_{\text{upper}} = \frac{3}{n} \text{ (with 95\% confidence)}$$
$$\lambda_{\text{upper}} = \frac{2.3}{n} \text{ (with 90\% confidence)} \qquad (2.19)$$

2.4 SAMPLE SIZE

For the conclusions to be meaningful, the sample must be of a certain size. We now explain how to determine the appropriate number of observations to meaningfully estimate the population parameters μ, σ, p, and λ.

2.4.1 Sample Size to Estimate μ

We want to determine the sample size n such that the unknown population mean μ is estimated with the desired degree of precision. The confidence interval for μ is the precision with which μ is being estimated. This confidence interval for μ is given by $\bar{x} \pm Z_{\alpha/2}\sigma/\sqrt{n}$. If we want to estimate μ within $\pm\Delta$ from the true value, then $Z_{\alpha/2}\sigma/\sqrt{n} = \Delta$ which leads to the estimate of sample size:

$$n = \left(\frac{Z_{\alpha/2}}{d}\right)^2 \quad \text{where } d = \frac{\Delta}{\sigma} \tag{2.20}$$

Note that n depends not on the individual values of Δ and σ but on their ratio. The procedure to determine n is as follows:

1. Assume a value for σ. This may be based upon prior data or judgment.
2. Specify Δ, the level of acceptable uncertainty regarding μ.
3. Calculate $d = \Delta/\sigma$.
4. Select the desired level of confidence and obtain the value of $Z_{\alpha/2}$.
5. Compute $n = (Z_{\alpha/2}/d)^2$.

For some practical values of d and % confidence, the number of observations to estimate μ are summarized in Table 2.1.

As an example, suppose we wish to estimate the population mean hardness μ of a product within ± 0.05 from the true value. The standard deviation of hardness is judged to be 0.15. How much data should be collected? Since $d = \Delta/\sigma = 0.33$, for 95% confidence, $Z_{\alpha/2} = 1.96$ and $n = (1.96/0.33)^2 = 36$.

How well should we know μ? Why was Δ selected to be 0.05 for the hardness example? The answers depend upon the consequence of misjudging μ. For example, if μ is not on target, what will be the consequence? Will it result in a significant loss in consumer satisfaction? Will it result in the product being outside specification? Considerations such as these drive the selection of Δ.

Table 2.1 Number of Observations to Estimate μ

	Sample Size n		
d	90% Confidence	95% Confidence	99.7% Confidence
1.00	3	4	9
0.75	5	7	16
0.50	11	16	36
0.33	25	36	83
0.25	44	62	144
0.20	68	97	225
0.10	271	385	900

2.4.2 Sample Size to Estimate σ

From Equation 2.15, the confidence interval for the standard deviation σ is given by

$$\sqrt{\frac{(n-1)s^2}{\chi^2_{\alpha/2,n-1}}} \leq \sigma \leq \sqrt{\frac{(n-1)s^2}{\chi^2_{1-\alpha/2,n-1}}}$$

which can be restated as the confidence interval for σ as a percentage of s as follows:

Table 2.2 95% Confidence Limits for 100 (σ/s)

Sample Size	Lower Limit	Upper Limit
3	52	626
4	57	373
6	62	245
8	66	204
10	69	183
16	74	155
21	77	144
31	80	134
41	82	128
51	84	124
71	86	120
101	88	116
200	90	110
500	94	106
1000	96	104
5000	98	102

$$100\sqrt{\frac{(n-1)}{\chi^2_{\alpha/2,n-1}}} \leq \frac{100\sigma}{s} \leq 100\sqrt{\frac{(n-1)}{\chi^2_{1-\alpha/2,n-1}}} \qquad (2.21)$$

For various values of n and 95% confidence ($\alpha = 0.05$), Table 2.2 shows these confidence limits for σ as a percentage of s. Note that the limits are not symmetric. Equation (2.21) was used for values of n up to 101. After that, the following approximation was used. The standard deviation of S as a percentage of σ is approximately given by $100/\sqrt{2n}$ (Box, Hunter, and Hunter). In constructing Table 2.2 for $n \geq 200$, approximate symmetric two-sigma limits were used to construct the 95% confidence interval.

Table 2.2 may be used to select the sample size to estimate σ. Suppose we want to know σ within 20% of the true value. Then for 95% confidence, the approximate sample size is 50. If we wish to know σ within 10% of the true value, the approximate sample size is 200, and so on. A formula could be used to approximately estimate the sample size. If we want to know σ within δ% of the true value, then, approximately, for 95% confidence

$$n = 2\left(\frac{100}{\delta}\right)^2 \qquad (2.22)$$

As examples, for δ = 20%, 10%, 5%, and 2% Equation (2.22) gives values of n of 50, 200, 800, and 5000, respectively. For 90% confidence, the sample sizes will be approximately 33% smaller. Knowing σ well requires very large number of observations.

2.4.3 Sample Size to Estimate p and λ

Binomial p. If x (> 0) is the number of defectives in n products, then $\bar{p} = x/n$ provides an estimate of p. The approximate confidence interval for p is $\bar{p} \pm Z_{\alpha/2}\sqrt{\bar{p}(1-\bar{p})/n}$. If we wish to estimate p within $\pm\,\Delta$, then $\Delta = Z_{\alpha/2}\sqrt{\bar{p}(1-\bar{p})/n}$ and, approximately,

$$n = \frac{(Z_{\alpha/2})^2\,\bar{p}(1-\bar{p})}{\Delta^2} \qquad (2.23)$$

Since the sample size is to be determined prior to collecting data, $\bar{p}$ is unknown and is replaced by our initial guess for p. Sample sizes to estimate p generally turn out to be large. For example, if the fraction defective p is expected to be around 5% and we wish to estimate it within ± 1% from the true value with 95% confidence then the sample size is

$$n = \frac{(2)^2(0.05)(0.95)}{(0.01)^2} = 1900$$

What if $x = 0$? It has been shown earlier that we can calculate an upper limit for p. For example, $p_{\text{upper}} = 3/n$ for 95% confidence and $p_{\text{upper}} = 2.3/n$ for 90% confidence. This gives the sample size as

$$n = \frac{3}{p_{\text{upper}}} \text{ (for 95\% confidence)}$$
$$n = \frac{2.3}{p_{\text{upper}}} \text{ (for 90\% confidence)} \tag{2.24}$$

If we wish to demonstrate with 95% confidence that the fraction defective $p \leq 1\%$, then, from Equation (2.24), $n = 3/0.01 = 300$. This means that if zero defectives are found in 300 randomly selected products, then we are 95% confident that $p \leq 1\%$. The zero defectives formula given by Equation (2.24) leads to the smallest possible sample size to demonstrate a certain level of p.

Poisson λ. The confidence interval for λ is $\bar{c} \pm Z_{\alpha/2}\sqrt{\bar{c}/n}$, where $\bar{c}$ is the average number of defects per product based upon examining n products. If we want to estimate λ within $\pm\,\Delta$, then $\Delta = Z_{\alpha/2}\sqrt{\bar{c}/n}$ and

$$n = \frac{(Z_{\alpha/2})^2\,\bar{c}}{\Delta^2} \tag{2.25}$$

Since the sample size is to be determined prior to collecting data, $\bar{c}$ is replaced by our initial estimate of λ.

What if $\bar{c} = 0$? It has been shown earlier that in this case $\lambda_{\text{upper}} = 3/n$ with 95% confidence and $\lambda_{\text{upper}} = 2.3/n$ with 90% confidence. This gives the sample size as

$$n = \frac{3}{\lambda_{\text{upper}}} \text{ (For 95\% confidence)}$$
$$n = \frac{2.3}{\lambda_{\text{upper}}} \text{ (For 90\% confidence)} \tag{2.26}$$

This zero defects formula for n leads to the smallest possible sample size to demonstrate a certain value of λ.

2.5 TOLERANCE INTERVALS

Tolerance intervals differ from confidence intervals. Whereas confidence intervals provide limits within which the parameters of a distribution are expected to be, tolerance intervals provide limits within which a certain proportion $(1 - p)$ of the population is expected to be. p is the fraction of the population outside the tolerance interval. For small values of p, if the tolerance limits are within specification limits, the process is expected to produce a large proportion of acceptable products. Otherwise, the fraction defective may be large. Thus, tolerance intervals are often used to validate the process. In this sense, they have a connection with process capability indices. Tolerance intervals may also be used to establish capability-based specifications, which we will explore in Chapter 5.

If the process is assumed to be normally distributed with known μ and σ, then an interval that contains $100(1 - p)\%$ of the population is given by $\mu \pm Z_{p/2}\sigma$. Such an interval may be called a $100{:}100(1 - p)$ tolerance interval meaning that we are 100% confident that the tolerance interval contains $100(1 - p)\%$ of the population. The value of p can be chosen to be suitably small to include a large proportion of the population within the tolerance interval.

With μ and σ unknown, we can no longer be 100% sure that any finite interval contains $100(1 - p)\%$ of the population. However, it is possible to construct a tolerance interval to contain $100(1 - p)\%$ of the population with $100(1 - \alpha)\%$ confidence. Such an interval is called a $100(1 - \alpha){:}100(1 - p)$ tolerance interval and is given by

$$100(1 - \alpha){:}100(1 - p) \text{ tolerance interval} = \bar{x} \pm ks \qquad (2.27)$$

The values of the constant k depend upon percentage confidence, percentage of the population contained within the tolerance interval, and the sample size. For various combinations of α, p, and n, values of k have been tabulated in Appendix D for both two-sided and one-sided tolerance intervals.

Continuing with the candy bar thickness example, if we assume that μ and σ are known to be 3 mm and 0.1 mm, respectively, then the 100:95 tolerance interval is 3 ± 0.2 mm. On the other hand, based upon 10 observations, if the observed $\bar{x} = 3$ mm and $s = 0.1$ mm then using Equation (2.27) and k values from Appendix D

$$95{:}95 \text{ tolerance interval} = 3 \pm 3.379\,(0.1) = 3 \pm 0.338$$

$$99{:}95 \text{ tolerance interval} = 3 \pm 4.265\,(0.1) = 3 \pm 0.426$$

The penalty for not knowing μ and σ is quite substantial and leads to a doubling of the width of the tolerance interval.

As the sample size increases, the value of k approaches $Z_{p/2}$. What should the sample size be? For the selected values of α and p, we can determine the percentage increase in the value of k over the value of k corresponding to $n = \infty$. This provides us with some guidance regarding n. For example, in constructing a 95:95 tolerance interval, if we do not want the value of k to be more than 20% larger than the absolute minimum, then $n = 50$ because from Appendix D the minimum k is approximately 2 and the k value corresponding to $n = 50$ is approximately 2.4.

2.6 NORMALITY, INDEPENDENCE, AND HOMOSCEDASTICITY

The statistical analyses considered in this book make three key assumptions. This section describes these assumptions, provides ways of checking them, and discusses approaches to take if the assumptions are not satisfied.

Consider the case in which data have been collected on the bond strength of a product manufactured under fixed conditions. Such data may be described by the following equation or model:

$$x = \mu + e$$

where x is the observed bond strength, μ is the population mean bond strength, and e is called error, or the deviation of the observed bond strength from the true mean μ. Statistical analyses considered in this book make the following key assumptions regarding the statistical properties of error. Note that the assumptions pertain to errors and not necessarily to the data.

1. The errors are normally distributed.
2. The errors are independent.
3. The errors have zero mean and constant standard deviation.

These assumptions are usually denoted by $e \sim NID(0, \sigma)$. The assumption of zero mean is easily satisfied by a proper estimation of μ so the three key assumptions are that the errors are normally, independently distributed with a constant variance σ^2. Constancy of variance is referred to as homoscedasticity.

For the simple case considered above, it follows that the observations (x) themselves must be independent drawings from a normal distribution with mean μ and variance σ^2. Such need not always be the case. For example, if bond strength depends upon temperature (T) then the relationship may be

$$x = \alpha_0 + \alpha_1 T + e$$

and even if $e \sim NID(0, \sigma)$, bond strength will not have a normal distribution if the data are collected for varying temperatures. So the three key assumptions apply to error and, depending upon the model, may or may not apply to the data.

Significant departures from these assumptions can cause large errors in some statistical analyses; other analyses may be influenced to a smaller extent. Examples of conclusions that may be seriously in error include computation of tolerance limits, computation of control limits for the chart of individuals, translation of the capability index to fraction defective, and significance levels in hypotheses testing. On the other hand, the design of an $\bar{X}$ chart will be influenced to a much smaller extent. It is best to check these assumptions and take appropriate actions when the assumptions are seriously violated.

2.6.1 Normality

Checking Normality. There are many tests to check normality. The chi-square test described below is based upon a comparison of the observed histogram frequencies to the corresponding theoretical normal distribution frequencies. The procedure is as follows:

1. Determine the mean and variance from the data. The theoretical distribution is assumed to be a normal distribution with this mean and variance.
2. Classify the data into k cells (like a histogram) such that the expected frequency based upon the fitted normal distribution is at least five for each cell.
3. Calculate

$$\text{chi square} = \frac{\sum_{i=1}^{k} (O_i - E_i)^2}{E_i} \tag{2.28}$$

Table 2.3 Chi-Square Test

Cell	Observed Frequency (O_i)	Expected Frequency (E_i)	$(O_i - E_i)^2/E_i$
≤ 0.59	4	5.1	0.2431
0.59–0.615	4	5.5	0.3965
0.615–0.64	11	7.2	1.9494
0.64–0.665	7	7.3	0.0113
0.665–0.690	6	5.6	0.0325
> 0.690	4	5.3	0.3224
		Computed chi square	2.9552

where
O_i = observed frequency in cell i
E_i = expected frequency in cell i

4. Compare the calculated value of chi square with the critical chi-square value for $k - 3$ degrees of freedom. Note that one degree of freedom each is lost due to mean, variance, and frequency. For the weight data shown below, Table 2.3 shows the computations for the chi-square test. The data are: 0.56, 0.57, 0.58, 0.59, 0.60, 0.60, 0.60, 0.61, 0.62, 0.63, 0.63, 0.63, 0.63, 0.64, 0.64, 0.64, 0.64, 0.64, 0.64, 0.65, 0.65, 0.65, 0.65, 0.65, 0.66, 0.66, 0.67, 0.67, 0.68, 0.68, 0.69, 0.69, 0.70, 0.70, 0.72, 0.72. Since there are six cells, the degrees of freedom are (6 – 3) = 3. From the chi-square table in Appendix C, the critical chi-square value for 95% confidence and three degrees of freedom is 9.35. The computed chi-square value of 2.95 is less than the critical value and the assumption of normal distribution cannot be rejected.

Measures of Nonnormality. The test for normality is often supplemented by two commonly used measures of nonnormality called skewness and kurtosis.

The lack of symmetry of the distribution is measured by skewness. It is calculated as follows:

$$\text{Skewness} = \frac{\Sigma(x_i - \bar{x})^3}{ns^3} \tag{2.29}$$

If the distribution is symmetric, the positive and negative cubed deviations cancel each other and the expected value of skewness is zero. For positively skewed distributions with the long tail going toward $+\infty$, skewness is a positive number. For negatively skewed distributions with the long tail going toward $-\infty$, skewness is a negative number.

A distribution may be symmetric but it may or may not be perfectly bell-shaped. For example, a uniform distribution is rectangular; it is symmetric, with zero skewness, but not bell-shaped. Kurtosis measures the departures from the perfect bell shape. It is calculated as follows:

$$\text{Kurtosis} = \frac{\Sigma(x_i - \bar{x})^4}{ns^4} - 3 \tag{2.30}$$

For a normal distribution the expected value of kurtosis is zero. If the distribution is more peaked than normal, kurtosis is positive. If the distribution is flatter than normal, kurtosis is negative.

Dealing with Nonnormality. If the data are nonnormal, what steps should be taken? These steps are briefly described below.

1. Wherever possible, we should try to identify the reason for nonnormality. Perhaps the characteristic being measured is not expected to be normally distributed. If the distribution is expected to be normal and turns out to be not so, then the section on histograms describes many potential causes to consider. If such a cause is found, it should be corrected.
2. The next step is to assess the impact of nonnormality on the analysis we intend to do. Not every analysis and conclusion is equally sensitive to departures from normality. In general, if the conclusions have to do with the middle of the distribution, such as an $\bar{X}$ chart, the conclusions are robust against departures from normality. If the conclusions have to do with the tail of the distribution, such as estimating the fraction defective, the conclusions are sensitive to departures from normality.
3. One approach is to fit an appropriate nonnormal distribution (Johnson and Kotz), such as an exponential, lognormal, or gamma distribution, to the data and then do the analysis using this nonnormal distribution. This approach is not always easy. Distribution-free analyses can also be done but such analyses can lead to wide uncertainties regarding conclusions and, consequently, large sample sizes.
4. A nonnormal distribution can be transformed into a normal distribution by an appropriately chosen transformation $Y = f(X)$. The analysis can then be done in terms of Y, which has a normal distribution. The final results can be reverse transformed and reported in terms of X. Such an approach is appealing when the transformation makes physical sense such as the log transformation in dealing with particle size data, which may be log-normally distributed.

The following illustrates other situations in which a transformation can correct departures from normality. Suppose that the equation for the observed values x is

$$x = \sqrt{\mu + e}$$

where $e \sim NID(0, \sigma)$. Clearly,

$$x^2 = \mu + e$$

and will have a normal distribution. Similarly, if x represents binomial proportion with mean μ based on samples of size n, then

$$y = \text{arc}\ \sin\sqrt{x}$$

is approximately normally distributed with mean arc $\sin\sqrt{\mu}$ and variance equal to $0.25/n$.

2.6.2 Independence

The errors are said to be independent if knowledge of some error values tells us nothing about the remaining error values. Lack of independence often occurs because of correlations in time or space. For example, if time-ordered observations are collected on the output of a chemical process, the cyclic nature of the process could cause successive observations to be highly positively correlated. If one observation is high, the next observation is also likely to be high. If the equation $x_t = \mu + e_t$ is used to describe the observations, where t represents time, the errors will also be highly positively correlated and the assumption of independence will fail.

This assumption can be tested by computing correlation coefficients (see Chapter 3) between successive observations, observations two lags apart, and so on. If the correlation coefficients are large, the assumption of independence is violated. For serially correlated data, techniques such as time series analysis may be used to obtain equations in which the errors are independent. In general, violation of independence is difficult to correct and precautions such as random sampling should be taken.

2.6.3 Homoscedasticity

There are several situations in which the assumption that error variance is constant (homoscedasticity) can be violated. For example, if we wish to compare two formulations, it is possible that the formulations differ in terms of their variability. Tests to compare two or more variances are described in Chapter 3.

If the observed data show evidence of a systematic relationship between mean and variance, then constant variance may be achieved through an appropriate transformation of the data. These variance-stabilizing transformations are known as the Box–Cox transformations (Box, Hunter, and Hunter) and are shown in Table 2.4.

As an example, suppose we measure bond strength at three different temperatures—low, medium, and high—and it turns out that the standard deviation of bond strength is proportional to the mean bond strength. If we build the functional relationship

$$\text{Bond Strength} = f_1(T) + e$$

Table 2.4 Variance Stabilizing Transformations

Transformation	Relationship between σ and μ
$Y = 1/X$	$\sigma \propto \mu^2$
$Y = 1/\sqrt{X}$	$\sigma \propto \mu^{1.5}$
$Y = \log X$	$\sigma \propto \mu$
$Y = \sqrt{X}$	$\sigma \propto \mu^{0.5}$
$Y = X$	$\sigma \propto$ constant
$Y = X^2$	$\sigma \propto \mu^{-1}$
$Y = \text{arc sin}\sqrt{X}$	X is fraction defective

where T represents temperature, then error variance will not be constant. From Table 2.4, if we build the relationship

$$\log(\text{Bond Strength}) = f_2(T) + e$$

then error variance will more nearly be constant.

3

Comparative Experiments and Regression Analysis

This chapter presents statistical tools to design and analyze comparative experiments and to build equations relating input and output factors. An understanding of these tools of comparative experimentation and regression analysis is very helpful in answering questions such as:

1. A process change was made. Is the variability after the change smaller than variability before the change? How much smaller?
2. A certain car is rated at 30 mpg. On five trips, each with a tank-full of gas, the miles per gallon were: 28.5, 29.3, 30.7, 29.8, and 28.9. Is the rating justified?
3. In 1954, the incidence of polio was three in 10,000. The Salk polio vaccine was expected to reduce the incidence rate by half. How many people should participate in a double-blind test to prove the assertion?
4. Data are available on the performance of multiple machines. Do the machines perform alike?
5. Oven temperature was thought to control product moisture. However, based upon data routinely collected in manufacturing, the correlation coefficient between temperature and moisture turned out to be –0.3. Does this mean that oven temperature has little effect on moisture?

Statistical Methods for Six Sigma by Anand M. Joglekar

ISBN 0-471-20342-4

6. How can we identify key input factors and build predictive equations relating these factors to outputs of interest?

The chapter begins with a discussion of the hypothesis-testing framework commonly used to design and analyze comparative experiments. Statistical tests, including the t-test, F-test, and ANOVA are presented to compare one, two, and multiple populations. The use of hypothesis tests based sample size formulae to design the experiment and the confidence interval approach to analyze the experiment is recommended in practice. The chapter also considers the topic of correlation and regression analysis, whose main purpose is to obtain an equation relating a dependent factor to independent factors. Data are often available on multiple input and output factors and regression analysis becomes a useful tool to unearth relationships between these factors.

3.1 HYPOTHESIS TESTING FRAMEWORK

The hypothesis-testing framework can be qualitatively explained by considering the American judicial system. If a person is accused of a crime, the initial hypothesis is that the person is not guilty. This is called the null hypothesis, H_0. The alternate hypothesis, H_1, is that the person is guilty. Between the two hypotheses, all possibilities are exhausted. In order to determine which hypothesis is more believable, we collect data by conducting a trial. Prior to data collection, decision rules have to be established: the trial will be a jury trial, a unanimous verdict is necessary to convict, certain evidence cannot be presented, and so on. Based upon the data and the decision rules, a decision is made and the person is declared either guilty or not guilty. There are two ways in which the decision may be wrong. The two probabilities of wrong decision are called the α and β risks defined as:

α = Probability that an innocent person goes to jail

β = Probability that a guilty person goes free

In this case the α-risk is generally considered to be the more critical one and should be minimized. It is possible to make any one of the two risks zero. For example, α risk will be zero if we do not have a trial at all! Of course, in this case, the β risk will be 100%. To make both risks small, we must have almost perfect knowledge of what really happened; namely, we need a large amount of data. So the sample size is determined based upon acceptable levels of risks. To summarize, the hypothesis testing framework is to set forth two contradictory and exhaustive hypotheses, establish a decision-making procedure, determine the amount of data to be collected based upon acceptable levels of the two risks of wrong decisions, collect data, and make a decision.

Let us now consider the hypothesis test procedure quantitatively. A certain raw material lot will be accepted if the average bond strength exceeds 5 lbs. If the average bond strength is much less than 5 lbs, we will reject the lot. We want to test the raw material lot to determine if it should be accepted or rejected.

Hypotheses. In order to formulate the hypotheses correctly, we must define the objectives of data collection precisely. For example, consider the following four objectives:

1. We want to estimate the true average bond strength μ within ± 0.1 lbs.
2. We want to know if μ is within 5 ± 0.1 lbs.
3. We want to compare a new material to the old material and assess whether μ for the new material exceeds that for the old material by 0.1 lbs.
4. We want to identify the raw material with the largest true bond strength from the collection of raw materials being tested.

While these objectives may look similar, they represent different hypotheses, require different sample sizes and analyses.

For the bond strength example under consideration, the null hypothesis is that the mean bond strength μ for the lot of raw material is greater than or equal to 5 lbs. The alternate hypothesis is that μ is less than 5 lbs.

$$H_0 : \mu \geq 5 \qquad \text{and} \qquad H_1 : \mu < 5$$

Test Procedure. A test procedure is a rule based upon collected data to decide whether or not to reject H_0. It consists of a test statistic and a rejection region. For example, from the lot of raw material under consideration, we may decide to take $n = 25$ random samples and measure their bond strength. From this data, the average bond strength $\bar{x}$ can be determined and constitutes the test statistic. If the true bond strength is 5 lbs, the computed $\bar{x}$ could well be somewhat less than 5 lbs because of variability. So we may set up a rejection region as follows:

If $\bar{x} < k = 4.5$, reject H_0

Otherwise, do not reject H_0

Values of n and k define the test procedure.

The Two Risks. Once the decision is made, it may be wrong in two ways. The associated risks are:

$$\alpha = \text{Probability of rejecting } H_0 \text{ when it is true.}$$
$$\beta = \text{Probability of accepting } H_0 \text{ when it is false.} \tag{3.1}$$

Presently, the two risks are:

α = Probability of rejecting a good lot

β = Probability of accepting a bad lot

How big should these risks be? The answer depends upon the consequences of the risks. If we reject a good raw material lot that is already paid for, it will cost us at least the price of the lot. If we accept a bad raw material lot, we may end up producing a bad lot of finished product, which will cost us much more. In this case, the β risk is more important than the α risk and we may decide to take a β risk of 1% and an α risk of 5%.

Computing α and β Risks. To compute the α and β risks, specific values of μ are selected for H_0 and H_1. For the bond strength example, we may select $\mu = 5$ for H_0 because an average bond strength of 5 lbs is acceptable. We may select $\mu = 4.2$ for H_1 if a bond strength of 4.2 lbs is sure to result in bad finished product. The current test procedure is to take 25 observations per lot and reject the lot if $\bar{x} < 4.5$ lbs. Let us assume that the individual bond strength measurements are normally distributed with a known standard deviation $\sigma = 1$. Then $\sigma_{\bar{x}} = 0.2$ and the α and β risks can be computed from the properties of normal distribution as follows. If P denotes probability, then

$$\alpha = P(\bar{x} < 4.5 \text{ when } \mu = 5) = 0.62\%$$

$$\beta = P(\bar{x} > 4.5 \text{ when } \mu = 4.2) = 6.68\%$$

The risks are graphically shown in Figure 3.1. It is easy to see from Figure 3.1 that if we change the value of k, the risks change. If k is changed from 4.5 to 4.7, α risk increases and β risk reduces. Changing k increases one risk and reduces the other.

For a fixed value of k, increasing the sample size n simultaneously

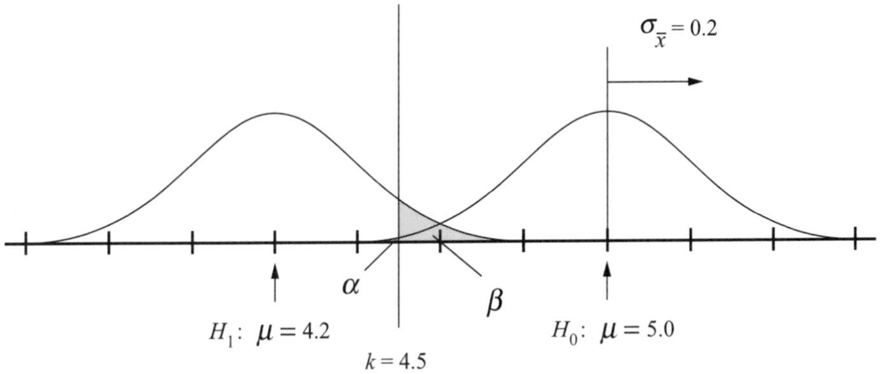

FIGURE 3.1
Computing α and β risks.

reduces both risks. For example, if n = 36, $\sigma_{\bar{X}}$ = 0.167, which leads to α = 0.13% and β = 3.6%. As shown in Figure 3.2, increasing n reduces both risks and changing k changes the balance between risks. Thus by a proper choice of n and k, any desired values of α and β risks can be obtained.

Determining Sample Size and Rejection Region. The procedure to determine n and k is illustrated below for the bond strength example. The two hypotheses are $H_0 : \mu \geq \mu_0$ and $H_1 : \mu = \mu_1 < \mu_0$. The test procedure is to reject H_0 when $\bar{x} < k$. The two risks are:

$$\alpha = P(\bar{x} < k \text{ when } \mu = \mu_0) = P\left(\frac{\bar{x} - \mu_0}{\sigma/\sqrt{n}} < \frac{k - \mu_0}{\sigma/\sqrt{n}}\right) = P\left(Z < \frac{k - \mu_0}{\sigma/\sqrt{n}}\right)$$

$$\beta = P(\bar{x} > k \text{ when } \mu = \mu_1) = P\left(\frac{\bar{x} - \mu_1}{\sigma/\sqrt{n}} > \frac{k - \mu_1}{\sigma/\sqrt{n}}\right) = P\left(Z > \frac{k - \mu_1}{\sigma/\sqrt{n}}\right)$$

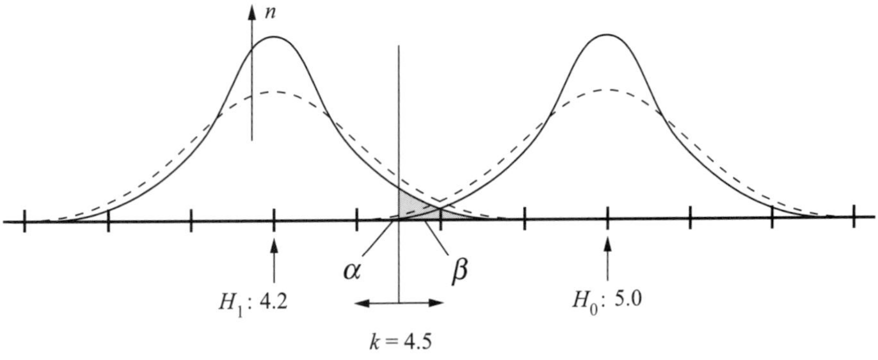

FIGURE 3.2
Effect of n and k on α and β.

It follows that

$$\frac{k - \mu_0}{\sigma/\sqrt{n}} = Z_\alpha \quad \text{and} \quad \frac{k - \mu_1}{\sigma/\sqrt{n}} = Z_\beta$$

The above equations can be simultaneously solved to obtain n and k.

$$n = \left[\frac{\sigma(Z_\alpha + Z_\beta)}{(\mu_0 - \mu_1)} \right]^2$$
$$k = \mu_1 + Z_\beta \sigma/\sqrt{n} = \mu_0 - Z_\alpha \sigma/\sqrt{n} \tag{3.2}$$

For the bond strength example, $\sigma = 1$, $\mu_0 = 5$, $\mu_1 = 4.2$, $\alpha = 5\%$, and $\beta = 1\%$. Therefore, $Z_\alpha = 1.64$, $Z_\beta = 2.33$ and from Equation (3.2), $n \approx 25$ and $k \approx 4.67$.

3.2 COMPARING SINGLE POPULATION

We begin by considering hypothesis tests to compare parameters of a single population, such as μ, σ, and fraction defective p, to specified values. For example, viscosity may be an important characteristic in a process validation experiment and we may want to determine if the population standard deviation of viscosity is less than a certain value or not. Additional examples of such comparisons are suggested by the following questions.

1. Is the process centered on target? Is the measurement bias acceptable?
2. Is the measurement standard deviation less than 5% of the specification width? Is the process standard deviation less than 10% of the specification width?
3. Let p denote the proportion of objects in a population that possess a certain property such as products that exceed a certain hardness, or cars that are domestically manufactured. Is this proportion p greater than a certain specified value? A local newspaper stated that less than 10% of the rental units did not allow renters with children. Is the newspaper claim justified?

3.2.1 Comparing Mean (Variance Known)

Z Test. If we wish to determine whether the population mean μ is equal to μ_0 or not, then the two hypotheses are $H_0 : \mu = \mu_0$ and $H_1 : \mu \neq \mu_0$. If the population standard deviation is assumed to be known, then from Chapter 2, under the null hypothesis, the test statistic

$$Z = \frac{\overline{X} - \mu_0}{\sigma/\sqrt{n}} \tag{3.3}$$

has a standard normal distribution with zero mean and a standard deviation of one. By taking random observations from the population, the Z value can be calculated. If the calculated Z value is close to zero, H_0 cannot be rejected. But if the calculated Z value exceeds the critical Z value, i.e., falls in the rejection region; the null hypothesis is rejected. This Z-test applies when the variance is known from prior data or when the sample size n exceeds 30, in which case the computed s is used as σ. The rejection region depends upon the specific hypotheses as stated below:

Hypotheses		Rejection Region
$H_0 : \mu = \mu_0$	$H_1 : \mu \neq \mu_0$	$Z_{\alpha/2} \leq Z \leq -Z_{\alpha/2}$
$H_0 : \mu \geq \mu_0$	$H_1 : \mu < \mu_0$	$Z < -Z_\alpha$
$H_0 : \mu \leq \mu_0$	$H_1 : \mu > \mu_0$	$Z > Z_\alpha$

If the computed value of Z falls in the rejection region, then the null hypothesis is rejected with 100 (1 – α)% confidence, and the difference is said to be statistically significant with 100(1 – α)% confidence.

Example. A new type of construction material will be used if the true average compressive strength exceeds 4000 psi. Based upon 36 observations, the average compressive strength $\bar{x} = 3990$ and standard deviation of compressive strength $s = 120$. Should we not use the material?

Let μ be the true average compressive strength. We want to test the hypotheses $H_0 : \mu \geq 4000$ versus $H_1 : \mu < 4000$. With $\mu_0 = 4000$, the computed value of Z is

$$Z = \frac{\bar{x} - \mu_0}{\sigma/\sqrt{n}} = \frac{3990 - 4000}{120/\sqrt{36}} = -0.5$$

Since the alternate hypothesis has a < sign, the rejection region is $Z < -Z_\alpha$. For an α risk of 5%, $-Z_\alpha = -1.64$. Since the computed $Z = -0.5$ is greater than –1.64, the null hypothesis that the average compressive strength μ is at least 4000, cannot be rejected with 95% confidence.

Confidence Interval. The 100 (1 – α)% confidence interval for μ is

$$\bar{x} \pm Z_{\alpha/2}\frac{\sigma}{\sqrt{n}} \tag{3.4}$$

For the above example, the 95% confidence interval for μ is

$$3990 \pm 2\frac{120}{\sqrt{36}} = 3990 \pm 40 = 3950 \text{ to } 4030$$

Since the confidence interval includes 4000, we cannot reject the null hypothesis with 95% confidence. While the confidence interval and the Z-test lead to the same conclusion with respect to accepting or rejecting the null hypothesis, the confidence interval provides greater information regarding whether the differences are practically meaningful or practically significant and also whether additional data should be collected. As shown in Figure 3.3, the confidence interval and the Z-test lead to the same practical conclusions for cases (A) and (D). However, for

Case (B): The confidence interval tells us that even though $\mu <$ 4000, the difference is very small and may be practically unimportant.

Case (C): The confidence interval suggests that more data should be collected because the uncertainty regarding μ is very large and μ could be substantially smaller than 4000.

Determining Sample Sizes. For the compressive strength example, what if the real average compressive strength is 3980 psi? With a sample size of 36, what is the probability that we will wrongly conclude that

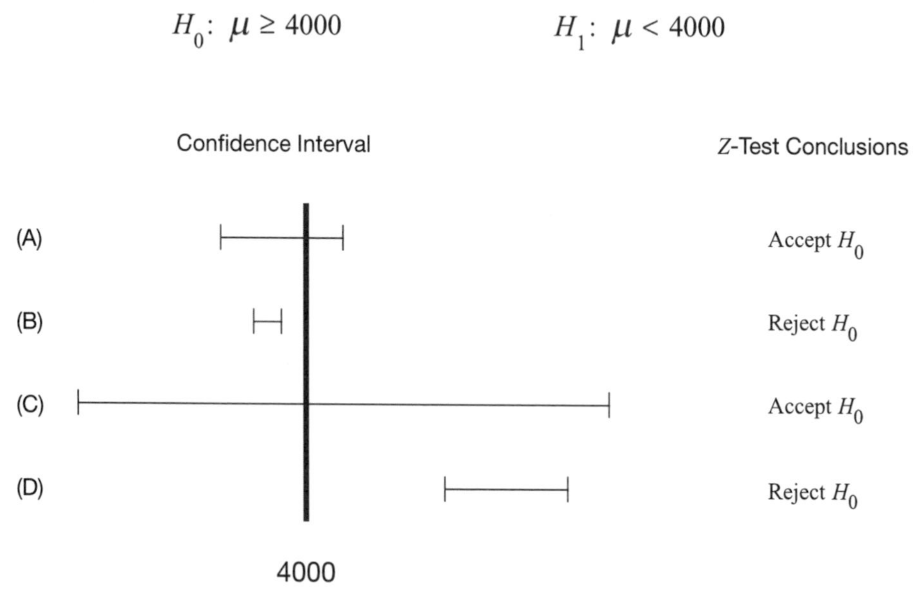

FIGURE 3.3 Confidence interval is better than a Z-test.

the compressive strength is at least $\mu_0 = 4000$ psi? This brings us to the β risk.

$$\beta = P(\text{accept } H_0 \text{ when } \mu = 3980)$$

H_0 is accepted whenever

$$Z = \frac{\bar{x} - \mu_0}{\sigma/\sqrt{n}} = \frac{\bar{x} - 4000}{20} > -1.64 \text{ or } \bar{x} > 3967$$

Hence, $\beta = P(\bar{x} > 3967 \text{ when } \mu = 3980)$

$$\beta = P\left(\frac{\bar{x} - 3980}{20} = Z > \frac{3967 - 3980}{20} = -0.64\right) = 0.74$$

This means that there is a 74% probability of concluding that the mean compressive strength is 4000 psi when it is actually 3980 psi if the sample size is 36. What is the appropriate sample size to use? The sample size formula given by Equation (3.2) may be written as

$$n = \frac{(Z_\alpha + Z_\beta)^2}{d^2} \quad \text{for a one-sided test}$$
$$n = \frac{(Z_{\alpha/2} + Z_\beta)^2}{d^2} \quad \text{for a two-sided test} \tag{3.5}$$

where $d = \Delta/\sigma$ and Δ is the smallest difference in the means $|\mu_0 - \mu_1|$ we wish to detect. The one-sided and two-sided tests refer to whether the alternate hypothesis is one- or two-sided.

For the compression test example, if we want to detect a difference in mean of $\Delta = 20$ with an α-risk of 10% and a β-risk of 5%, then $\mu_0 = 4000$, $\mu_1 = 3980$, $Z_\alpha = 1.28$, $Z_\beta = 1.64$, $\sigma = 120$, and $d = 20/120 = 0.167$. Hence,

$$n = \frac{(Z_\alpha + Z_\beta)^2}{d^2} = \frac{(1.28 + 1.64)^2}{(0.167)^2} = 307$$

The sample size will have to be 307 to reduce the β-risk from 74% to 5%.

In making comparisons, the test should be properly designed by computing the necessary sample size, and once the data are collected decisions should be made on the basis of confidence intervals.

3.2.2 Comparing Mean (Variance Unknown)

***t*-Test.** When the population variance is unknown, σ is replaced by the sample standard deviation s and the test statistic

$$t = \frac{\bar{X} - \mu_0}{S/\sqrt{n}} \tag{3.6}$$

has a t-distribution with $(n - 1)$ degrees of freedom.

The t-distribution looks like a normal distribution but has thicker tails when compared with the normal distribution. Critical values of t-distribution, denoted by $t_{\alpha,n-1}$, are values such that the probability of a t_{n-1} distribution exceeding the critical value is α. These are tabulated in Appendix B for various values of α and degrees if freedom $\nu = n - 1$. The thicker tails mean that the critical value $t_{\alpha,n-1}$ is larger than Z_α and when the variance is unknown, the difference between the sample mean and the specified value has to be greater than in the case of known variance to be considered statistically significant. With sample size $n > 30$, the t values approach the Z value (see Appendix B) and the difference between the t-test and Z-test becomes practically small. So it is important to use this t-test when $n < 30$.

Example. A certain car is rated at 30 mpg. On five trips, each with a tank-full of gas, the miles per gallon were 28.4, 29.2, 30.9, 29.8, and 28.6. Is the rating justified? From the data, $n = 5$, $\bar{x} = 29.4$, and $s = 1.0$. The hypotheses being tested are $H_0 : \mu = 30.0$ versus $H_1 : \mu \neq 30.0$. The computed t value is

$$t = \frac{29.4 - 30.0}{1/\sqrt{5}} = -1.34$$

For $\alpha = 0.05$, from the t-table in Appendix B, $t_{\alpha/2,n-1} = 2.776$. Since the computed value of t is between -2.776 and $+ 2.776$, the rating of 30 mpg cannot be rejected.

Confidence Interval. The 100 $(1 - \alpha)$% confidence interval for μ is

$$\bar{x} \pm t_{\alpha/2,n-1} \frac{s}{\sqrt{n}} \tag{3.7}$$

For the car mpg example, the 95% confidence interval for μ is

$$29.4 \pm 2.776 \frac{1.0}{\sqrt{5}} = 28.16 \text{ to } 30.64$$

The confidence interval includes 30; hence, the rating cannot be rejected at the 95% confidence level. However, the confidence interval is relatively wide, suggesting that more data should be collected. Once again, the confidence interval provides more practically useful information than the t-test.

Sample Size. The sample size formula remains the same as given by Equation (3.2). This means that σ must be approximately known to calculate the sample size. Such an approximate estimate of σ may be obtained from prior experience or relevant data. Otherwise, a small amount of data may be initially collected and an improved estimate of sample size obtained by using the observed standard deviation or its upper confidence limit.

3.2.3 Comparing Standard Deviations

Chi-Square Test. If the true population standard deviation is σ_0, then the statistic

$$\frac{(n-1)S^2}{\sigma_0^2} \tag{3.8}$$

has a χ^2_{n-1} distribution. The critical chi-square values are tabulated in Appendix C.

Example. The standard deviation of the fat content of ground beef was claimed to be 1.0 (% fat). On the basis of 16 samples, the observed standard deviation was 1.5 (% fat). Is the claim justified? In this case, $H_0 : \sigma_0 = 1$ and $H_1 : \sigma \neq 1$. The calculated chi-square value is

$$\chi^2 = \frac{(16-1)(1.5)^2}{(1.0)^2} = 33.75$$

From the χ^2 table in Appendix C, for $\alpha = 0.05$, $\chi^2_{\alpha/2,n-1} = 27.5$ and $\chi^2_{1-\alpha/2,n-1} = 6.26$. Since the computed chi-square is greater than 27.5, the claim is not justified.

Confidence Interval. The 95% confidence interval for σ^2 is given by

$$\frac{(n-1)s^2}{\chi^2_{\alpha/2,n-1}} \quad \text{to} \quad \frac{(n-1)s^2}{\chi^2_{1-\alpha/2,n-1}} \tag{3.9}$$

For the fat content example, the confidence interval for σ^2 is

$$\frac{15(1.5)^2}{27.5} \quad \text{to} \quad \frac{15(1.5)^2}{6.26} = 1.2 \text{ to } 5.4$$

The confidence interval for σ is calculated by taking square roots and is 1.1 to 2.32. Clearly, the claim is not included in the confidence interval and is deemed to be unjustified. If σ needs to be known more precisely, larger numbers of observations will be necessary.

Sample Size. Suppose we want to test the hypotheses $H_0 : \sigma = \sigma_0$ versus $H_1 : \sigma \neq \sigma_0$. To select a sample size, we first need to select a value σ_1 such

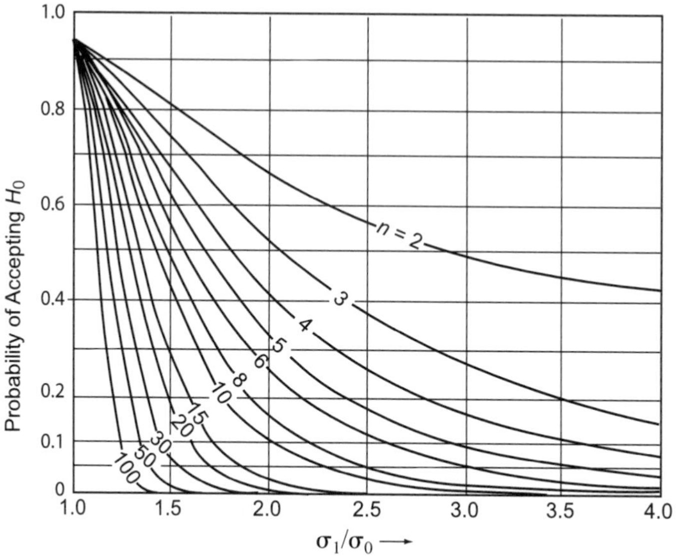

FIGURE 3.4 Sample size to compare standard deviation σ_1 to a target σ_0 ($\alpha = 5\%$).

that if $\sigma = \sigma_1$ we want to reject H_0 with high probability. Then the sample size can be determined from the operating characteristic curves in Figure 3.4, whose x-axis is the ratio σ_1/σ_0 and the y-axis shows the probability of accepting H_0 for various sample sizes. For $\sigma_1/\sigma_0 = 1$, the probability of accepting H_0 is 95% for all curves, meaning that the α-risk is 5%. For all other values of σ_1/σ_0, the y-axis corresponds to the β-risk. Figures 3.4 and 3.6 are reproduced from Ferris, Grubbs, and Weaver, "Operating Characteristics of the Common Statistical Tests of Significance," *Annals of Mathematical Statistics,* June 1946, with permission from the Institute of Mathematical Statistics.

Suppose we want to compare $\sigma_0 = 10$ versus $\sigma_1 = 15$. This is a difference of 50% in standard deviation. Let $\alpha = 5\%$ and $\beta = 10\%$. Then $\sigma_1/\sigma_0 = 1.5$ and from Figure 3.4, for a β-risk of 10%, the sample size is obtained corresponding to the curve that goes through the point of intersection of a vertical line at 1.5 and a horizontal line at 0.1. Presently, $n = 25$.

3.2.4 Comparing Proportion

Z Test. Let p denote the proportion of objects in a population that possess a certain attribute. For example, p may be the proportion of products that are outside the specification. Each of the n sample products is classified as being either within specification or outside specification, so the primary data are of the discrete (0, 1) type. Then

$$\bar{p} = \frac{\text{number of products outside specification}}{n}$$

To test the hypothesis $H_0 : p = p_0$ versus $H_1 : p \neq p_0$, under H_0, $\bar{p}$ has a mean equal to p_0 and standard deviation equal to $\sqrt{p_0(1 - p_0)/n}$. Consequently, the test statistic

$$Z = \frac{\bar{p} - p_0}{\sqrt{p_0(1 - p_0)/n}} \tag{3.10}$$

has a standard normal Z distribution. If the computed value of Z exceeds the critical value, H_0 is rejected.

Example. A local newspaper stated that less than 10% of the rental properties did not allow renters with children. The city council conducted a random sample of 100 units and found 13 units that excluded children. Is the newspaper statement wrong based upon this data? In this case $H_0 : p \leq 0.1$ and $H_1 : p > 0.1$. In this case $p_0 = 0.1$ and the computed Z value is

$$Z = \frac{0.13 - 0.1}{\sqrt{0.1(0.9)/100}} = 1.0$$

For $\alpha = 0.05$, $Z_\alpha = 1.64$ and the newspaper statement cannot be rejected based upon this data at the 95% level of confidence.

Confidence Interval. The approximate 100 $(1 - \alpha)$% confidence interval for p is

$$\bar{p} \pm Z_{\alpha/2}\sqrt{\bar{p}(1 - \bar{p})/n} \tag{3.11}$$

The 95% confidence interval for p for the rental properties example is

$$0.13 \pm 2\sqrt{0.13(0.87)/100} = 0.06 \text{ to } 0.20$$

Although the newspaper statement might be correct, this confidence interval is very wide, suggesting the need for a larger sample size if we wish to understand the situation better.

Sample Size. To test the hypotheses $H_0 : p = p_0$ versus $H_1 : p = p_1$, the necessary sample size may be determined as follows.

Let $\sigma_0 = \sqrt{p_0(1 - p_0)}$ and $\sigma_1 = \sqrt{p_1(1 - p_1)}$. Then

$$n = \left[\frac{Z_\alpha\sigma_0 + Z_\beta\sigma_1}{p_1 - p_0}\right]^2 \quad \text{for a one-sided test}$$

$$n = \left[\frac{Z_{\alpha/2}\sigma_0 + Z_\beta\sigma_1}{p_1 - p_0}\right]^2 \quad \text{for a two-sided test} \tag{3.12}$$

In the case of rental units, to test the hypotheses $H_0: p = p_0 = 0.1$ versus $H_1: p = p_1 = 0.12$, what is the required sample size for $\alpha = 5\%$ and $\beta = 10\%$? The sample size n, for this one-sided test, may be computed as follows:

$$\sigma_0 = \sqrt{0.1(0.9)} = 0.3 \qquad \sigma_1 = \sqrt{0.12(0.88)} = 0.325$$

$$n = \left[\frac{1.64 * 0.3 + 1.28 * 0.325}{0.12 - 0.10}\right]^2 = 2060$$

3.3 COMPARING TWO POPULATIONS

This section considers hypothesis tests to compare parameters of two populations with each other. For example, we may want to know if after a process change the process is different from the way it was before the change. The data after the change constitute one population to be compared with the data prior to change, which constitute the other population. Some specific comparative questions are: Has the process mean changed? Has the process variability reduced? If the collected data are discrete, such as defectives and nondefectives, has percent defective changed?

3.3.1 Comparing Two Means (Variance Known)

Z Test. The following test applies when we want to compare two population means and the variance of each population is either known or the sample size is large ($n > 30$). Let μ_1, n_1, $\bar{x}_1$, and σ_1 denote the population mean, sample size, sample average and population standard deviation for the first population and let μ_2, n_2, $\bar{x}_2$, and σ_2 represent the same quantities for the second population. The hypotheses being compared are $H_0: \mu_1 = \mu_2$ and $H_1: \mu_1 \neq \mu_2$. Under the null hypothesis

$$\text{Mean } (\bar{X}_1 - \bar{X}_2) = 0 \qquad \text{Variance } (\bar{X}_1 - \bar{X}_2) = \sigma_1^2/n_1 + \sigma_2^2/n_2$$

Therefore, the test statistic

$$Z = \frac{\bar{X}_1 - \bar{X}_2}{\sqrt{\sigma_1^2/n_1 + \sigma_2^2/n_2}} \qquad (3.13)$$

has a standard normal distribution. If the computed value of Z exceeds the critical value, the null hypothesis is rejected.

Example. We want to determine whether the tensile strength of products from two suppliers are the same. Thirty samples were tested from each supplier with the following results: $\bar{x}_1 = 51$, $s_1^2 = 10$, $\bar{x}_2 = 48$, $s_2^2 = 15$, and

$$Z = \frac{51 - 48}{\sqrt{10/30 + 15/30}} = 3.28$$

The $Z_{\alpha/2}$ value for $\alpha = 0.001$ is 3.27; hence, the two means are different with 99.9% confidence.

Confidence Interval. The 100 $(1 - \alpha)$% confidence interval for $(\mu_1 - \mu_2)$ is

$$(\bar{x}_1 - \bar{x}_2) \pm Z_{\alpha/2} \sqrt{\frac{\sigma_1^2}{n_1} + \frac{\sigma_2^2}{n_2}} \tag{3.14}$$

For the tensile strength example, the 99.9% confidence interval for $(\mu_1 - \mu_2)$ is

$$(51 - 48) \pm 3.27 \sqrt{\frac{10}{30} + \frac{15}{30}} = 0.01 \text{ to } 5.98$$

Since the confidence interval does not include zero, the two means are statistically different with 99.9% confidence.

Sample Size. The sample size n is given by

$$n = n_1 = n_2 = \frac{(\sigma_1^2 + \sigma_2^2)(Z_{\alpha/2} + Z_\beta)^2}{\Delta^2} \tag{3.15a}$$

for a two-sided test, where Δ is the smallest difference in means $(\mu_1 - \mu_2)$ to be detected. For a one-sided test, $Z_{\alpha/2}$ is replaced by Z_α. If the standard deviations of the two populations are assumed equal to σ then

$$n = n_1 = n_2 = \frac{2(Z_{\alpha/2} + Z_\beta)^2}{d^2} \tag{3.15b}$$

for a two-sided test, where $d = \Delta/\sigma$. As an example, if we want to detect a difference of two units in the tensile strength and if we assume $\sigma_1^2 = \sigma_2^2 = 12$, then for $\alpha = \beta = 5\%$, the sample size is

$$n = \frac{(12 + 12)(1.96 + 1.64)^2}{(2)^2} = 78 \text{ observations for each supplier}$$

For values of α and β risks in the neighborhood of 5 to 10 percent, the sample size formula may be further simplified.

$$n = n_1 = n_2 = \frac{20}{d^2} \text{ for a two-sided test} \tag{3.15c}$$

Consider the following example. In a study involving animals, 1 μg and 5 μg dosages of a certain drug were given to five different animals each, and the following peak responses were observed:

Peak response (1 μg dose) = 80, 5, 10, 30, 120

Peak response (5 μg dose) = 990, 240, 80, 240, 480

When responses change by orders of magnitude, it is often best to log-transform the data prior to analysis:

log (peak response for 1 μg dose): 1.9, 0.7, 1.0, 1.5, 2.1

log (peak response for 5 μg dose): 3.0, 2.4, 1.9, 2.4, 2.7

For the log-transformed data, the estimated standard deviation is 0.5 and based upon this standard deviation and Equation (3.15c), the following conclusions may be drawn regarding the sample size:

1. For $n = 5$, $d = 2$ and $\Delta = 2\sigma = 1$. This is the difference in mean that can be detected for the log-transformed data and implies that with a sample size of five, a tenfold change in peak response can be detected.
2. For $n = 10$, $\Delta = 0.7$ and a fivefold change in peak response can be detected.
3. For $n = 20$, $\Delta = 0.5$ and a threefold change in peak response can be detected.

3.3.2 Comparing Two Means (Variance Unknown but Equal)

Independent *t*-Test. This test is used to compare two population means when the sample sizes are small, the population variances are unknown but may be assumed to be equal. In this situation, a pooled estimate of the standard deviation is used to conduct the t-test. Prior to using this test, it is necessary to demonstrate that the two variances are not different, which can be done by using the F-test in Section 3.3.4. Remember that F comes before t! The hypotheses being tested are $H_0 : \mu_1 = \mu_2$ versus $H_1 : \mu_1 \neq \mu_2$. A pooled estimate of variance is obtained by weighting the two variances in proportion to their degrees of freedom as follows:

$$s^2_{\text{pooled}} = \frac{(n_1 - 1)s_1^2 + (n_2 - 1)s_2^2}{(n_1 + n_2 - 2)}$$

The test statistic

$$t = \frac{\bar{X}_1 - \bar{X}_2}{S_{\text{pooled}}\sqrt{\dfrac{1}{n_1} + \dfrac{1}{n_2}}} \tag{3.16}$$

has a $t_{n_1+n_2-2}$ distribution. If the computed value of t exceeds the critical value, H_0 is rejected and the difference is said to be statistically significant.

Example. The following results were obtained in comparing surface soil pH at two different locations:

$$n_1 = 8 \qquad n_2 = 12$$
$$\bar{x}_1 = 8.1 \qquad \bar{x}_2 = 7.8$$
$$s_1 = 0.3 \qquad s_2 = 0.2$$

Do the two locations have the same pH?

Assuming that the two variances are equal, we first obtain a pooled estimate of variance:

$$s^2_{\text{pooled}} = \frac{(n_1 - 1)s_1^2 + (n_2 - 1)s_2^2}{(n_1 + n_2 - 2)} = \frac{7(0.3)^2 + 11(0.2)^2}{8 + 12 - 2} = 0.06$$

$$s_{\text{pooled}} = 0.24$$

Then the t statistic is computed:

$$t = \frac{8.1 - 7.8}{0.24\sqrt{\dfrac{1}{8} + \dfrac{1}{12}}} = 2.74$$

For a two-sided test with $\alpha = 0.05$ and $(n_1 + n_2 - 2) = 18$ degrees of freedom, the critical value of t is $t_{0.025,18} = 2.1$. Since computed value of t exceeds the critical value 2.1, the hypothesis that the two locations have the same pH is rejected.

Confidence Interval. The 100 $(1 - \alpha)$% confidence interval for $(\mu_1 - \mu_2)$ is

$$(\bar{x}_1 - \bar{x}_2) \pm t_{\alpha/2, n_1+n_2-2}\, s_{\text{pooled}} \sqrt{\frac{1}{n_1} + \frac{1}{n_2}} \tag{3.17}$$

For the pH example, the 95% confidence interval for $(\mu_1 - \mu_2)$ is

$$(8.1 - 7.8) \pm 2.1(0.24)\sqrt{\frac{1}{8} + \frac{1}{12}} = 0.07 \text{ to } 0.53$$

As an analysis tool, a confidence interval is better than conducting a t-test. Both lead to the same conclusions with respect to accepting or rejecting the null hypothesis. However, the confidence interval provides greater practical information regarding the importance of differences and also whether additional data are necessary. With reference to Figure 3.5, where Δ represents a practically important difference:

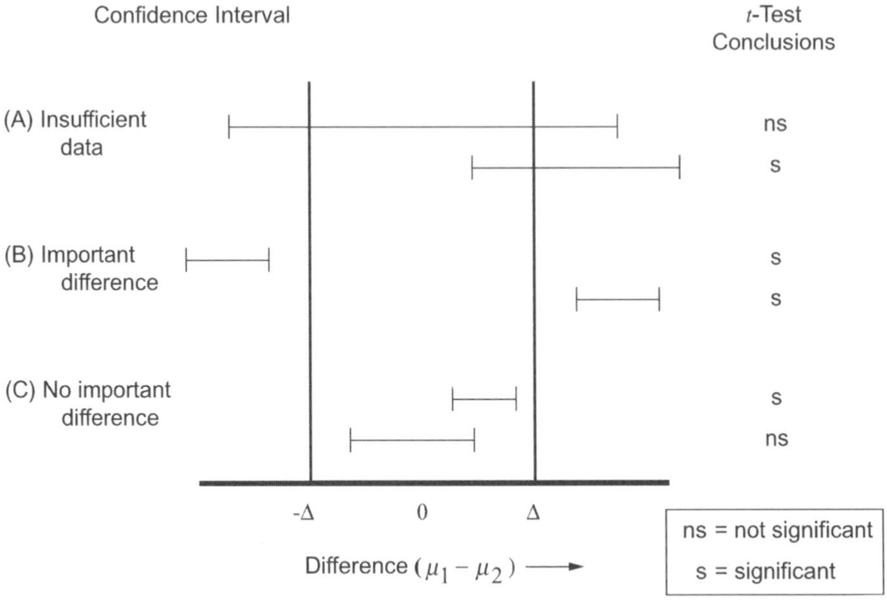

FIGURE 3.5 Confidence interval is better than a t-test.

Case (A): The confidence interval straddles Δ, indicating that we need more data to decide if the difference is smaller or bigger than Δ. The t-test concludes differently.

Case (B): The confidence interval and the t-test lead to similar conclusions.

Case (C): The difference between the two means is not important for both confidence intervals shown because the absolute difference is less than Δ. The t-test concludes differently for one of the confidence intervals.

The conclusion is that we should use the hypothesis-test-based sample size formulae to determine how much data to collect but then analyze the collected data using the confidence interval approach. The use of confidence intervals rather than hypothesis tests as an analysis tool is even more important in those cases where the sample size is either too small or too large because it was arbitrarily chosen.

3.3.3 Comparing Two Means (Variance Unknown and Unequal)

Independent *t*-Test. This test is used to compare two population means when the sample sizes are small ($n < 30$), the variance is unknown, and the two population variances are not equal, which should first be demonstrated by conducting the F-test to compare two variances. If the two variances are not different, the test in the previous section will prove more sensitive. To test the hypothesis $\mu = \mu_0$ against $\mu \neq \mu_0$, the test statistic t and the degrees of freedom v are

$$t = \frac{\bar{x}_1 - \bar{x}_2}{\dfrac{s_1^2}{n_1} + \dfrac{s_2^2}{n_2}} \qquad v = \frac{\left(\dfrac{s_1^2}{n_1} + \dfrac{s_2^2}{n_2}\right)^2}{\dfrac{(s_1^2/n_1)^2}{n_1 - 1} + \dfrac{(s_2^2/n_2)^2}{n_2 - 1}} \tag{3.18}$$

If the computed t exceeds the critical t, the null hypothesis is rejected.

Example. The following data were obtained on the life of light bulbs made by two manufacturers:

$$n_1 = 6 \qquad n_2 = 10$$

$$\bar{x}_1 = 2050 \text{ hours} \qquad \bar{x}_2 = 1890 \text{ hours}$$

$$s_1 = 90 \text{ hours} \qquad s_2 = 200 \text{ hours}$$

Is there a difference in the mean life of light bulbs made by the two manufacturers? Note that the F-test in Section 3.3.5 shows that the two standard deviations are not equal. The computed t and v are:

$$t = \frac{2050 - 1890}{\sqrt{\dfrac{90^2}{6} + \dfrac{200^2}{10}}} = 2.18 \qquad \text{and} \qquad v = \frac{(90^2/60 + 200^2/20)^2}{\dfrac{(90^2/6)^2}{5} + \dfrac{(200^2/10)^2}{9}} = 13$$

For $\alpha = 0.05$, $t_{\alpha/2,v} = 2.16$. Since the computed value exceeds the critical t value, we are 95% sure that the mean life of the light bulbs from the two manufacturers is different.

Confidence Interval. The 100 $(1 - \alpha)$% confidence interval for $(\mu_1 - \mu_2)$ is

$$(\bar{x}_1 - \bar{x}_2) \pm t_{\alpha/2,v}\sqrt{\frac{s_1^2}{n_1} + \frac{s_2^2}{n_2}} \tag{3.19}$$

For the light bulb example, the 95% confidence interval for $(\mu_1 - \mu_2)$ is computed below:

$$(2050 - 1890) \pm 2.16\sqrt{\frac{90^2}{6} + \frac{200^2}{10}} = 2 \text{ to } 318$$

3.3.4 Comparing Two Means (Paired *t*-test)

This test is used to compare two population means when there is a physical reason to pair the data and the two sample sizes are equal. A paired test is more sensitive in detecting differences when the population standard deviation is large. To test $\mu = \mu_0$ versus $\mu \neq \mu_0$, the test statistic is

$$t_{\text{paired}} = \frac{\bar{d}}{s_d/\sqrt{n}} \tag{3.20}$$

where
d = difference between each pair of values
$\bar{d}$ = observed mean difference
s_d = standard deviation of d

Example. Two operators conducted simultaneous measurements on percentage of ammonia in a plant gas on nine successive days to find the extent of bias in their measurements. The data are shown in Table 3.1 as reported in (Davies). Since the day-to-day differences in gas composition were larger than the expected bias, the tests were designed to permit paired comparison.

For the data in Table 3.1,

$$\bar{d} = 5.3,\ s_d = 7.9$$

and

$$t_{\text{paired}} = \frac{5.3}{7.9/\sqrt{9}} = 2.04$$

For $\alpha = 0.05$, $t_{0.025,8} = 2.31$. Since the computed t value is less than the critical t value, the results do not conclusively indicate that a bias exists.

TABLE 3.1. Data on Percent Ammonia

Day	Operator A	Operator B	d = (B – A)
1	4	18	14
2	37	37	0
3	35	38	3
4	43	36	–7
5	34	47	13
6	36	48	12
7	48	57	9
8	33	28	–5
9	33	42	9

Confidence Interval. The 100 $(1 - \alpha)\%$ confidence interval for $(\mu_1 - \mu_2)$ is

$$\bar{d} \pm t_{\alpha/2,n-1}(s_d/\sqrt{n}) \tag{3.21}$$

Presently, the 95% confidence interval for the difference in means (bias) is

$$5.3 \pm 2.31(7.9/\sqrt{9}) = -0.8 \text{ to } 11.4$$

If the potential bias and the uncertainty regarding the bias is considered large, then more data should be collected to quantify the bias precisely.

3.3.5 Comparing Two Standard Deviations

***F*-Test.** This test is used to compare two standard deviations and applies for all sample sizes. To test $\sigma_1 = \sigma_2$ versus $\sigma_1 \neq \sigma_2$, the ratio

$$F = \frac{S_1^2/\sigma_1^2}{S_2^2/\sigma_2^2} \tag{3.22}$$

has an F distribution, which is a skewed distribution and is characterized by the degrees of freedom used to estimate S_1 and S_2, called the numerator degrees of freedom $(n_1 - 1)$ and denominator degrees of freedom $(n_2 - 1)$, respectively. Under the null hypothesis, the F statistic becomes S_1^2/S_2^2. The percentile points of the F distribution are shown in Appendix E. In calculating the F ratio, the larger variance is in the numerator, so that the calculated value of F is greater than one. If the computed value of F exceeds the critical value $F_{\alpha/2,n_1-1,n_2-1}$, tabulated in Appendix E, the null hypothesis is rejected.

Example. Precision is important when manufacturing sporting ammunition. For a certain design of ammunition, the results were $n_1 = 21$ and $s_1 = 0.8$. For an improved design, the results were $n_2 = 16$ and $s_2 = 0.6$. Has the variability been reduced? If σ_1 and σ_2 denote the standard deviations for the old and the new design, then the null hypothesis is $\sigma_1 \leq \sigma_2$. The alternate hypothesis is $\sigma_1 > \sigma_2$. The computed value of the F statistic is

$$F = (0.8/0.6)^2 = 1.78$$

The critical F value for $\alpha = 5\%$ is $F_{0.05,20,15} = 2.33$. Since the computed F value is less than 2.33, there is no conclusive evidence that the new design has reduced variability.

Confidence Interval. The 100 $(1 - \alpha)\%$ confidence interval for σ_1^2/σ_2^2 is

$$\frac{(s_1^2/s_2^2)}{F_{\alpha/2,n_2-1,n_1-1}} \quad \text{to} \quad \frac{(s_1^2/s_2^2)}{F_{\alpha/2,n_1-1,n_2-1}} \tag{3.23}$$

Confidence interval for σ_1/σ_2 is computed by taking square roots. For the ammunition example, $n_1 = 21$ and $n_2 = 16$; and for $\alpha = 10\%$, the confidence interval for σ_1^2/σ_2^2 is

$$\frac{(0.8^2/0.6^2)}{2.2} \quad \text{to} \quad (0.8^2/0.6^2) * 2.33 = 0.8 \text{ to } 4.14$$

The 90% confidence interval for σ_1/σ_2 is 0.89 to 2.03, obtained by taking square roots. Since the confidence interval for σ_1/σ_2 includes 1, there is no conclusive evidence that the new design has reduced variability. Also, the confidence interval is very wide so that, with the number of tests conducted, a rather large change in standard deviation will go undetected.

Sample Size. The sample size necessary to detect a certain ratio of two standard deviations σ_1/σ_2 can be obtained from the operating characteristic curves in Figure 3.6, which apply for an α-risk of 5%. The y-axis gives the β-risk corresponding to a selected ratio σ_1/σ_2 and a selected sample size. For the ammunition example, if the new design reduces the standard deviation by 33% compared to the prior design, it would be considered a significant improvement. What should be the sample size? The standard deviation ratio is $1/0.66 \approx 1.5$. From Figure 3.6, for $\alpha = 5\%$ and $\beta = 10\%$, the sample size is 50. This means that 50 observations for each design are necessary to detect a 33% difference in standard deviations.

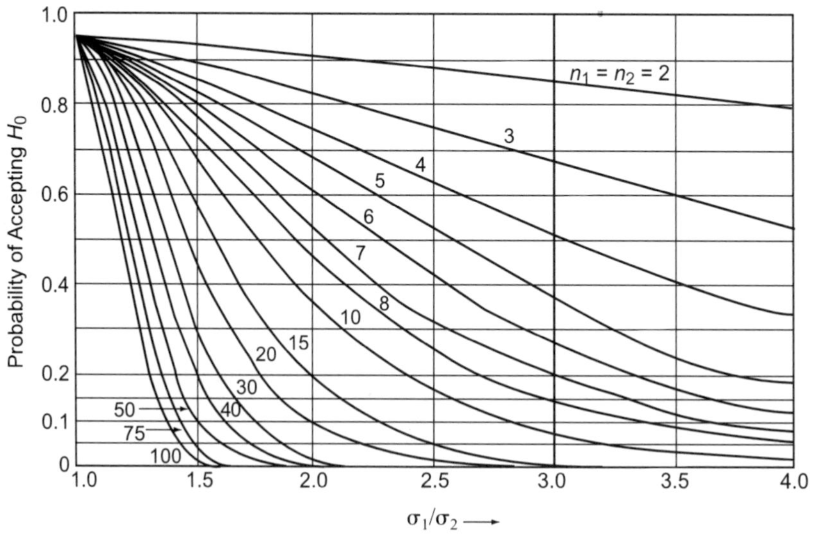

FIGURE 3.6 Sample size to compare two standard deviations ($\alpha = 5\%$).

3.3.6 Comparing Two Proportions

Z Test. This test is used to compare the proportion (p_1) of items in one population to the proportion (p_2) of items in another population when the primary data are discrete (0,1 type). To compare $p_1 = p_2$ against $p_1 \neq p_2$, the test statistic is

$$Z = \frac{\bar{p}_1 - \bar{p}_2}{\sqrt{\bar{p}(1-\bar{p})(1/n_1 + 1/n_2)}} \tag{3.24}$$

where

$$\bar{p} = \frac{\text{all items with the specified characteristic}}{(n_1 + n_2)}$$

The null hypothesis is rejected if Z exceeds the critical value.

Example. If one is found to be guilty, does it pay to have pleaded guilty to avoid a prison term? Following data were obtained on defendants with similar prior history and accused of similar crimes:

	Plea	
	Not Guilty	Guilty
Number judged guilty	82	206
Number sentenced to prison	68	110
Sample proportion	0.829	0.534

Here, $n_1 = 82$, $\bar{p}_1 = 0.829$, $n_2 = 206$, and $\bar{p}_2 = 0.534$. Hence,

$$\bar{p} = \frac{68 + 110}{82 + 206} = 0.618$$

$$Z = \frac{0.829 - 0.534}{\sqrt{0.618(1 - 0.618)(1/82 + 1/206)}} = 4.65$$

The value of Z is highly significant, indicating that if one strongly expects to be found guilty, initially pleading guilty may be a good strategy to avoid prison.

Confidence Interval. The approximate 100 $(1 - \alpha)$% confidence interval for $(p_1 - p_2)$ is

$$(\bar{p}_1 - \bar{p}_2) \pm Z_{\alpha/2}\sqrt{\bar{p}(1-\bar{p})(1/n_1 + 1/n_2)} \tag{3.25}$$

For the above example, the 95% confidence interval for $(p_1 - p_2)$ is

$$(0.829 - 0.534) \pm 2\sqrt{0.618(1 - 0.618)(1/82 + 1/206)} = 0.17 \text{ to } 0.42$$

Sample Size. The sample size formula for a one-sided test is given below. For a two-sided test, α is replaced by $\alpha/2$. The two sample sizes are assumed to be equal, i.e., $n = n_1 = n_2$.

$$n = \frac{[Z_\alpha\sqrt{(p_1 + p_2)(q_1 + q_2)/2} + Z_\beta\sqrt{p_1q_1 + p_2q_2}]^2}{d^2} \tag{3.26a}$$

where $q_1 = 1 - p_1$, $q_2 = 1 - p_2$ and $d = p_1 - p_2$. Approximately, the formula may be simplified to

$$n = \frac{(Z_\alpha + Z_\beta)^2(p_1 + p_2)(q_1 + q_2)}{d^2} \tag{3.26b}$$

As an example, consider the Salk polio vaccine experiment conducted in 1954. Prior to the vaccine, the probability of getting polio was $p_1 = 0.0003$. It would have been considered a significant improvement if the vaccine reduced this incident rate by half, i.e., $p_2 = 0.00015$. What was the required sample size for $\alpha = 0.05$ and $\beta = 0.10$?

$$n = \frac{[1.64\sqrt{(0.00045)(1.99955)/2} + 1.28\sqrt{(0.00015)(0.99985) + (0.0003)(0.9997)}]^2}{(0.0003 - 0.00015)^2} = 170{,}000$$

3.4 COMPARING MULTIPLE POPULATIONS

We now consider tests to compare multiple populations. In comparing multiple means, the null hypothesis being tested is whether all means may be considered to be equal. This is answered by conducting analysis of variance (ANOVA), which is a generalization of the t-test to compare two means. If the ANOVA suggests that the means are different, then we want to know which particular means are different. ANOVA does not provide a direct answer, and multiple comparison procedures are necessary. The two issues involved in making these multiple comparisons may be explained as follows. Suppose we have k treatments to compare. Then there are $k(k - 1)/2$ possible pair-wise comparisons. However, there are only $(k - 1)$ degrees of freedom between treatments. Hence, all possible pair-wise comparisons are not independent of each other since there could be at most $(k - 1)$ independent comparisons. The second concern is that if we wish to make k_1 independent pair-wise comparisons, each at a significance level α, then the probability of false rejection of the null hypothesis is $1 - (1 - \alpha)^{k_1}$. For $\alpha = 0.05$ and $k_1 = 10$, this probability is 40%, which is very high. These two issues are addressed by first conducting an overall ANOVA in which the total experiment-wise risk of false rejection of the null hypothesis is fixed at α. This is followed by multiple comparison tests to assess differences between means. Similar considerations apply in comparing multiple variances.

Two commonly used experimental strategies to compare multiple means are called completely randomized design and randomized block design. These are presented below, followed by some multiple comparison procedures to assess differences between means. This section ends with a brief discussion of methods to compare multiple variances. The computations for some of these methods are best done using a computer program, and the focus here is on the interpretation of results.

3.4.1 Completely Randomized Design

For a completely randomized design, random samples are taken from multiple populations and the averages are compared to each other. This is an extension of the independent t-test. Consider the following example.

We want to compare the moisture protection afforded a food product by four different packaging films. Six packages were made from each film. The moisture pickup of the packaging contents at the end of 30-day storage at 85°F and 90% relative humidity was measured. Inadvertently, one of the units for film A was lost. The data are shown in the Table 3.2.

From the average moisture pickup for each film, there appear to be differences between films. Are these differences statistically significant? Analysis of variance answers this question. In principle, the variability between averages is compared to the variability within a film to answer the question: given the variability within a film, could the observed averages have occurred if there were no differences between the films? The

TABLE 3.2. Data on Moisture Pickup

	Film A	Film B	Film C	Film D
	1.43	1.88	1.90	2.10
	1.88	1.82	2.05	2.28
	1.66	1.88	1.95	2.03
	1.67	1.85	2.17	2.14
	1.04	1.59	1.90	2.25
		1.62	2.29	2.50
Average	1.536	1.773	2.043	2.217
			Grand Average	1.907

hypotheses being compared are: the means are equal versus they are not equal. The analysis of variance (ANOVA) constitutes the hypothesis test and is shown in Table 3.3.

The results may be interpreted as follows. Source refers to the sources of variability in the data. Part of the variability is due to differences in films. The remainder is due to replicates or random variation within a film. To assess whether the differences between films are significant when compared to variation within films, ANOVA partitions the total sum of squares (total SS) into film sum of squares (film SS) and replicate sum of squares (replicate SS) as follows:

1. Total SS is the sum of (observation – grand average)^2. One degree of freedom is lost in computing the grand average. Therefore, the total degrees of freedom are (23 – 1) = 22.
2. Film SS is that portion of the total SS that is due to differences between films. This is computed by comparing film averages to the grand average. One of the four degrees of freedom is lost in computing grand average and the degrees of freedom associated with films are (4 – 1) = 3.
3. Replicate SS is that portion of total SS that is due to package-to-package variation within each film. It is computed by comparing the observations for each film to the film average under the assumption that the variance within film is the same for all films. Four degrees of freedom are lost in computing four film averages and the degrees of freedom for replicates are (23 – 4) = 19.

The following relationship holds true for the sum of squares:

$$\text{Total SS} = \text{Film SS} + \text{Replicate SS}$$

The mean square is the sum of squares divided by the corresponding degrees of freedom. The F ratio is the film mean square divided by replicate mean square and has 3 and 19 degrees of freedom in this case. The last column shows the percent confidence corresponding to the computed F ratio. We are more than 99.9% sure that the films do not have the same

TABLE 3.3. ANOVA for Moisture Pickup

Source	Degrees of Freedom	Sum of Squares	Mean Square	F	% Confidence
Film	3	1.482	0.494	12.3	>99.9%
Replicate	19	0.763	0.040		
Total	22	2.246	0.102		

moisture pickup. In the ANOVA table, this percent confidence number is the most important number to focus on. All other numbers essentially represent intermediate calculation steps. Another important number is the replicate (or error) degrees of freedom, which should be 20 or more for a reasonable estimate of replicate (or error) variance given by the replicate (or error) mean square.

ANOVA suffers from the same criticism that a t-test does; namely, it focuses on statistical significance but does not tell us how large the differences between the films are or could be and whether the sample size is adequate. An approximate answer to these questions can be obtained by constructing confidence intervals for the mean moisture pickup for each film or by conducting analysis of means.

What about the necessary sample size? Can we use the same formula as for a t-test to compare two means? The key issue with multiple comparisons is that k comparisons are being simultaneously made and with some approximation, if each comparison has a probability α of producing an incorrect conclusion, then k comparisons have probability $k\alpha$ that at least one of the conclusions is wrong. To make the total experiment-wise risk equal to α, the risk for each pair-wise comparison needs to be α/k. In the sample size formula for the t-test to compare two populations, if we substitute α/k for α, we will obtain an approximate sample size for this multiple comparison. Other more exact and complex methods are also available (Sahai and Ageel).

3.4.2 Randomized Block Design

This approach is an extension of the paired t-test and is used to compare the means of several populations when the comparisons are constructed to block out the effect of an extraneous factor. Consider the example of developing a pickling solution to produce crisper pickles. Seven solutions were studied, as shown in Table 3.4, the control being the current pickling solution. Three lots of cucumbers were used and the experiment was structured to eliminate the extraneous cucumber lot-to-lot variation. The cucumber lots are "blocks."

The ANOVA is shown in Table 3.5. The difference between solutions is significant at the 97% confidence level. The cucumber lots are not significantly different from each other, indicating that blocking in this case was unnecessary. The same data could also be analyzed as an unblocked (completely randomized) design.

3.4.3 Multiple Comparison Procedures

LSD Approach. The purpose of multiple comparison procedures is to identify the specific pairs of means that are different from each other by conducting pair-wise comparisons. A commonly used procedure is called

TABLE 3.4. Data on Crispness of Pickles

Cucumber Lot	Control	2% Lactose	4% Lactose	6% Lactose	2% Sucrose	4% Sucrose	6% Sucrose
1	19.2	20.6	24.2	25.1	21.2	24.3	22.3
2	21.7	23.4	22.1	25.7	22.3	23.4	23.6
3	22.6	21.4	24.4	22.9	21.2	24.9	24.2

TABLE 3.5. ANOVA for Crispness of Pickles

Source	Degrees of Freedom	Sum of Squares	Mean Square	F	% Confidence
Solution	6	33.4	5.6	3.44	97%
Lot	2	2.4	1.2	0.74	50%
Residual	12	19.4	1.6		
Total	20	55.2			

the least significant difference (LSD) procedure. If the ANOVA shows differences, then pair-wise comparisons are performed using a t-test at level α. The procedure is now illustrated using the film moisture pickup example.

Assuming a two-sided alternate hypothesis, two means are said to be statistically different if

$$|\bar{x}_i - \bar{x}_j| > LSD$$

where,

$$LSD = t_{\alpha/2,(N-k)} s_{\text{pooled}} \sqrt{\left(\frac{1}{n_i} + \frac{1}{n_j}\right)} \tag{3.27a}$$

N is the total number of observations, k is the number of treatments (films) to be compared, s_{pooled} is the pooled error standard deviation based upon $(N - k)$ degrees of freedom, and n_i and n_j are sample sizes for treatments i and j. If all sample sizes are equal, then

$$LSD = t_{\alpha/2,k(n-1)} s_{\text{pooled}} \sqrt{2/n} \tag{3.27b}$$

for the film example, the sample sizes are not all equal. We have $N = 23$, $k = 4$, and for $\alpha = 0.05$, the t-value is 2.093. s_{pooled} is $\sqrt{0.04} = 0.2$. For $n_i = n_j = 6$,

$$LSD = 0.242$$

If two means differ by more than 0.242, then those two films are statistically different. Thus films B and D are different but C and D are not. The magnitude of LSD is half the width of the confidence interval for difference in two means (with appropriate degrees of freedom). It thus provides a measure of practical significance as discussed previously in Section 3.3.2. The prior use of ANOVA in this LSD procedure ensures that the overall probability of falsely rejecting the null hypothesis is limited to α, however, the LSD approach does not protect against falsely finding chance differences among a pair of treatments to be statistically significant.

Bonferroni Approach. We now consider a second approach to multiple comparisons called the Bonferroni approach. The basic idea is that if k pair-wise comparisons are to be made and we want the overall risk to be α, then each comparison should be made, or equivalently, each confidence interval for difference in a pair of means should be constructed, at level α/k. The method is simple and works well for small values of k. For the film example, there are a total of six possible pair-wise comparisons among four films. For $\alpha = 0.05$ and $k = 6$, $\alpha/k \approx 0.01$ and the corresponding $t_{19,0.005} = 2.861$, which gives the confidence interval for difference in means to be approximately $(\bar{x}_i - \bar{x}_j) \pm 0.33$. So films B and D are different but C and D are not.

There are many other multiple comparison procedures such as those due to Tukey and Kramer, Duncan, Dunnett, Newman and Keuls, and so on. For brief discussions of these methods and further references, see (Sahai and Ageel).

3.4.4 Comparing Multiple Standard Deviations

$F_{\max}$ Test. A simple test to compare multiple variances is due to Hartley and represents an extension of the F-test to compare two variances. It applies when all sample sizes are equal to n. If there are k variances to be compared, then the test statistic is

$$F_{\max} = \frac{\max(s_i^2)}{\min(s_i^2)} \tag{3.28}$$

where $\max(s_i^2)$ and $\min(s_i^2)$ represent the largest and smallest sample variances, respectively. If all variances are equal, the value of $F_{\max}$ is expected to be close to 1. If the computed value of $F_{\max}$ exceeds the critical $F_{\max}[k, n-1, \alpha]$, then the variances are statistically deemed to be different. The percentage points of the $F_{\max}$ distribution are given in Appendix F.

As an illustrative example, consider the following situation in which the same product was measured ten times by four different operators in a blind test. The variance for each operator was computed and is shown below. Do the operators differ in terms of variability at $\alpha = 0.05$?

$$s_i^2 : 0.8, 0.3, 1.2, 0.9$$

$$F_{\max} = \frac{1.2}{0.3} = 4$$

Appendix F shows that the critical $F_{\max}[4, 9, 0.05] = 6.31$. Since the computed value of $F_{\max}$ is less than the critical value, the null hypothesis that all operators have equal variance is not rejected.

There are many other tests to compare multiple variances, including those by Cochran, Bartlett, Levene, Brown and Forsythe, O'Brien, and so on. See (Sahai and Ageel) for additional references.

3.5 CORRELATION

We now consider the subject of establishing relationships between two or more factors. For example, we may be interested in the relationship between various ingredients and product hardness, or the relationship between various attributes of a product such as moisture, hardness, and composition. Correlation and regression analyses are useful techniques to establish such relationships. Correlation analysis is a graphical and computational technique used to measure the nature and strength of the relationship between two factors. Regression analysis is a method used to obtain an equation relating a dependent factor (an effect) to independent factors (the causes).

3.5.1 Scatter Diagram

Suppose we are interested in understanding the relationship between thread moisture content and elongation. We may obtain paired data (moisture content, % elongation) as follows: (1.2, 8.0), (1.6, 8.3), (2.1, 8.8), (1.5, 8.0), etc. The plot of the paired data, called a scatter diagram, may appear as shown in Figure 3.7. This scatter diagram indicates that the greater the moisture, the greater the elongation. The visually observed linear relationship is referred to as correlation. There are many potential pitfalls in interpreting scatter diagrams.

Correlation without Causation. In interpreting such relationships, what we can say is that there appears to be a relationship between X

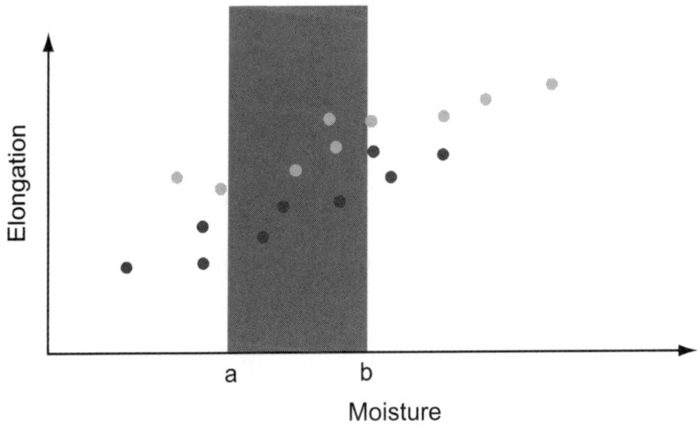

FIGURE 3.7 Scatter diagram.

and Y of a certain nature. Whether this relationship is causal or not is a separate question, which may have to be answered on the basis of technical knowledge or by using active planned experiments. For example, if we plot the salary of accountants against the sale of microwave ovens over a period of several years, we may observe a relationship similar to that in Figure 3.7. This does not mean that an accountant should buy a microwave oven to get a salary increase. A possible explanation of this spurious correlation is that as economy expands, the sale of microwave ovens increases and so does the salary of accountants. Such nonsense correlation can arise, particularly when passively collected data are plotted against one another.

Inadequate Range of *X*. The range of data values along the x-axis should be large, otherwise the relationship may not be evident. In Figure 3.7, if the range of moisture in the data is constrained to be between a and b, the relationship will not be evident, thus there may be causation without correlation. With passively collected data, the range of data values along the x-axis may be small, either by chance or by design.

Stratification. In establishing a relationship between thread moisture content and elongation, we may want to stratify (separate) the data by the type of thread, as shown in Figure 3.7. The relationship may vary with the type of thread. Also, the unstratified data may show no relationship, whereas the stratified data may show a relationship. The reverse can also be true.

3.5.2 Correlation Coefficient

Scatter diagrams are interpreted in terms of the nature of suggested variation between the two factors. With reference to Figure 3.8 (a), when X is large, Y is large, and when X is small, Y is small. In this case, we say that there is a positive correlation between X and Y. Figure 3.8 (b) shows a negative correlation between X and Y. As X increases, Y decreases. Figure 3.8 (c) shows no correlation between X and Y. No apparent relationship exists. Figure 3.8 (d) shows that the relationship between X and Y is not linear, but curvilinear.

Correlation coefficient R measures the strength of the linear relationship between X and Y. Consider the following quantity, known as covariance.

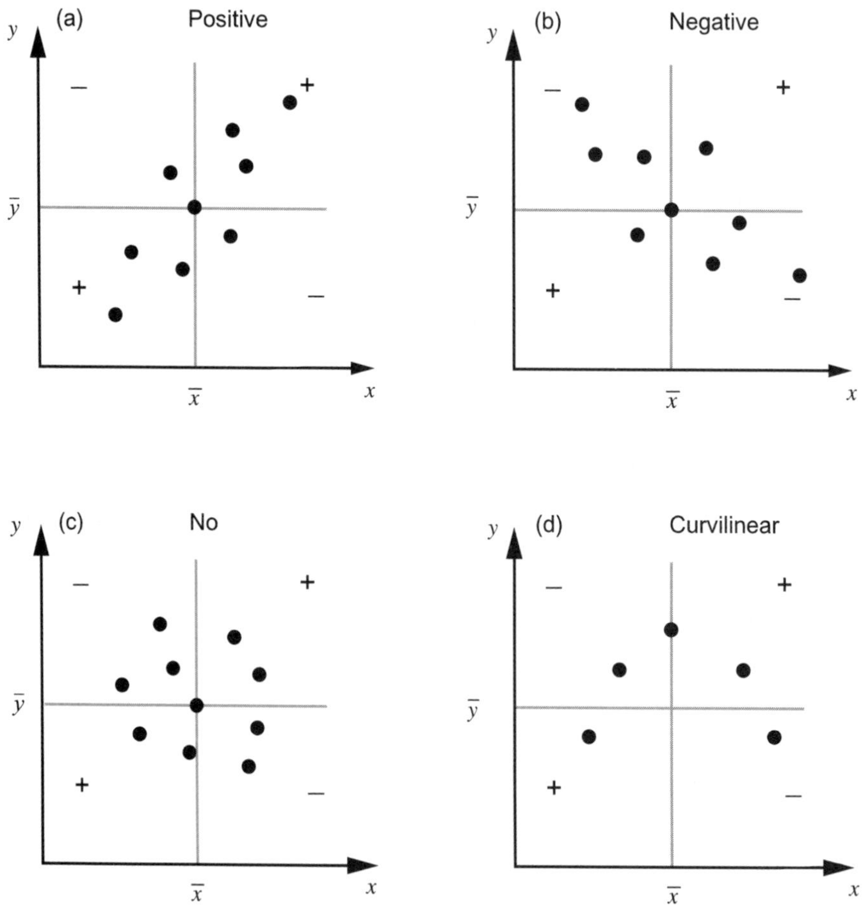

FIGURE 3.8 Correlation coefficient.

$$s_{xy} = \frac{1}{(n-1)}\Sigma(x_i - \bar{x})(y_i - \bar{y}) \qquad (3.29)$$

If the relationship is strongly positive, as in Figure 3.8 (a), then an x_i value above the mean $\bar{x}$ will often be paired with a y_i value above the mean $\bar{y}$, so that $(x_i - \bar{x})(y_i - \bar{y}) > 0$. Similarly when x_i is below $\bar{x}$, y_i will often be below $\bar{y}$, and $(x_i - \bar{x})(y_i - \bar{y})$ will still be positive. Figure 3.8 shows the quadrants in which $(x_i - \bar{x})(y_i - \bar{y})$ is positive or negative. A positive s_{xy} means positive correlation, a negative s_{xy} means a negative correlation, and s_{xy} close to zero means no correlation.

As a measure of the strength of relationship between X and Y, s_{xy} has a serious defect. By changing the scale of measurement of X or Y, the value of s_{xy} can be made arbitrarily large or small. Therefore, the correlation coefficient R is obtained by standardizing the value of s_{xy} as follows:

$$R = \frac{s_{xy}}{s_x \cdot s_y} \qquad (3.30)$$

where s_x and s_y are the calculated standard deviations of X and Y, respectively.

R measures the degree of linear relationship between the two factors. It does not measure the degree of curvilinear relationship. For example, for Figure 3.8 (d), $R = 0$! Yet there is a strong curvilinear relationship. A pitfall of the correlation coefficient is that a value of R near zero is not a guarantee of no relationship, but only demonstrates the absence of a linear relationship between the two factors.

The correlation coefficient has the following properties. The value of R is independent of the units in which X and Y are measured and lies between -1 and $+1$. It does not depend upon which variable is labeled X and which is labeled Y. If $R = 1$, all (x, y) pairs lie on a straight line with positive slope. If $R = -1$, they lie on a straight line with a negative slope.

Example. Let us suppose that we are interested in reducing the abrasion loss of rubber. For various rubber specimen, data are available on abrasion loss and the corresponding compounding material properties, as well as processing conditions. Let

X_1 = raw material property A

X_2 = raw material property B

X_3 = curing temperature

Y_1 = hardness of rubber

Y_2 = abrasion loss

The correlation coefficient for each pair of variables can be determined from the data. For example, given three pairs of values for Y_1 and Y_2: (40, 400), (50, 350) and (60, 150), we have

$$s_{y_1y_2} = \frac{1}{2}\Sigma(y_1 - \bar{y})(y_2 - \bar{y}) = -1250 \quad \text{and} \quad s_{y_1} = 10, s_{y_2} = 132$$

The correlation coefficient R turns out to be

$$R = -\frac{1250}{10(132)} = -0.95$$

This strong negative correlation suggests that the greater the hardness less the abrasion loss.

Correlation Matrix. The complete correlation matrix is shown in Table 3.6. From the correlation matrix, there appears to be a strong linear relationship between abrasion loss and hardness, between abrasion loss and raw material property B, and between hardness and raw material property B.

When data are available on a large number of input and output factors, the correlation matrix is helpful in identifying potential key relationships. In interpreting the correlation matrix, the various pitfalls of scatter diagrams and correlation coefficient must be kept in mind.

Statistical Significance. The true correlation coefficient is denoted by ρ. The estimated value is R. The correlation coefficient is statistically significant, i.e., $\rho \neq 0$, if

$$t = \frac{R\sqrt{n-2}}{\sqrt{1-R^2}} \tag{3.31}$$

exceeds the critical value for t_{n-2}.

For example, based upon a sample size $n = 16$, the correlation coeffi-

TABLE 3.6. Correlation Matrix

	X_1	X_2	X_3	Y_1	Y_2
X_1	1				
X_2	0.60	1			
X_3	–0.15	–0.10	1		
Y_1	–0.35	–0.82	0.08	1	
Y_2	0.49	0.88	–0.20	–0.95	1

cient between oven temperature and crack size was $R = 0.716$. Is the correlation statistically significant? The computed t value is

$$t = \frac{0.716\sqrt{14}}{\sqrt{1-(0.716)^2}} = 3.84$$

The critical value of t for 99% confidence and 14 degrees of freedom is 2.624. Since the computed t is greater than the critical value, we are more than 99% sure that the true correlation coefficient between crack size and temperature is greater than zero. Had the correlation coefficient R turned out to be 0.5, it would have been statistically insignificant in this case.

Practical Significance. The rule of thumb to assess whether the correlation coefficient is practically significant or not is as follows:

$$|R| < 0.5 \text{ Weak correlation}$$

$$0.5 < |R| < 0.8 \text{ Moderate correlation}$$

$$|R| > 0.8 \text{ Strong correlation}$$

This rule is based upon the fact that the square of the correlation coefficient R^2 measures the percent variation explained by that factor (see Section 3.6.1). If $R = 0.5$, then X explains 25% of the variation of Y.

3.6 REGRESSION ANALYSIS

The usual purpose of regression analysis is to build an equation relating output (Y) to one or more input factors (X). The relationship between Y and X may be linear, polynomial, or some other multivariate general function. This section assumes that regression analysis software is available to fit equations and the discussion is limited to the interpretation of results (Draper and Smith).

3.6.1 Fitting Equations to Data

In order to determine the relationship between abrasion loss of rubber and hardness of rubber, fifteen different rubber specimen were obtained. For each specimen, hardness and abrasion loss were determined using standard methods. It was felt that abrasion loss may also be related to the tensile strength of rubber, so tensile strength data were collected as well. Table 3.7 summarizes the results, which are a portion of the data reported in (Davies).

TABLE 3.7. Physical Properties of Rubber

Hardness	Tensile Strength	Abrasion Loss
45	162	372
55	233	206
61	232	175
66	231	154
71	231	136
71	237	112
81	224	55
86	219	45
53	203	221
60	189	166
64	210	164
68	210	113
79	196	82
81	180	32
56	200	228

Linear Regression. Figure 3.9 shows a plot of the relationship between the abrasion loss (y) and hardness (x). There appears to be a linear relationship between abrasion loss and hardness given by

$$y = a + bx + e$$

where a is the intercept, b is the slope, and e denotes the errors or residuals, namely, the unexplained deviation of the observed data from the straight line. Regression analysis computes the values of a and b such

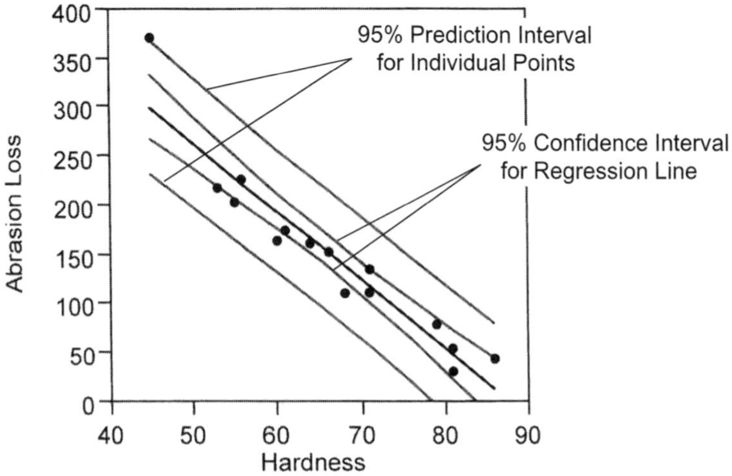

FIGURE 3.9 Linear regression.

TABLE 3.8. Parameter Estimates

Term	Estimate	Standard Error	% Confidence
a	615.6	42.4	>99.99
b	–7.0	0.6	>99.99

that the sum of squares of errors Σe_i^2 is minimized. This approach to estimate the coefficients in an equation is called the least squares approach and produces estimates of a and b having the smallest variance, assuming the observations to be independent with equal variance. The estimated parameters are in Table 3.8.

From Table 3.8, the fitted straight line is

$$y = 615.6 - 7.0x + e$$

and shows how abrasion loss reduces as hardness increases. The percent confidence column indicates that both coefficients are highly statistically significant, meaning that zero is not a likely value for either coefficient. Standard error refers to the standard deviation associated with the estimated coefficient so that the estimated coefficient ± 2 (standard error) provides an approximate 95% confidence interval for the true value of the coefficient. This uncertainty regarding the two parameters is graphically captured by the 95% confidence interval for the regression line shown by the two inside lines around the fitted line in Figure 3.9. The interpretation is that we are 95% sure that the true straight line could be anywhere inside this interval. The outer 95% prediction interval lines in Figure 3.9 are for individual points, meaning that 95% of the population is expected to be inside these outer lines. The analysis of variance (ANOVA) is shown in Table 3.9. "Source" refers to the source for sum of squares. The corrected total sum of squares is $\Sigma(y_i - \bar{y})^2$. With 15 observations, it has 14 degrees of freedom because one degree of freedom is lost in estimating $\bar{y}$. The error sum of squares is Σe_i^2. This is the portion of the corrected total sum of squares that is not explained by the model (fitted equation). The model sum of squares is the difference between the corrected total and the error sums of squares. The model sum of squares

TABLE 3.9. ANOVA for Abrasion Loss of Rubber

Source	Degrees of Freedom	Sum of Squares	Mean Square	F Ratio	% Confidence
Model	1	96153	96153	124	>99.99
Error	13	10084	776		
Corrected Total	14	106237	7588		

has one degree of freedom, one less than the number of parameters. Hence, error sum of squares has 13 degrees of freedom. The computed sum of squares are as shown. The mean square is the sum of squares divided by the corresponding degrees of freedom. The error mean square estimates σ_e^2 and the corrected total mean square estimates σ_Y^2. The F ratio is the ratio of the model mean square divided by the error mean square and has one and 13 degrees of freedom. The percent confidence corresponding to the computed F ratio is 99.99%, meaning that we are 99.99% sure that the fitted equation is statistically significant. We judge the goodness of the fitted equation in the following three ways.

1. **Statistical significance.** We look to see if the fitted coefficients are statistically significant as discussed before in connection with Table 3.8. Those coefficients that are statistically insignificant are deleted from the model, starting with the least significant coefficient. The remaining coefficients are then reestimated. This process, called model reduction, is continued until an equation is obtained containing only the statistically significant coefficients. The practical significance of the coefficients is then judged to assess if the equation provides practically useful information.

2. ***R*² and Adjusted *R*².** The R^2 value, known as the coefficient of determination, is computed as follows:

$$R^2 = \frac{\text{Model Sum of Squares}}{\text{Corrected Total Sum of Squares}}$$

$$R^2 = 1 - \frac{\text{Error Sum of Squares}}{\text{Corrected Total Sum of Squares}}$$

R^2 may be interpreted as the fraction of the variation (sum of squares) in Y, which is explained by the model. In the case of linear regression, R^2 is also mathematically equal to the square of the correlation coefficient. Presently, R^2 can be calculated from Table 3.9 and is 0.9 (= 96153/106237), meaning that the linear equation explains 90% of the variation in Y. R^2 can take values from 0 to 1. We want R^2 to be high for the model to be useful.

If υ_e and υ_Y denote the error and the corrected total degrees of freedom, respectively, then

$$R^2 \text{ estimates } 1 - \frac{\upsilon_e \sigma_e^2}{\upsilon_Y \sigma_Y^2} \tag{3.32b}$$

If υ_e and υ_R are essentially equal, then R^2 may be interpreted as the fraction of the variance of Y explained by the model. Adjusted R^2 adjusts for the degrees of freedom as follows:

$$\text{Adjusted } R^2 = 1 - \frac{\text{Mean Square Error}}{\text{Mean Square Corrected}} \tag{3.33a}$$

$$\text{Adjusted } R^2 \text{ estimates } 1 - \frac{\sigma_e^2}{\sigma_Y^2} \quad (3.33b)$$

thus providing a direct estimate of the fraction of the variance of Y explained by the model. By adjusting for the degrees of freedom, it also becomes a better measure to compare the goodness of alternate models, particularly with small data sets. The adjusted R^2 can take values from 0 to 1. We want the adjusted R^2 to be high for the model to be useful.

The following points should be remembered in connection with R^2 and the adjusted R^2.

(a) R^2 can be arbitrarily increased by adding terms in the model. For example, if we have four pairs of data relating X and Y, we could fit a cubic equation that will go through each data point and R^2 will be 1! The correct approach is to keep only those terms in the model that are statistically significant and then evaluate the resultant R^2 for goodness.

(b) R^2 depends upon the range of input factors. The larger the range, the larger the R^2 is likely to be, assuming error variance to be constant. This may be seen by realizing that as the range of X increases, σ_Y^2 will increase while σ_e^2 will stay the same.

(c) A low value of R^2, say $R^2 = 0.2$, implies that the model explains only 20% of the variance of Y. This low value may be due to two reasons: either σ_e^2 is too large, or σ_Y^2 is too small, or both. σ_e^2 may be large because the measurement error is large or there are one or more key input factors that varied during data collection but have not been included in the model, perhaps because they were unknown to us. σ_Y^2 may be small if the range of X is too small compared to what it should have been. A low R^2 suggests the need to review the range of X, the measurement error variance, and the existence of other key factors influencing the output.

3. **Standard Deviation of Error.** The standard deviation of error is obtained as the square root of the error mean square. From Table 3.9, an estimate of σ_e is $\sqrt{776} = 27.8$. This may be approximately interpreted to mean that $\pm 2\sigma_e$ is the 95% prediction interval on any predictions made from the fitted equation. The 95% prediction interval for predicted abrasion loss is approximately ±56. We want σ_e to be low.

Curvilinear Regression. Note that the linear regression above is unlikely to represent the true relationship between abrasion loss and hardness because with a linear relationship abrasion loss will become negative for large values of hardness and this is not possible. In general, an empirically fitted equation only represents the approximate relationship within the range of data. A careful examination of the data in Figure 3.9

suggests a curvilinear relationship. The following quadratic model was obtained:

$$y = 1175 - 24.3x + 0.13x^2 + e$$

Figure 3.10 shows the fitted equation along with the confidence and prediction intervals. All model coefficients were statistically significant: the R^2 was 0.95, which is an improvement on the linear model, and the standard deviation of error reduced to 21.1. As we add more and more terms in the equation, R^2 will automatically increase and the standard deviation of error will automatically decrease. So the idea is not to keep adding terms in the equation until R^2 becomes 100% but to only include statistically significant coefficients in the equation and then evaluate the resultant R^2 for goodness.

Multiple Regression. When output is expressed as a function of more than one input factor, it is called multiple regression. When the tensile test data in Table 3.7 were also included, the following equation was obtained:

$$\text{Abrasion Loss} = 1190 - 25.27 \text{ Hardness} + 0.14 \text{ Hardness}^2 + 0.08 \text{ Tensile Strength} + \text{Error}$$

The coefficient corresponding to tensile strength was statistically insignificant, indicating that the model was not improved by the addition of tensile strength.

Setting Specifications. The results of regression analysis can also be used to set specifications. Let us suppose that the specification for abrasion loss is a maximum loss of 300. What should be the specification for

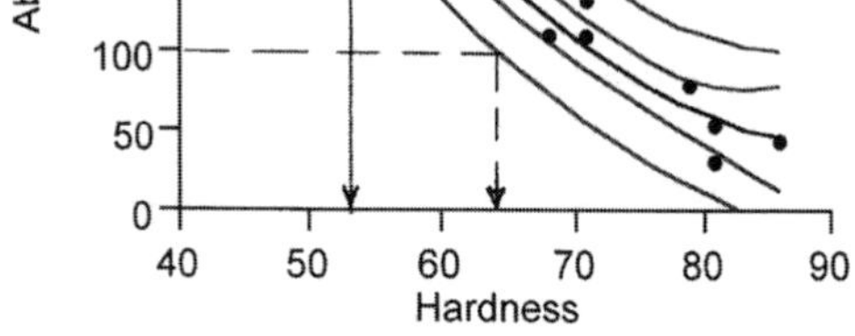

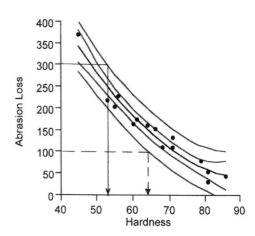

FIGURE 3.10 Curvilinear regression.

hardness? It may be tempting to use the fitted equation and find the value of hardness (x) corresponding to $y = 300$. This turns out to be incorrect because of the uncertainty associated with the model. The correct approach, as shown in Figure 3.10, is to draw a horizontal line from $y = 300$ to the upper individual prediction limit and then read the corresponding value of hardness. From Figure 3.10, if hardness exceeds 53, we are 95% (actually 97.5%) sure that abrasion loss will be less than 300.

What if the specification is two-sided? The specification for abrasion loss is unlikely to be two-sided but just as an example, if there were also a lower specification limit of 100, what would be the specification for hardness? As shown in Figure 3.10, the specification for hardness would be 53 to 64. The variability associated with the fitted equation causes the hardness specification to become much narrower than what it could have been.

3.6.2 Accelerated Stability Tests

This section describes the application of regression analysis to accelerated battery stability tests. Similar approach may be taken to test drug stability or the shelf life of a food product. The response of interest was the internal resistance of the battery, which increased upon storage. The specification required the internal resistance to be less than 500 ohms for a two-year storage at room temperature (25° C). The purpose of conducting accelerated stability tests was to establish acceleration factors so that appropriate short-duration accelerated tests could be conducted to evaluate battery performance.

Data from the accelerated stability tests are shown in Table 3.10. The initial resistance of a large number of batteries was measured and

TABLE 3.10. Accelerated Battery Stability Test

Temperature (T) °K	Time (t) (months)	Resistance y_t (ohms)	$\ln\left[\frac{y_t - y_0}{t}\right]$	$\ln\left[\frac{\ln(y_t/y_0)}{t}\right]$	$1/T$	Sample Size
298	6	13.27	–0.281	–2.665	0.003356	12
298	9	21.73	0.367	–2.291	0.003356	12
298	12	36.30	0.831	–2.131	0.003356	12
313	3	20.52	1.368	–1.257	0.003195	12
313	6	62.98	2.202	–1.111	0.003195	12
313	9	115.15	2.470	–1.250	0.003195	12
323	1	15.47	1.907	–0.560	0.003096	12
323	2	35.03	2.576	–0.365	0.003096	12
323	2.5	44.58	2.663	–0.428	0.003096	12
323	3	70.07	3.018	–0.365	0.003096	12

the average was found to be 8.74 ohms. Batteries were stored at three temperatures—25° C, 40° C, and 50° C—that respectively translate to 298° K, 313° K, and 323° K. Twelve batteries were periodically withdrawn and the observed average resistance values are shown in Table 3.10. It was anticipated that the change in resistance would be described by either a zero order or a first order reaction.

Zero Order Reaction. If y_t denotes the resistance at time t, then for a zero order reaction,

$$\frac{dy_t}{dt} = k_0$$

That is, the reaction rate is constant. Upon integration,

$$y_t = y_0 + k_0 t$$

where y_0 is the initial resistance of 8.74 ohms. The rate constant k_0 increases with temperature and it was assumed that the relationship between k_0 and absolute temperature can be expressed by the following Arrhenius equation:

$$k_0 = Ae^{-(E/RT)}$$

where T is the absolute temperature in degrees Kelvin, E is the activation energy, R is the gas constant, and A is a constant. Hence,

$$y_t = y_0 + Ae^{-(E/RT)}t \tag{3.34a}$$

The parameters to be estimated are A and (E/R). The above equation may be linearized as follows.

$$\ln\left[\frac{y_t - y_0}{t}\right] = \ln A - \frac{E}{RT} \tag{3.34b}$$

This is an equation for a straight line if we plot $\ln[(y_t - y_0)/t]$ against $(1/T)$ as shown in Figure 3.11. The fitted equation is

$$\ln\frac{y_t - y_0}{t} = 29.51 - \frac{8678}{T}$$

Both the intercept and slope are highly statistically significant and the equation explains approximately 80% of the variance of data.

First Order Reaction. For a first order reaction,

$$\frac{dy_t}{dt} = k_1 y_t$$

That is, the reaction rate is proportional to y_t. Upon integration,

$$\ln y_t = \ln y_0 + k_1 t$$

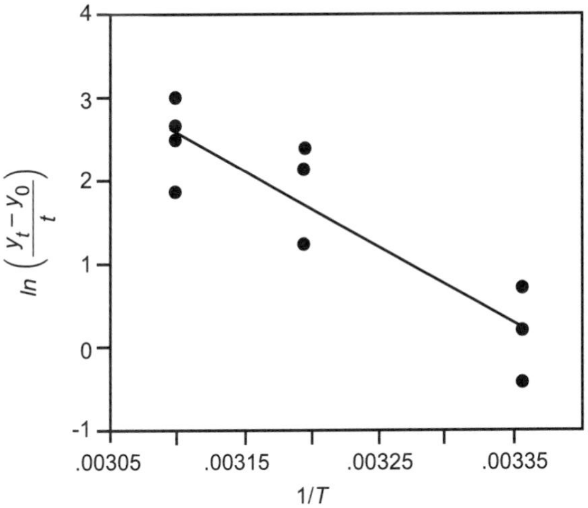

FIGURE 3.11 Zero order reaction.

Using the Arrhenius equation for k_1,

$$\ln y_t = \ln y_0 + Ae^{-(E/RT)}t \tag{3.35a}$$

Rearranging the terms, we get the following equation for a straight line:

$$\ln\left[\frac{\ln(y_t/y_0)}{t}\right] = \ln A - \frac{E}{RT} \tag{3.35b}$$

The fitted equation, shown in Figure 3.12, is

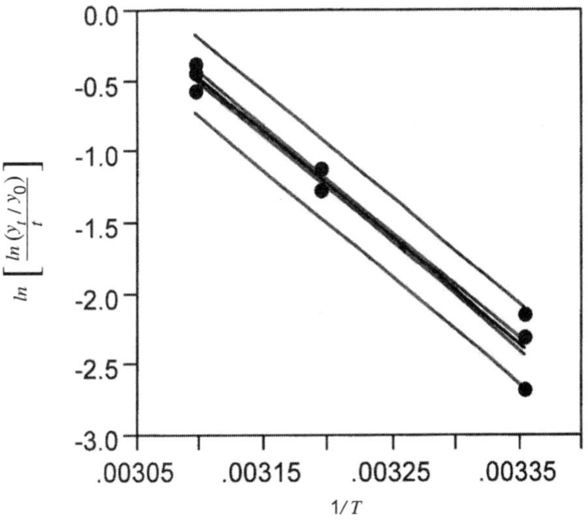

FIGURE 3.12 First order reaction.

$$\ln\left[\frac{\ln(y_t/y_0)}{t}\right] = 22.57 - \frac{7432}{T}$$

This first order model provides a significantly better fit to the data and the equation explains 97% of the variance of data.

Acceleration Factor. If t_{25} and t_{50} denote the time to reach 500 ohms at 25° C and 50° C, respectively, then the acceleration factor (*AF*) may be defined as

$$AF = \frac{t_{25}}{t_{50}}$$

Using the fitted first order reaction equation, we get the following two equations at 25° C (298° K) and 50° C (323° K):

$$\ln \ln(500/8.47) - \ln t_{25} = 22.57 - \frac{7432}{298}$$

$$\ln \ln(500/8.47) - \ln t_{50} = 22.57 - \frac{7432}{323}$$

By subtracting the first equation from the second,

$$\ln\left(\frac{t_{25}}{t_{50}}\right) = \frac{7432}{298} - \frac{7432}{323} = 1.93$$

Hence,

$$AF = \frac{t_{25}}{t_{50}} = e^{1.93} = 6.9$$

This means that 50° C accelerates the degradation rate by 6.9 times compared to 25° C so that two years storage at 25° C is equivalent to 3.5 months storage at 50° C.

Nonlinear Regression. It should be pointed out that Equation (3.35a) may be properly written as

$$\ln y_t = \ln y_0 + Ae^{-(E/RT)}t + e_t \qquad (3.36)$$

where e_t represents error and has $NID(o, \sigma)$ distribution. A similar situation occurs with the zero order model. Equation (3.36) is a nonlinear equation in parameters A and E/R. In regression analysis, what matters is whether the equation is linear or nonlinear in parameters, not in input factors. If the derivative of y_t with respect to any parameter is a function of any parameter, then the equation is nonlinear, as is the case here. On the other hand, Equation (3.35b) is linear. Equation (3.36) cannot be linearized by using the ln transformation. Nonlinear equations require iterative solutions so the parameter estimates obtained above can only be considered as starting guesses for the nonlinear estimation procedures.

On the other hand, it could also be that Equation (3.35a) should be written as

$$\ln y_t = \ln y_0 + Ae^{-(E/RT)}te_t \tag{3.37}$$

which could be linearized as

$$\ln\left[\frac{\ln(y_t/y_0)}{t}\right] = \ln A - \frac{E}{RT} + \ln e_t$$

If $\ln e_t$ has $NID(o, \sigma)$ distribution, then the linearization approach is entirely satisfactory.

4 Control Charts

Data are often collected over time. The descriptive data summaries considered in Chapter 2, such as the mean, standard deviation, and histogram, do not preserve the information along the time dimension. Thus, if there is a time trend in the data, information regarding the nature of this trend is lost in these three summaries. Therefore, we need an additional plot of the data over time as the fourth summary. The control chart is one way to plot data over time.

This chapter starts by defining the role of control charts. The basic principles behind determining control limits are then described. Formulae to design the most commonly used variable and attribute control charts are presented with examples. The out-of-control rules to detect special causes are explained along with the rationale for these rules. Finally, the key success factors for implementing effective charts are discussed.

4.1 ROLE OF CONTROL CHARTS

No two products are exactly alike. This is so because the process that produces these products has many causes of variability. Many of these are common causes. They are a part of the normal operation of the process and produce product variation that is stable and predictable over time. Others are special causes. They are not a part of the normal operation of the process. When they occur, they cause the product variation to become unstable and unpredictable over time.

To achieve product uniformity, it is necessary to either reduce the special and common causes of variation or reduce their effects on product variability. Reducing common causes or their effects requires product and process redesign, whereas reducing special causes requires action to ensure that the process operates in the way it is intended to. Thus, the corrective actions for the two ways to improve the process are fundamentally different. Confusion between common and special causes of variation is expensive and leads to counterproductive corrective actions. Shewhart developed control charts as a graphical method to distinguish between common and special causes of variation. This development brought about a new way of thinking regarding variation and improvements (Shewhart).

The old way of thinking was to classify variation as either meeting specifications or not meeting specifications. Thus, a product was either good (met specifications) or bad (did not meet specifications). A report was either on time or not on time. This thinking led to a strategy of detection, in which the product was inspected and reinspected in an attempt to sort good products from bad products. This strategy is wasteful because it permits resources to be invested in unacceptable products and services. Furthermore, by defining quality as meeting specifications, it permits complacence by not requiring continuous improvement toward meeting the ideal product targets. A better approach is to avoid the production of bad products in the first place by a strategy of prevention. This prevention strategy is based upon an understanding of the process, the causes of variability, and the nature of actions necessary to reduce variation and achieve consistent, on-target performance.

The Process. A process is a way of doing things. All products and services are a result of some process. As shown in Figure 4.1, a process includes the entire combination of customers, suppliers, and producers, involving people, machines, materials, methods, and environment that work together to produce a product or service.

Process quality is measured by the degree to which the process output is exactly as desired by the customers, namely by the degree to which product and service performance characteristics are consistently on target. The performance characteristics and their targets are selected on the basis of customer expectations.

The process output is not consistently on target because the process is affected by many causes of variation. A product varies because of people (operators, training, and experience), machines (machine-to-machine

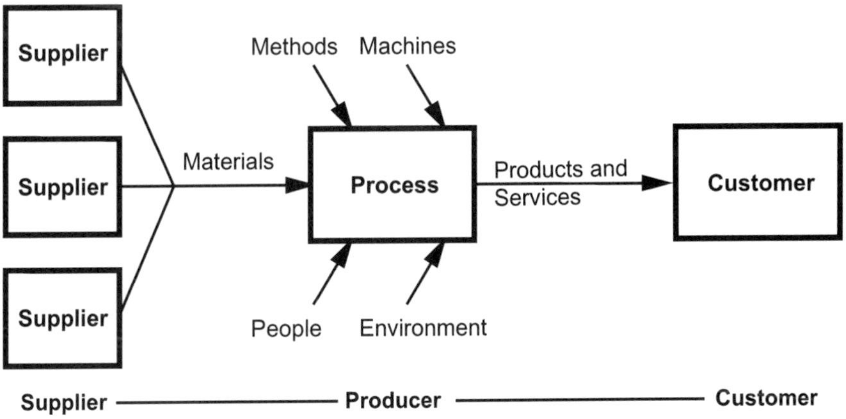

FIGURE 4.1 The process.

differences, wear, and maintenance), methods (temperature control), materials (lot-to-lot and within-lot differences), and environment (ambient temperature and humidity). The time to place a purchase order varies due to people performing the various steps, availability of people, the accuracy of the original request, the reliability of equipment used, and the procedures followed. The collected data additionally vary due to the variability of the measurement system. These causes of variation can be classified as common and special causes.

Common Causes of Variation. These causes of variation are a part of the normal operation of the process and are constantly present. Their effect manifests itself in short-term variability. They are usually large in number and the effect of any one of the causes is relatively small. Some examples are the small changes in process factors, raw materials, ambient conditions, and measurements that occur constantly. Due to the central limit theorem, the cumulative effect of common causes is variation in output characteristic that is usually normally distributed and is stable and repeatable over time. In this situation, shown in Figure 4.2, the process is said to be in a state of statistical control or in control or stable, and the output of the process is predictable within limits. A stable process has constant mean, standard deviation, and distribution over time.

Special Causes of Variation. These refer to causes of variation that are either not always present or are not always present to the same degree. They produce a large change in the output characteristic. Their effect manifests itself in the long term by making the long-term variability larger than the short-term variability. Special causes occur due to the introduction of a new cause that was previously absent or be-

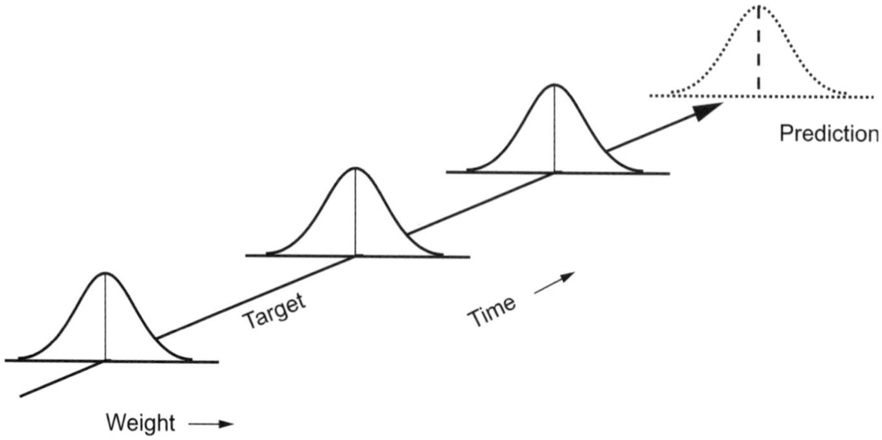

FIGURE 4.2 Common cause variation.

cause of a larger than usual change in a key common cause. Use of the wrong ingredient, wrong process setting, or an untrained operator are examples of special causes. When they occur, they cause an unpredicted change in the mean, variance, or shape of the distribution of output characteristic, as shown in Figure 4.3. Therefore, predictions regarding the future distribution of output characteristics cannot be made. The process is said to be out of control or unstable.

Improvement Actions. Since variation is due to either common or special causes, it follows that there are two ways to reduce variation. One

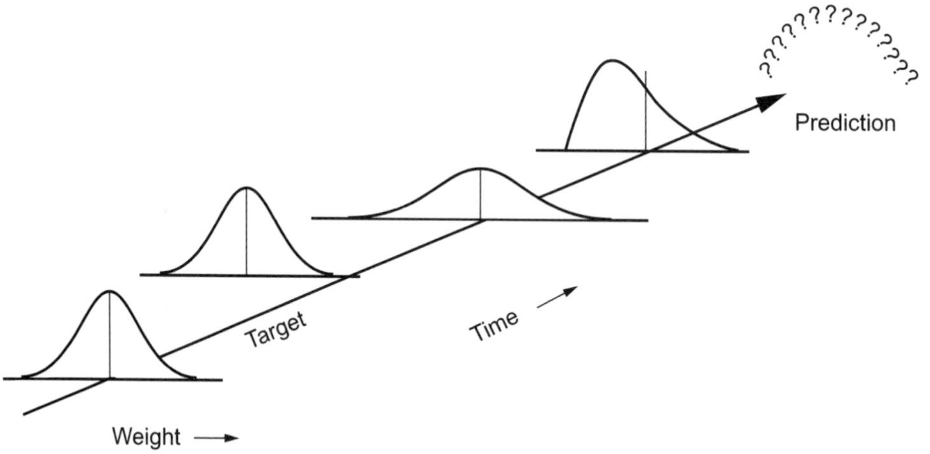

FIGURE 4.3 Special cause variation.

way is to reduce common causes or their effects; the other is to reduce special causes or their effects.

Common causes are inherent to the process. Therefore, to reduce common cause variation, the process itself must be changed. For example, if the raw material variability is one common cause of variation, then the process needs to be changed to either procure the raw material to tighter specifications or make product design and manufacturing less sensitive to raw material variation. If the purchase order cycle time variation is caused by the frequent absence of some personnel to sign purchase orders, then the process needs to be changed to either have alternate signers or a reduced number of signers. Reducing common cause variation requires a change in the process.

On the other hand, if the variation is due to special causes, then it has nothing to do with the way the process was intended to operate. If the use of a wrong tool is the special cause, the process need not be changed; rather, it needs to be executed as designed. Reducing special cause variation requires the identification and removal of special causes.

Two Mistakes. Confusion between common and special causes is costly and leads to counterproductive corrective actions. This is so because the two corrective actions are fundamentally different. The following two mistakes result from the confusion between common and special causes:

1. Mistake 1 is to ascribe variation to a special cause when it is the result of a common cause.
2. Mistake 2 is to ascribe variation to a common cause when it is the result of a special cause.

Overadjustment is a common example of mistake 1. Never doing anything to find and remove a special cause is a common example of mistake 2. Mistake 1 can be thought of as reacting to noise and mistake 2 as ignoring a signal. Both of these mistakes lead to increased variability.

Figure 4.4 shows the output of a stable process (A.I.A.G.), with the

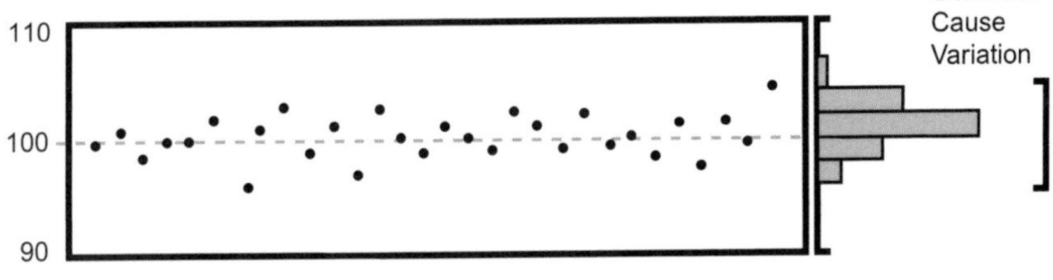

FIGURE 4.4 Results of a stable process with no adjustment.

process mean being at the target value of 100. There are no special causes of variation. However, if each deviation from target is considered to be due to a special cause and the mean is adjusted for this deviation, then the adjustment becomes an additional source of variation. This is called tampering and leads to higher variability, as shown in Figure 4.5. This is mistake 1.

The results of mistake 2 are shown in Figure 4.6. At certain point in time, a special cause shifts the process mean from 100 to 102. In this case, it would have been correct to adjust the mean but perhaps because this shift was not detected, no corrective action is taken. Again, this leads to higher variability.

It is therefore necessary to be able to distinguish between the two types of variation so that appropriate corrective actions may be taken. A control chart is a graphical method used to distinguish between common cause variation and special cause variation. The role of a control chart is to help us identify the presence and nature of special causes.

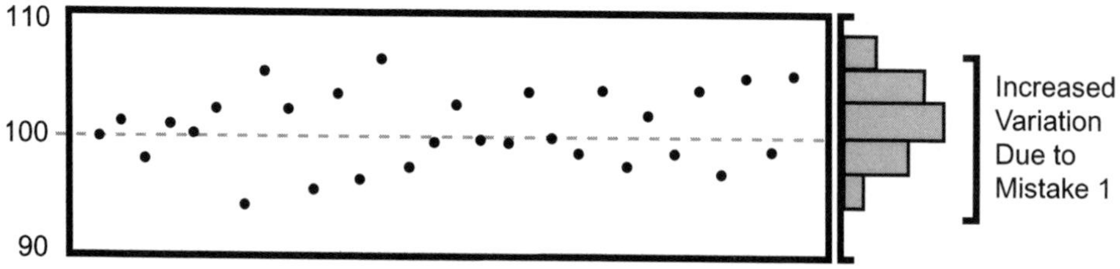

FIGURE 4.5 Results of a stable process with adjustments (mistake 1).

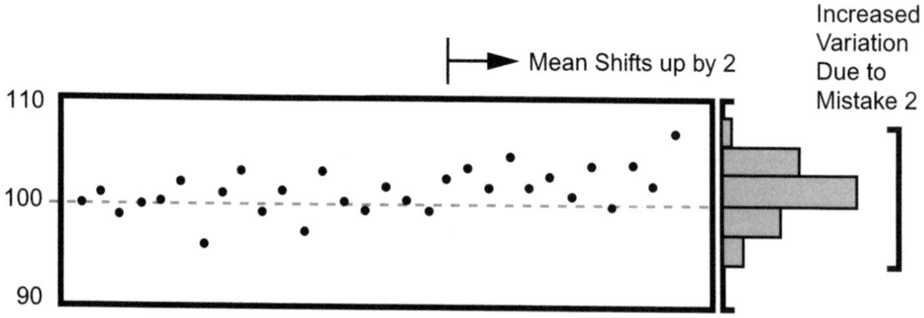

FIGURE 4.6 Results of an unstable process with no adjustments (mistake 2).

4.2 LOGIC OF CONTROL LIMITS

What distinguishes a control chart from a simple plot of data over time is the presence of control limits. Control limits provide the boundary to distinguish between common and special cause variation. If a plotted point falls outside the control limits, the presence of a special cause is detected. What is the logic used to calculate control limits? Let us consider an example.

Product weight is often a characteristic of interest. We wish to find out whether our process is stable with respect to this characteristic. For this purpose, every hour, we measure the weights of n consecutive products. The data collected every hour is known as a subgroup, the subgroup size being n. The sampling interval is one hour. For each subgroup, the subgroup average, standard deviation, and histogram may be constructed and plotted over time, as shown in Figure 4.7.

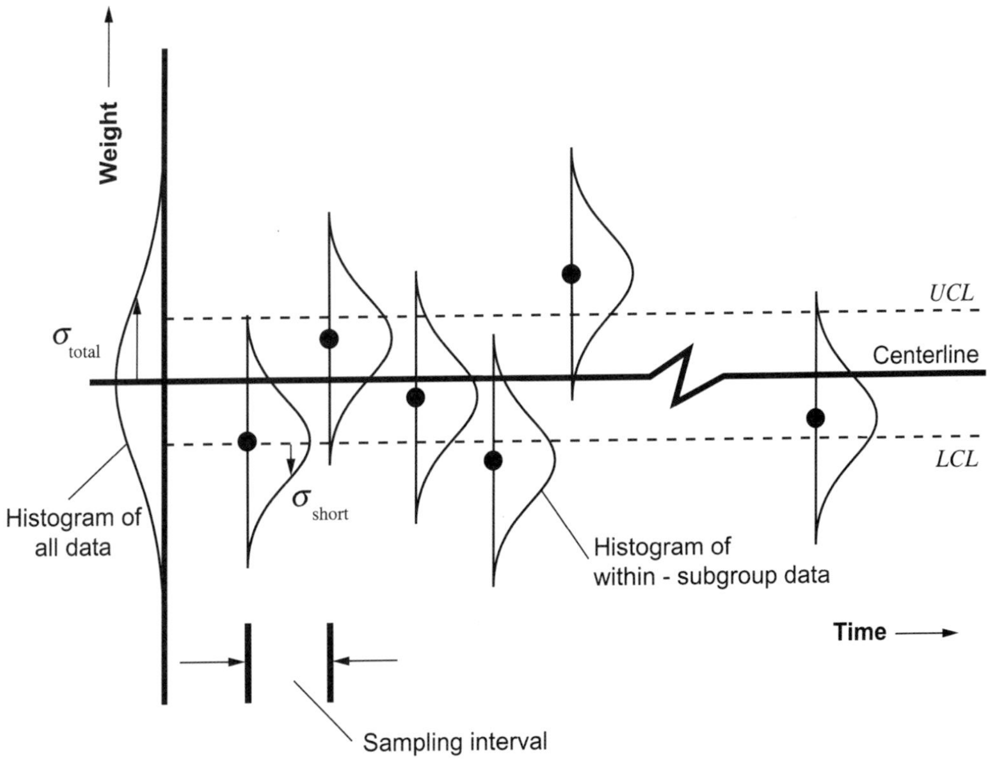

FIGURE 4.7 Logic of control limits.

Is the process mean constant over time? If it is not, then the process is unstable with respect to the mean. This question is answered by constructing a chart of averages known as an $\bar{X}$ chart. The individual values within a subgroup are denoted by x and $\bar{x}$ denotes the subgroup average. An $\bar{X}$ chart is characterized by a centerline, upper and lower control limits, subgroup size, and sampling interval. So to design an $\bar{X}$ control chart, we need to answer four questions:

1. Where should the centerline be?
2. How far away from the centerline should the control limits be?
3. What should be the subgroup size n?
4. What should be the sampling interval?

We will answer all of these questions in this chapter but for now, let us focus only on the first two questions.

Where should the centerline be? If the purpose of the control chart is to determine whether the process is stable, then it seems reasonable to draw the centerline at the grand mean of all data. The grand mean is usually denoted by $\bar{\bar{x}}$. On the other hand, if the purpose is to bring the mean on target, then the control chart should be centered on target. In Figure 4.7, the subgroup averages are shown by large dots and the centerline is drawn at $\bar{\bar{x}}$.

How far away from the centerline should the control limits be? The function of the control limits is to help us identify when the process is out of control. The interpretation will be that as long as the plotted $\bar{x}$ values fall randomly within the control limits, the process is in control. If a point falls outside the control limits, the process will be deemed to be out of control at that point. The logic to compute the control limits is as follows.

1. The upper control limit (*UCL*) and the lower control limit (*LCL*) should be symmetrically positioned around the centerline. This is so because $\bar{x}$ is likely to be normally distributed and a normal distribution is symmetric.
2. The distance between each control limit and the centerline should be 3 sigma. This is so because if the process is in control, namely, all the $\bar{x}$ values come from a normal distribution with a constant mean equal to $\bar{\bar{x}}$, then the probability that a specific $\bar{x}$ will fall outside the control limits by pure chance will be less than 0.3% or 3 in 1000. If a plotted $\bar{x}$ does fall outside the control limits, we will interpret it as an out of control signal rather than saying that the

3 in a 1000 event has just occurred. Since 3 out of 1000 plotted $\bar{x}$ values are likely to fall outside the control limits by pure chance when the process is in control, a false alarm will occur on the average once every 350 points or so. If this frequency of false alarm is deemed to be excessive, we can use (say) 4 sigma limits. The false alarm frequency will drop dramatically. However, it will simultaneously become more difficult to identify shifts in the process. Consequently, 3 sigma limits have now become a standard for all control charts.

3. What is sigma? Since we are plotting $\bar{x}$, sigma must be the standard deviation of $\bar{X}$. We know that $\sigma_{\bar{X}} = \sigma/\sqrt{n}$, where σ is the standard deviation of individual values X and n is the subgroup size. So the distance between the *UCL* and the centerline becomes $3\sigma/\sqrt{n}$.

4. How should σ be computed? It needs to be computed to only reflect the common cause variability of the process. Only then will the control limits correctly separate common cause variation from special cause variation. Therefore, it would be wrong to compute σ as the standard deviation of all data, denoted in Figure 4.7 by σ_{total}. This is so because the process may be unstable, in which case σ_{total} will include both common and special cause variability.

   The correct procedure is to compute σ based upon data within a subgroup. This is shown in Figure 4.7 as σ_{short}, indicating short-term variation. The actual computation involves first calculating the within-subgroup variance for each subgroup and then calculating σ_{short} as the square root of the average within-subgroup variance. This short-term standard deviation correctly reflects the common cause standard deviation of the process. By definition, special causes occur less frequently and could not possibly influence every within-subgroup variance. Therefore,

$$\text{Control limits for } \bar{X} \text{ chart} = \bar{\bar{x}} \pm 3\frac{\sigma_{\text{short}}}{\sqrt{n}} \qquad (4.1)$$

It is easy to see some consequences of this formula. As subgroup size increases, control limits narrow inversely proportional to $\sqrt{n}$. Also, the mere fact that the $\bar{X}$ chart control limits are inside the specification limits does not mean that the process is good. Specification limits usually are for individual values and $\bar{X}$ chart control limits are for averages. The control limits can be made narrow to any extent by simply increasing the subgroup size. It is usually wrong or counterproductive to draw control limits and specification limits on the same chart. Specifications are in no way used in setting these control limits.

4.3 VARIABLE CONTROL CHARTS

There are two types of control charts: variable control charts and attribute control charts. Variable control charts apply when the characteristic of interest is measured on a continuous scale. Attribute control charts deal with count data. This section introduces three commonly used variable control charts: the average and range ($\bar{X}$ and R), the average and standard deviation ($\bar{X}$ and S), and the individual and moving range (X and mR) charts.

4.3.1 Average and Range Charts

Product Weight Example. Product weight control is an important strategy in the food industry, the pharmaceutical industry, and in many other industries. If weight variability could be reduced, the average weight can be targeted closer to the labeled weight, resulting in significant cost reductions. With this in mind, the weight of a certain food product was the characteristic of interest. Weights of five consecutive products were measured every hour. Thus, the subgroup size is five and the sampling interval is one hour. Table 4.1 shows the collected data for 22 subgroups for the amount, in grams, by which the weight exceeded 250 grams. At the time the data were collected, no control charts were kept. As a result, an improvement opportunity was missed.

The $\bar{X}$ and R charts assume that, under stability, $X \sim NID(\mu, \sigma)$. For a normal distribution, μ and σ are independent of each other and two charts need to be designed: the $\bar{X}$ chart to monitor process mean and the R chart to monitor process variability. The necessary intermediate calculations are shown in Table 4.1, and two procedures are illustrated below to design the $\bar{X}$ chart.

Procedure 1. This procedure follows the method outlined in the previous section and uses Equation (4.1) to determine control limits. From Table 4.1, $\bar{\bar{x}} = 3.66$ and $s^2_{\text{spooled}} = 1.067$. Since the subgroup size ($n = 5$) is constant, s^2_{spooled} is simply the average of within-subgroup variance estimates. s_{spooled} estimates σ_{short} as 1.033 and the centerline and the lower and upper control limits (LCL and UCL) for the $\bar{X}$ chart are:

$$\bar{\bar{x}} \pm 3\frac{\sigma_{\text{short}}}{\sqrt{n}} = 3.66 \pm 3\left(\frac{1.033}{\sqrt{5}}\right) = 3.66 \pm 1.39$$

$$\text{Centerline} = 3.66$$

$$LCL = 2.27$$

$$UCL = 5.05$$

TABLE 4.1 Product Weight Data (x = weight − 250)

Subgroup	Data (x)					Average $\bar{x}$	Range (R)	Std. Dev. (s)	Variance $(s)^2$
1	4.2	1.9	4.1	2.5	3.8	3.30	2.3	1.04	1.08
2	2.4	5.1	4.4	3.7	5.9	4.30	3.5	1.34	1.80
3	3.1	2.4	3.6	3.9	3.1	3.22	1.5	0.57	0.33
4	2.4	3.2	5.4	4.8	3.9	3.94	3.0	1.20	1.45
5	5.1	2.8	3.8	4.2	4.9	4.16	2.3	0.92	0.85
6	3.1	2.8	6.1	3.4	3.9	3.86	3.3	1.32	1.73
7	2.2	3.8	5.4	3.6	4.4	3.88	3.2	1.17	1.37
8	2.1	1.8	1.1	1.3	1.7	1.60	1.0	0.40	0.16
9	3.3	2.8	4.1	3.9	3.2	3.46	1.3	0.53	0.28
10	4.1	4.5	2.8	5.1	4.4	4.18	2.3	0.85	0.73
11	1.9	4.8	3.7	2.7	3.8	3.38	2.9	1.11	1.24
12	2.8	3.4	3.1	4.8	3.4	3.50	2.0	0.77	0.59
13	3.8	3.2	4.9	3.7	4.1	3.94	1.7	0.63	0.39
14	2.8	3.1	3.1	2.4	4.4	3.16	2.0	0.75	0.56
15	4.5	5.9	1.8	5.4	4.6	4.44	4.1	1.58	2.51
16	4.4	4.1	1.8	4.5	4.2	3.80	2.7	1.13	1.28
17	3.1	3.8	3.9	2.7	3.3	3.36	1.2	0.50	0.25
18	5.9	4.1	1.9	4.9	4.7	4.30	4.0	1.49	2.22
19	3.9	3.4	3.7	5.2	4.4	4.12	1.8	0.70	0.50
20	1.9	2.8	5.2	4.3	4.1	3.66	3.3	1.30	1.70
21	2.9	2.2	4.8	3.1	4.5	3.50	2.6	1.11	1.23
22	3.5	5.4	3.1	2.6	2.9	3.50	2.8	1.11	1.24
						$\bar{\bar{x}} = 3.66$	$\bar{R} = 2.50$	$\bar{s} = 0.978$	$s^2_{spooled} = 1.067$
									$s_{spooled} = 1.033$

Procedure 2. A computationally simple method has been devised to approximately calculate σ_{short} and control limits without having to compute $s_{spooled}$. The control limit formulae are summarized in Table 4.2, and the required constants are given in Appendix G. The procedure is illustrated below.

$$\sigma_{short} = \frac{\bar{R}}{d_2} \tag{4.2}$$

For the product weight data, $\bar{R} = 2.50$, and for $n = 5$, from Appendix G, $d_2 = 2.326$. Hence $\sigma_{short} = 2.50/2.326 = 1.074$, which is slightly different from the value calculated before. The control limits for $\bar{X}$ chart are:

$$\bar{\bar{x}} \pm 3\frac{\sigma_{short}}{\sqrt{n}} = \bar{\bar{x}} \pm 3\frac{\bar{R}}{d_2/\sqrt{n}}$$

TABLE 4.2 Designing Variable Control Charts

Chart	Centerline	*LCL*	*UCL*
Average ($\bar{X}$)	$\bar{\bar{x}}$	$\bar{\bar{x}} - A_2\bar{R}$	$\bar{\bar{x}} + A_2\bar{R}$
Range (R)	$\bar{R}$	$D_3\bar{R}$	$D_4\bar{R}$
Average ($\bar{X}$)	$\bar{\bar{x}}$	$\bar{\bar{x}} - A_3\bar{s}$	$\bar{\bar{x}} + A_3\bar{s}$
Standard Deviation (S)	$\bar{s}$	$B_3\bar{s}$	$B_4\bar{s}$
Individuals (X)	$\bar{x}$	$\bar{x} - 2.66\ m\bar{R}$	$\bar{x} + 2.66\ m\bar{R}$
Moving Range (mR)	$m\bar{R}$	0	$3.268\ m\bar{R}$

If we let $A_2 = 3/d_2\sqrt{n}$ then, as shown in Table 4.2,

$$\text{Control limits for } \bar{X} \text{ chart} = \bar{\bar{x}} \pm 3A_2\bar{R} \tag{4.3}$$

For the product weight data, $\bar{\bar{x}} = 3.66$, $\bar{R} = 2.50$, and for $n = 5$, from Appendix G, $A_2 = 0.577$. Substituting in Equation (4.3),

$$\text{Control limits for } \bar{X} \text{ chart} = 3.66 \pm 0.577\ (2.50) = 3.66 \pm 1.44$$

$$\text{Centerline} = 3.66$$

$$LCL = 2.22$$

$$UCL = 5.10$$

These limits differ slightly from those calculated using the pooled estimate of variance because $\sigma_{\text{short}} = \bar{R}/d_2$ is an approximation. One advantage of this approximate procedure is that it is less sensitive to outliers. It is also simpler and we will use it for control chart design.

The centerline and control limits for the range chart are $\bar{R} \pm 3\sigma_R$ where σ_R is the standard deviation of range and equals $d_3\sigma_{\text{short}}$. Values of d_3 are tabulated in Appendix G. Hence,

$$\text{Control limits for } R \text{ chart} = \bar{R} \pm 3d_3\frac{\bar{R}}{d_2} = \bar{R}\left(1 \pm 3\frac{d_3}{d_2}\right)$$

If we let

$$D_3 = 1 - 3\frac{d_3}{d_2}$$

and

$$D_4 = 1 + 3\frac{d_3}{d_2}$$

Then, as shown in Table 4.2, the range chart control limits are computed as:

$$\text{Centerline} = \overline{R}$$
$$LCL = D_3\overline{R} \qquad (4.4)$$
$$UCL = D_4\overline{R}$$

The values of D_3 and D_4 are given in Appendix G. For the product weight data, $\overline{R} = 2.50$, $n = 5$, and from Appendix G, $D_3 = 0$ and $D_4 = 2.114$. Hence, the centerline and control limits for the R chart are:

$$\text{Centerline} = 2.50$$
$$LCL = 0$$
$$UCL = 2.114\,(2.50) = 5.29$$

The charts are shown in Figure 4.8. Note that the R chart is in control, indicating that the within-subgroup variability is constant over time. The $\overline{X}$ chart shows one subgroup mean below the lower control limit, indicating the presence of a special cause affecting the process average. If this process had been monitored in real time using a control chart, perhaps the special cause would have been identified, leading to an improvement.

4.3.2 Average and Standard Deviation Charts

Standard deviation is a better estimate of within-subgroup variability than range, and the average of within-subgroup standard deviations $\bar{s}$ leads to a better estimate of σ_{short} as compared to that obtained from $\overline{R}/d_2$. Therefore, $\overline{X}$ and S charts are preferred over $\overline{X}$ and R charts. However, the advantage is not substantial, particularly for subgroup sizes around 5, and since R is easier to compute, $\overline{X}$ and R charts are satisfactory, especially for hand computations. σ_{short} and control limits for $\overline{X}$ and S charts are calculated as follows:

$$\sigma_{\text{short}} = \frac{\bar{s}}{c_4} \qquad (4.5)$$

For the product weight data, $\bar{s} = 0.978$, and from Appendix G, $c_4 = 0.94$. Hence, $\sigma_{\text{short}} = 1.04$, which is slightly different from the prior estimates. From Table 4.2,

$$\text{Control limits for the } \overline{X} \text{ chart} = \bar{\bar{x}} \pm A_3\bar{s} \qquad (4.6)$$

For the product weight data, $\bar{\bar{x}} = 3.66$, $\bar{s} = 0.978$ and from Appendix G, $A_3 = 1.427$ Hence, from Equation (4.6),

$$\text{Control limits for the } \overline{X} \text{ chart} = 3.66 \pm 1.40$$
$$\text{Centerline} = 3.66$$

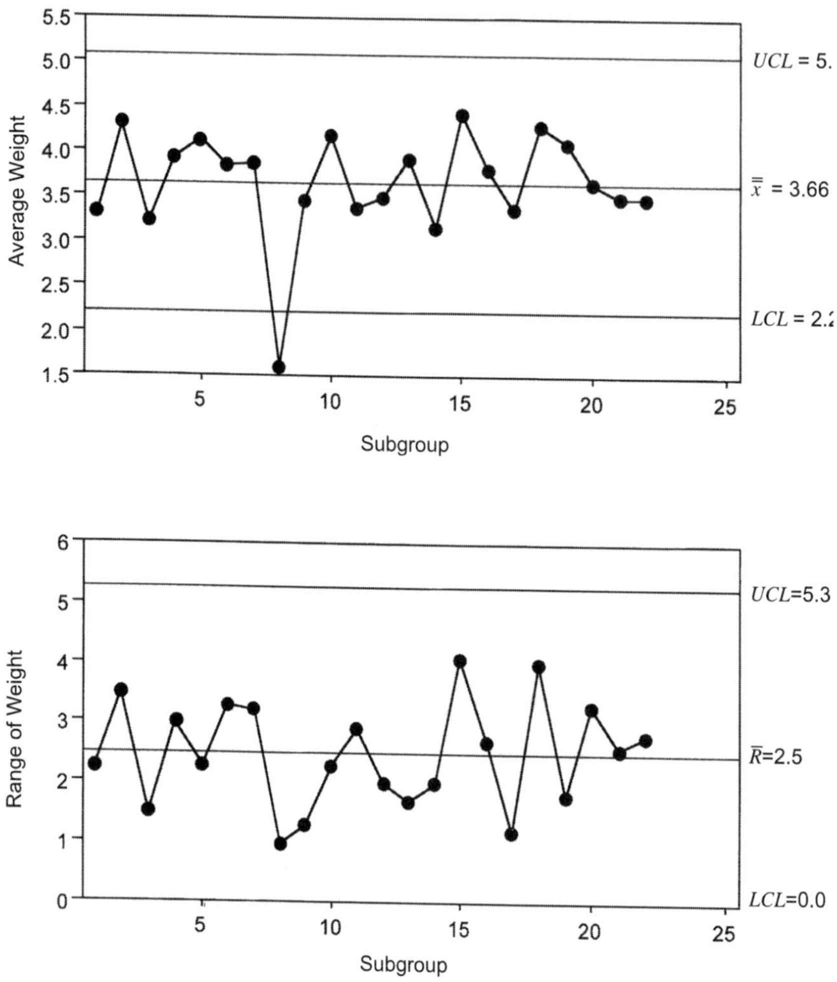

FIGURE 4.8 $\overline{X}$ and R charts.

$$LCL = 2.26$$
$$UCL = 5.06$$

Also from Table 4.2, the centerline and control limits for the S chart are:

$$\text{Centerline} = \bar{s}$$
$$LCL = B_3\bar{s} \qquad (4.7)$$
$$UCL = B_4\bar{s}$$

For the product weight data, $\bar{s} = 0.978$, and from Appendix G, $B_3 = 0$ and $B_4 = 2.089$. Hence, for the S chart,

$$\text{Centerline} = 0.978$$

$$LCL = 0$$

$$UCL = 2.04$$

The $\bar{X}$ and S charts are shown in Figure 4.9. The conclusions are the same as those obtained from the $\bar{X}$ and R charts.

4.3.3 Individual and Moving Range Charts

In many situations, the subgroup size n is equal to one and control charts need to be designed based upon individual values. This typically

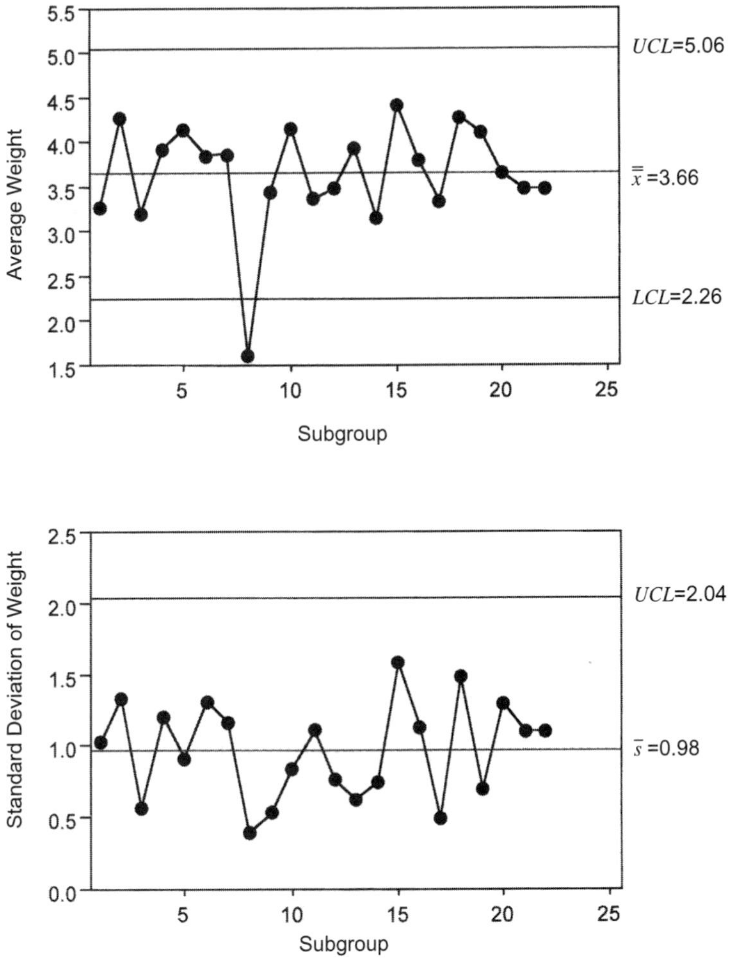

FIGURE 4.9 $\bar{X}$ and S charts.

occurs when one measurement is taken per batch, per day, per week, or per month; when measurements are expensive, as with destructive tests; or when the within-subgroup variance is small and multiple within-subgroup measurements are unnecessary. With only one measurement per subgroup, the within-subgroup variance cannot be calculated and the procedure to calculate control limits has to be modified.

The within-subgroup variance is an estimate of the short-term process variance. For a chart of individual values, short-term variation is represented by the difference between successive values. The absolute difference, or range, between successive values is called the moving range (mR), and the average of the moving range values, denoted by $m\bar{R}$, is used to estimate short-term variance. The procedure is illustrated with an example.

Production Rate Example. The daily production data, shown in Table 4.3, was displayed inside the entrance to a manufacturing plant. The daily production numbers refer to the number of containers of product produced per day. Alongside some of the numbers there were smiling faces. It turned out that the smiling faces were placed there when the daily production volume exceeded the goal of 3000 containers per day. When the production workers were asked about the message conveyed by the smiling faces, their response was that management was trying to communicate that if everybody worked as hard every day as they had worked on the smiling face days, there would be smiling faces everywhere. What can we learn from the data?

The calculated moving range values are also shown in Table 4.3. There will always be one less moving range value compared to the number of individual values. The average daily production $\bar{x}$ is 2860 containers and the average moving range $m\bar{R}$ is 279 containers. A measure of short-term variability and the control limits are obtained as follows, where d_2 is obtained from Appendix G for a subgroup size of 2:

$$\sigma_{\text{short}} = \frac{m\bar{R}}{d_2} = \frac{m\bar{R}}{1.128}$$

$$X \text{ chart control limits} = \bar{x} \pm 3\sigma_{\text{short}} = \bar{x} \pm 2.66\, m\bar{R} \qquad (4.8)$$

For the daily production example, $\sigma_{\text{short}} = 279/1.128 = 247$ and the X chart centerline and control limits are:

$$\text{Centerline} = 2860$$

$$LCL = \bar{x} - 2.66\, m\bar{R} = 2860 - 2.66(279) = 2118$$

$$UCL = \bar{x} + 2.66\, m\bar{R} = 2860 + 2.66(279) = 3602$$

Again, from Table 4.2, the centerline and control limits for $m\bar{R}$ chart are:

TABLE 4.3 Daily Production Data

Day	Daily Production	mR
1	2899	
2	3028—☺	129
3	2774	254
4	2969	195
5	3313—☺	344
6	3139—☺	174
7	2784	355
8	2544	240
9	2941	397
10	2762	179
11	2484	278
12	2328	156
13	2713	385
14	3215—☺	502
15	2995	220
16	2854	141
17	3349—☺	495
18	2845	504
19	3128—☺	283
20	2557	571
21	2845	288
22	3028—☺	183
23	2749	279
24	2660	89
25	2604	56

$\bar{x} = 2860$ $m\bar{R} = 279$

$$\text{Centerline} = m\bar{R}$$
$$LCL = 0 \tag{4.9}$$
$$UCL = 3.268\, m\bar{R}$$

For the production rate example, $m\bar{R} = 279$ and

$$\text{Centerline} = 279$$
$$LCL = 0$$
$$UCL = 912$$

The X chart is shown in Figure 4.10. In general, the mR chart does not contain any additional information and is not shown. The X chart shows the process to be stable. This means that the daily production rate could

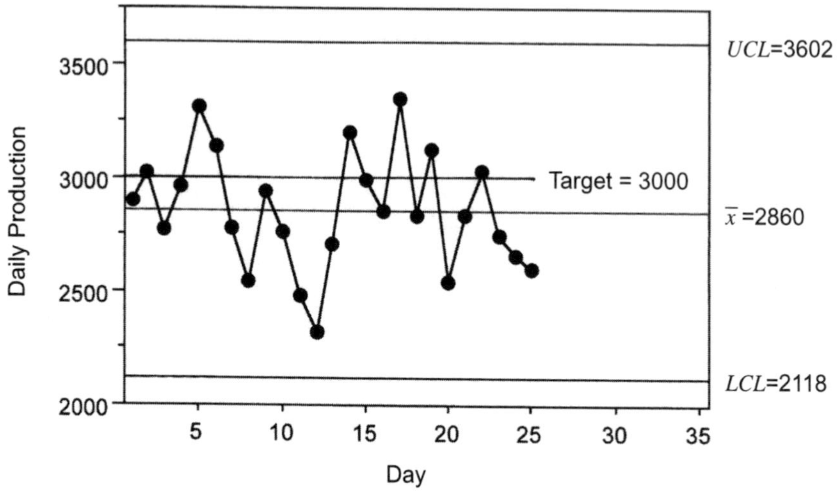

FIGURE 4.10 *X* chart for daily production.

fluctuate between 2118 and 3602 containers per day purely due to common cause variability. On some days, the goal of 3000 containers is met. This happens about 30% (7 smiling faces out of 25) of the time. There are no special causes. There is nothing special about the smiling face days. What is needed is an analysis of the entire process, including suppliers, to identify the causes of variability, and specific management actions to improve the process.

Product Weight Example (continued). We can treat the $\bar{x}$ data in Table 4.1 as if they were individual values and construct $\bar{X} - mR$ charts. The calculated value of $m\bar{R}$ = 0.69 and the control limits for $\bar{X}$ chart = 3.66 ± 1.83.

The $\bar{X}$ chart for the product weight example has now been designed in four different ways and the results are summarized in Table 4.4. The control limits obtained in these four ways are not exactly equal. The first

TABLE 4.4 Four Ways to Design the $\bar{X}$ Chart

Method	Control Limits
1. Using Pooled σ^2_{short}	3.66 ± 1.39
2. Using $\bar{X} - R$	3.66 ± 1.44
3. Using $\bar{X} - S$	3.66 ± 1.40
4. Using $\bar{X} - mR$	3.66 ± 1.83

three methods use σ_{short} to compute the control limits. The calculated limits are close to each other. The small differences are due to the varying approximations in calculating short-term σ.

The $\bar{X} - mR$ method is fundamentally different. It uses a combination of both within-subgroup ($\sigma_w = \sigma_{\text{short}}$) and between-subgroup ($\sigma_b$) standard deviations. Specifically, it uses an estimate of $3\sqrt{\sigma_w^2/n + \sigma_b^2}$ as the distance between the centerline and the control limits, n being the subgroup size. Consequently, the limits are wider. We will see later that this method is useful when $\sigma_w \ll \sigma_b$ and the production operator cannot take corrective action against the usual level of σ_b. Using the techniques of variance components described in Chapter 7, for the product weight data, $\sigma_w^2 = 1.0669$, $\sigma_b^2 = 0.145$, and $3\sqrt{\sigma_w^2/5 + \sigma_b^2} = 1.80$. The control limits are 3.66 ± 1.80, close to those reported in Table 4.4 for the $\bar{X} - mR$ approach.

4.4 ATTRIBUTE CONTROL CHARTS

Variable control charts dealt with characteristics measured on a continuous scale. A second category of data is one in which the characteristic is measured on a discrete scale in terms of the number of occurrences of some attribute. Such count data are called attribute data. Number of defects per product and classifying a characteristic as acceptable or not acceptable, present or absent, are examples of such data.

In some instances, attribute data occur naturally, as in the case of number of accidents per month. In other cases, measurements could be made on a continuous scale but attribute data are used because they are easy to understand, inexpensive to measure, or already available. The diameter of a shaft can be measured on a continuous scale, but a go/no-go gauge is used to classify the shaft as acceptable or not acceptable because it is easy to do so. The major disadvantage of attribute data is that attribute data do not provide detailed information regarding the characteristic. The resultant loss in discrimination increases the subgroup size requirements for control charts and reduces the ability for continuous improvement. As an example, if products are classified as within and outside specifications, the data do not have the ability to discriminate between products that are all within specification but at different distances from the target value. Therefore, every effort should be made to collect variable data whenever it is practical to do so.

Attribute data are count data. Generally, what is counted are defectives or defects. The product is either defective or not. There are only two classifications. This is (0, 1) type data. The number of defects per product may also be counted. This results in integer data of (0, 1, 2, 3, 4 . . .) type. For these counts to be compared, they must be based upon a fixed sample size. If the sample size varies, then the counts are divided by the sample size to make meaningful comparison possible.

The four commonly used control charts are based upon the type of count data (0,1 or integer) and whether the subgroup size is constant or changing. This is shown in Figure 4.11. The *np* and *p* charts are used for (0,1) data and the *c* and *u* charts are used for the integer data. The *np* and *p* charts are based upon the binomial distribution and the *c* and *u* charts use Poisson distribution. Since all count data consist of individual values, the chart of the individuals can be used as an alternative if the binomial or Poisson assumptions discussed below are not satisfied (Wheeler and Chambers).

Since the *p* chart can handle changing subgroup size, it can obviously work for a fixed subgroup size. So we do not need the *np* chart. Similarly, the *u* chart eliminates the need for the *c* chart. Only the *p* and *u* charts are discussed below.

4.4.1 Fraction Defective (*p*) Chart

Under certain conditions, number of defectives has a binomial distribution that forms the basis for a *p*-chart.

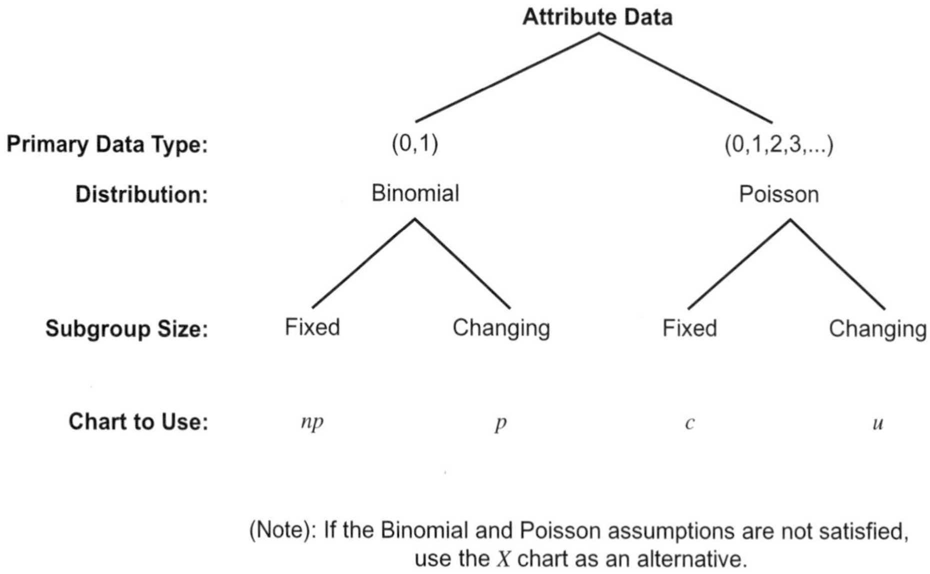

FIGURE 4.11 Selecting attribute control charts.

Inventory Accuracy Example. In order to assess inventory accuracy, a simple data collection scheme was devised. The intent was to check 100 inventory items each week and record the number of misplaced items. In reality, the number of items checked per week varied. The collected data are shown in Table 4.5.

For the binomial distribution to apply, the following assumptions have to be true.

1. Each item is classified as either having or not having some attribute (0, 1 data).
2. When the process is stable, there is a constant probability p of finding an item with the attribute.
3. The probability of finding an item with the attribute does not change as a function of another item having the attribute.

For the inventory accuracy example, the second assumption is somewhat questionable because the likelihood of misplacing small and big items may be different. If the assumptions are deemed satisfactory, then the variable subgroup size, which is the number of items checked each week, suggests a p chart.

TABLE 4.5 Inventory Accuracy Data

Week	# Checked (n_i)	# Misplaced (x_i)	Fraction Misplaced (p_i)
1	100	10	0.100
2	60	4	0.067
3	84	7	0.083
4	122	12	0.098
5	100	6	0.060
6	50	4	0.080
7	67	5	0.075
8	100	5	0.050
9	115	9	0.078
10	75	3	0.040
11	82	6	0.073
12	100	7	0.070
13	130	7	0.054
14	67	5	0.075
15	45	2	0.044
16	100	4	0.040
17	134	8	0.060
Total	1531	104	

The procedure to design the p chart is as follows:

1. Obtain the values of number misplaced (x_i) and number checked (n_i).
2. Calculate the fraction misplaced, $p_i = x_i/n_i$.
3. Calculate the centerline to be

$$\bar{p} = \frac{\text{total \# misplaced}}{\text{total \# checked}} = \frac{104}{1531} = 0.068$$

   Note that $\bar{p}$ is not calculated as the average of p_i values because the sample size is not constant. Under the assumption of stability, $\bar{p}$ estimates p, the probability of any randomly selected item being misplaced.

4. Under binomial assumptions, the standard deviation of p_i is estimated as $\sqrt{\bar{p}(1-\bar{p})/n}$. Hence,

$$\text{Control limits for } p \text{ chart} = \bar{p} \pm 3\sqrt{\bar{p}(1-\bar{p})/n} \tag{4.10}$$

The three-sigma control limits for all attribute charts are similarly derived and are summarized in Table 4.6. Substituting $\bar{p} = 0.068$ in Equation (4.10),

$$\text{Centerline} = 0.068$$

$$LCL = 0.068 - \frac{0.755}{\sqrt{n}}$$

$$UCL = 0.068 + \frac{0.755}{\sqrt{n}}$$

Since the subgroup size n changes from week to week, the control limits fluctuate as shown in Figure 4.12. In the case of the $\bar{X}$ chart or any other chart, if the subgroup size varies, the control limits will fluctuate as well.

TABLE 4.6 Control Limits for Attribute Charts

Chart	Centerline	Control Limits
np	$n\bar{p}$	$n\bar{p} \pm 3\sqrt{n\bar{p}(1-\bar{p})}$
p	$\bar{p}$	$\bar{p} \pm 3\sqrt{\bar{p}(1-\bar{p})/n}$
c	$\bar{c}$	$\bar{c} \pm 3\sqrt{\bar{c}}$
u	$\bar{u}$	$\bar{u} \pm 3\sqrt{\bar{u}/n}$

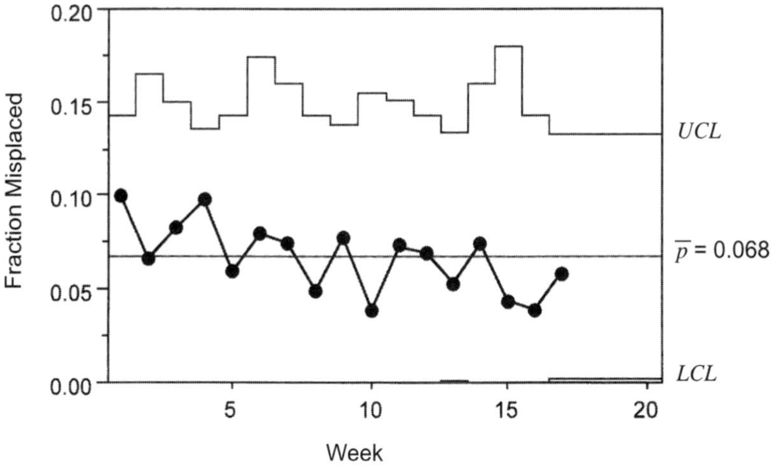

FIGURE 4.12 *p* chart for inventory accuracy.

None of the plotted points are outside the control limits. In fact, none of the out-of-control tests in Section 4.5 is broken, suggesting a stable process. If true, this would mean that no improvements in inventory accuracy have taken place over time. However, visually it is easy to see an overall downward trend. Regression analysis shows that this downward trend is statistically significant, suggesting that improvements have occurred. What this means is that there can be trends that the standard out-of-control rules in the next section do not detect and, on occasion, the rules may have to be supplemented by other statistical methods such as regression analysis.

We also notice that the lower control limit is essentially zero (the computed limit is negative) and the upper control limit is around 0.15. This means that while the average % misplaced (inventory inaccuracy) is 6.8%, it can fluctuate between 0% to 15% due to common causes. For a point to fall outside the control limits, inventory accuracy would either have to be perfect or very bad. The limits are too wide to detect meaningful changes. The limits can be narrowed by increasing the subgroup size. The current average subgroup size of about 90 is large. How much larger would it have to be? We will answer this question in Chapter 6.

The conventional 3 sigma limits in Table 4.6 assume a normal approximation to the binomial. Binomial distribution is a positively skewed distribution and the exact binomial control limits can also be determined for a false alarm probability of 0.13% below the lower limit and 0.13% above the upper limit. These are shown in Figure 4.13. There is now a nonzero lower limit; however, the overall conclusions remain essentially the same.

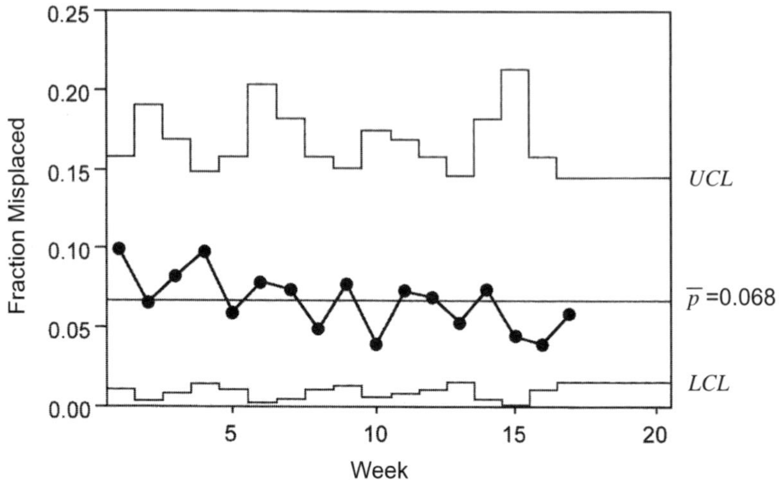

FIGURE 4.13 *p* chart for inventory accuracy (exact binomial limits).

4.4.2 Defects per Product (*u*) Chart

A *c* chart is used for number of defects per product when the sample size is constant. A *u* chart is used for number of defects per product with fixed or variable sample size. Under certain conditions, number of defects per product has a Poisson distribution, which forms the basis for these charts.

Number of Accidents per Month. Beginning in 1990, the number of accidents per month were reported at a certain manufacturing facility. Every month, a management presentation was made that compared that month's accidents to the previous month's accidents. Conclusions were drawn and decisions were made on the basis of these two point comparisons. When accidents were down, there was a feeling that things were working well. When accidents were up, the feeling was that people were not being careful, despite being told to do so. In August of 1991, it was felt that the number of accidents had gone up. In October, 1991, control chart procedures and appropriate corrective actions were implemented based upon team recommendations. The data for the monthly number of accidents are shown in Table 4.7.

For accidents or defects to have a Poisson distribution, the following assumptions have to be met. These assumptions are based upon the Poisson approximation to the binomial distribution discussed in Chapter 2.

TABLE 4.7 Number of Accidents per Month

Month	Number of Accidents	Month	Number of Accidents	Month	Number of Accidents
January 1990	9	November 1990	8	September 1991	9
February 1990	7	December 1990	3	October 1991	15
March 1990	10	January 1991	4	November 1991	11
April 1990	11	February 1991	14	December 1991	8
May 1990	7	March 1991	10	January 1992	4
June 1990	5	April 1991	12	February 1992	2
July 1990	9	May 1991	15	March 1992	8
August 1990	10	June 1991	9	April 1992	5
September 1990	8	July 1991	6	May 1992	3
October 1990	13	August 1991	14	June 1992	2

1. For the sample under consideration, each defect can be discretely counted.
2. If the sample is subdivided such that only one defect could occur within a subdivision, then the probability of finding a defect within any subdivision is constant and small when the process is in control.
3. This probability of finding a defect within a subdivision does not depend upon another subdivision having a defect.
4. The sample consists of a very large number of such subdivisions.

The accident data generally satisfy these assumptions. The employee population and month define the sample. The employee population was essentially the same over the time period considered. Each accident can be separately counted. Each employee is a natural subdivision of the sample and the probability of any one employee having an accident is small. It is assumed that the accidents are independent. There were a large number of employees.

If we ignore the small differences between the employee population from month to month and also ignore the difference in the number of working days each month, then the employees–month sample size may be considered to be constant and arbitrarily set equal to one. In this case, the c and u charts will coincide. For more precision, we could determine the actual sample size and standardize the number of accidents by dividing by the sample size.

The procedure to design a u chart is as follows.

1. Obtain the values of defects (number of accidents) x_i and the corresponding sample sizes (employees–month) n_i.
2. Calculate $u_i = x_i/n_i$. For this example, since $n_i = 1$, $u_i = x_i$.
3. Calculate the centerline,

$$\bar{u} = \frac{\text{Total observed defects (accidents)}}{\text{Total sample size}}$$

For example, the u chart was designed using 22 months of data up to October, 1991. The total sample size was 22 employees–months and the total number of accidents were 208. Hence, $\bar{u} = 208/22 = 9.46$.

4. For the Poisson distribution, the standard deviation of u_i is $\sqrt{\bar{u}/n_i}$. Hence,

$$\text{Control limits for the u chart} = \bar{u} \pm 3\sqrt{\frac{\bar{u}}{n_i}} \quad (4.11)$$

Substituting $\bar{u} = 9.46$ in Equation (4.11) gives

$$\text{Centerline} = 9.46$$

$$LCL = 0.20$$

$$UCL = 18.7$$

The control chart is shown in Figure 4.14. The control limits suggest that monthly accidents could vary between 0 and 19 due to com-

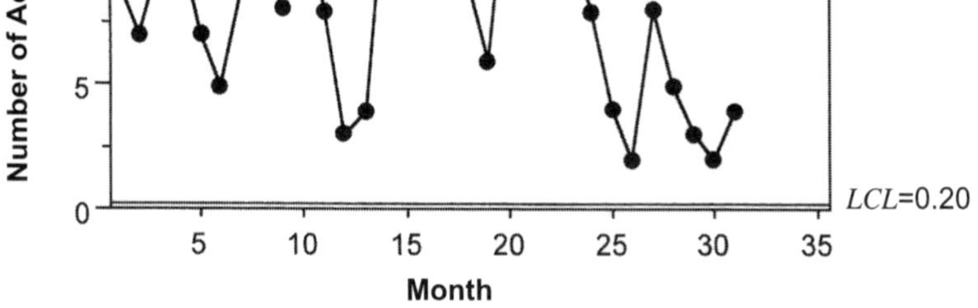

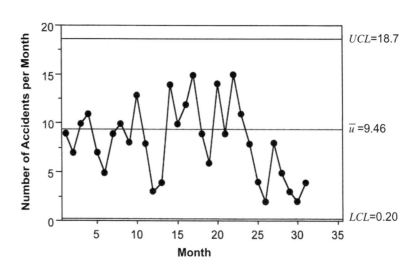

FIGURE 4.14 u chart for number of accidents.

mon causes. The last several points after October 1991 consistently fall below the centerline, suggesting that the implemented corrective actions are reducing the number of accidents.

4.5 INTERPRETING CONTROL CHARTS

Control charts provide more information regarding process instability than a thermometer does regarding fever. Control charts not only identify the presence of special causes but also provide information regarding the nature of special causes. If the process is stable, then the successive points on the control chart would be like random drawings from the expected distribution of what is being plotted. Significant deviations from this expected random behavior signal special causes, and the nature of these deviations provide clues regarding the nature of special causes.

As an example, for a stable process, the plotted values of $\bar{x}$ will be like random drawings from a normal distribution with $\bar{\bar{x}}$ and $\bar{R}/d_2\sqrt{n}$ as the estimated mean and standard deviation. For a normal distribution, the expectation is that the plotted points will be randomly distributed around the mean, that approximately 68% of the plotted points will be within one standard deviation from the mean, 0.3% of points will be greater than three standard deviations away from the mean, and so on. If, out of 30 plotted points, one is greater than three standard deviations away from the mean, or none are within one standard deviation from the mean, or there is a definite time trend, then these results would be unexpected and would signal the presence of special causes.

4.5.1 Tests for the Chart of Averages

The basic approach is that if a pattern of points has a very low probability of occurrence under the assumption of statistical control and can be meaningfully interpreted as a special cause, then that pattern constitutes a test for special cause. Eight commonly used tests for detecting special causes of variation for normally distributed averages or individual values are described below and graphically shown in Figure 4.15. Tests 1, 2, 3, and 4 are applied to the upper and lower halves of the chart separately. Tests 5, 6, 7, and 8 are applied to the whole chart.

For the purposes of these tests, the ± 3 sigma control limits are divided into six zones, each one sigma wide, and are labeled A, B, and C, as shown in Figure 4.15.

Test 1. Special cause is indicated when a single point falls outside 3 sigma control limits (beyond zone A). This suggests a sporadic shift or the beginning of a sustained shift.

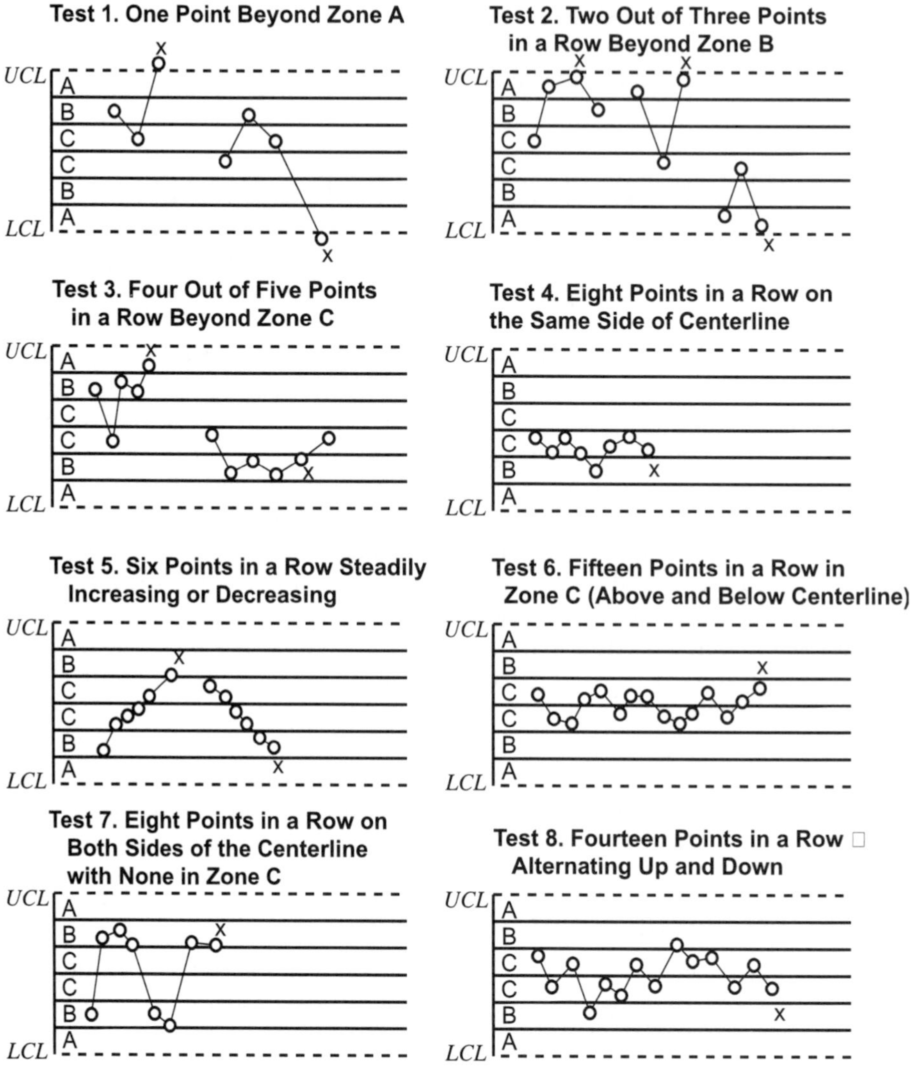

FIGURE 4.15 Tests for special causes.

Test 2. Special cause is indicated when at least two out of three successive points fall more than two sigma away (beyond zone B) on the same side of the centerline. This test provides an early warning of a shift.

Test 3. Special cause is indicated when at least four out of five successive points fall more than one sigma away (beyond zone C) on the same side of the centerline. This test provides an early warning of a sustained shift.

Test 4. Special cause is indicated when at least eight successive points fall on the same side of the centerline. This pattern suggests a sustained shift.

Test 5. Special cause is indicated when at least six consecutive points are steadily increasing or decreasing. The pattern suggests an upward or downward time trend.

Test 6. Special cause is indicated when there are at least 15 points in a row within ± one sigma from centerline (zone C) on both sides of the centerline. Such a pattern could result if the gauge is faulty, if the within-subgroup data come from a systematic sampling from two or more distributions, if process variability has dramatically decreased, or if wrong control limits are used.

Test 7. Special cause is indicated when there are eight or more points in a row on both sides of the centerline with none in zone C. Such a pattern indicates that observations within a subgroup come from a single distribution but each subgroup comes from one of two or more distributions.

Test 8. Special cause is indicated when there are at least 14 points in a row alternating up and down. This pattern indicates the presence of a systematic factor, such as two alternately used machines, suppliers, or operators.

Additionally, if the structure of the data suggests the need to investigate other meaningful patterns, the data should be so examined. For example, if the same raw material is supplied by two suppliers, the data may be segregated to see if there is a difference between suppliers. Similarly, the data may be examined for differences between shifts, operators, test equipment, and so on. Comparative experiments discussed in Chapter 3 are useful in this regard.

Probability Basis for Tests. The above tests are constructed so that when the process is in statistical control, the probability of incorrectly concluding that there is a special cause, namely, the probability of false alarm, is approximately less than five in a thousand for each test.

Table 4.8 shows the probability of a point falling in various zones for $\bar{X}$ and R charts (AT&T). Note that since the distribution of range is not symmetric, the probabilities in the upper and lower half are unequal.

As an illustration, the probability of false alarm for the first four

TABLE 4.8 Probabilities for Distribution of $\overline{X}$ and R

Zone	Normal Distribution	Distribution of Range (n = 5)	Distribution of Range (n = 2)
Upper half			
Above *UCL*	0.00135	0.0046	0.0094
A	0.02135	0.0294	0.0360
B	0.1360	0.1231	0.1162
C	0.3413	0.3120	0.2622
Lower half			
C	0.3413	0.3396	0.1724
B	0.1360	0.1735	0.1910
A	0.02135	0.0178	0.2128
Below *LCL*	0.00135	0	0

tests for normally distributed averages or individual values may be computed as follows from Table 4.8:

Test 1. The probability of a point falling outside the upper three sigma limit is 0.00135.

Test 2. The probability of a point falling above the upper two sigma value is (0.00135 + 0.02135) = 0.0227. Therefore, the probability of two out of three successive points beyond zone B on the same side of the centerline is $\binom{3}{2}(0.0227)^2(1 - 0.0227) = 0.0015$.

Test 3. The probability of a point falling above the upper one sigma value is (0.00135 + 0.02135 + 0.1360) = 0.1587. Therefore, the probability of four out of five successive points beyond Zone C on the same side of the centerline is $\binom{5}{4}(0.1587)^4(1 - 0.1587) = 0.0027$.

Test 4. The probability of a point being above the centerline is 0.5. Therefore, the probability of eight successive points being above the centerline is $(0.5)^8 = 0.0039$.

The α Risk. The α risk is the total probability of false alarm. For a given control chart, the α risk is a function of the specific tests used to detect special causes. For example, if the first four tests are used then, from Table 4.9, the α risk for normally distributed averages is (0.0094 + 0.0094) = 0.0188. This means that approximately two out of a hundred points will violate one or more of the first four tests purely by chance. If only the first test is used, the α risk is (0.00135 + 0.00135) = 0.0027 or three out of a thousand plotted points will violate the first test purely by chance. If all eight tests are used, the false alarm probability becomes 3% to 4%. Thus, indiscriminate use of a large number

TABLE 4.9 Probability of Getting a Reaction to the Tests

Test	Normal Distribution	Distribution of Range (n = 5)	Distribution of Range (n = 2)
Upper half of chart			
Single point out	0.0013	0.0046	0.0094
2 out of 3	0.0015	0.0033	0.0059
4 out of 5	0.0027	0.0026	0.0029
8 in a row	0.0039	0.0023	0.0010
Total	0.0094	0.0128	0.0192
Lower half of chart			
8 in a row	0.0039	0.0063	0.0121
4 out of 5	0.0027	0.0054	0.0792
2 out of 3	0.0015	0.0009	0.1069
Single point out	0.0013	—	—
Total	0.0094	0.0126	0.1982

of tests for special causes will lead to an unacceptable level of α risk. One approach is to start by using tests 1, 4, and 5 and then add or remove tests based upon practical experience with the process.

4.5.2 Tests for Other Charts

The tests for the chart of averages may be used for other control charts. However, the α risks are somewhat different and, occasionally, special tests may be necessary. These changes in α risk occur because the range and attribute counts do not have a normal distribution. Also, the distribution of range and the binomial and Poisson distributions for counts are not symmetrical. Therefore, the probabilities of getting a reaction to the tests are different in the upper and lower halves of the control chart. As an example, for ranges based upon a subgroup of size five, from Table 4.9, the α risk for the range chart using first four tests is 0.0254. This is slightly higher than the α risk of 0.0188 for the chart of averages, but the difference is practically not significant.

Constructing Special Tests. If necessary, special tests can be constructed from the appropriate distribution of what is being plotted using the probability principles illustrated above. For example, from Table 4.9, if we used all four tests, the α risk for a range chart with subgroup size two is (0.0192 + 0.1982) = 0.2174, which is extremely high. Many special tests can be constructed (AT&T). One such test is to only use Test 1 and Test 4, which reduces the risk to (0.0094 + 0.0131) = 0.0225.

Joint Interpretation of Charts. Control charts for variable data come in pairs (e.g., $\bar{X}$ and R) but there is only one control chart for each type of attribute data. This is because for variable data, the mean and the standard deviation are independent of each other and two separate charts are needed to monitor the mean and the standard deviation. For attribute data, standard deviation is a deterministic function of the mean and only one chart is necessary. As an example, for the u chart, mean = λ and standard deviation = $\sqrt{\lambda}$ so that the standard deviation is completely determined by the mean.

The pair of charts for variable data need to be jointly interpreted for special causes. The R and S charts are only affected by changes in variability, whereas the $\bar{X}$ chart is affected by changes in both the mean and the variability. Therefore, the R and S charts are evaluated first for special causes and the $\bar{X}$ chart is evaluated next in the context of any special causes found on the R and S charts. Also, if the points on the $\bar{X}$ chart are correlated with those on the R and S charts, then this indicates that the underlying distribution is skewed.

4.6 KEY SUCCESS FACTORS

The computation of control limits is essentially mechanical; they follow from Tables 4.2 and 4.6. What then are the key factors for successful implementation of control charts? These factors include identification of key characteristics to control; selection of rational subgroups; selection of a proper control chart, subgroup size, and sampling interval; the design and redesign of charts based upon adequate data; and taking corrective actions when the control chart produces a signal. These key success factors are briefly described below.

Selecting Key Characteristics. The objective is to chart those vital few characteristics that control the final product performance cost effectively. Figure 4.16 shows the cascading set of characteristics. Performance characteristics are those that directly measure customer preference. Clarity of picture on a TV screen, sensory characteristics of a food product, and time taken by a car to accelerate from 0 to 60 mph, are examples of performance characteristics. These performance characteristics are controlled by measurable features of the product called product characteristics. Examples include crust thickness and amount of cheese on pizza, and viscosity of salad dressing. Product characteristics are often further classified as critical, major, and minor. Critical characteristics are those which, if outside specifications, could cause safety concern. Major characteristics are those which, if outside specifications, could re-

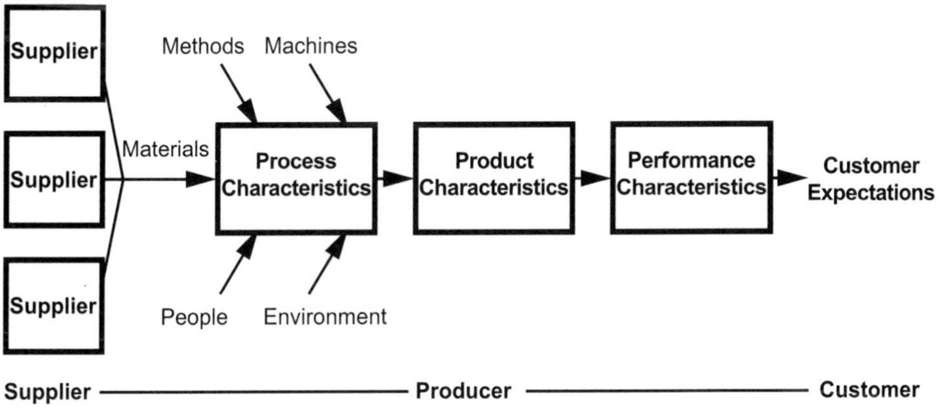

FIGURE 4.16 Cascading set of characteristics.

sult in unacceptable product performance. Minor characteristics do not impair product performance to any significant degree. These product characteristics are controlled by process characteristics, typically consisting of machines, methods, materials, environment, and people. Examples of process characteristics include mix speed, mix time, frying temperature, ambient humidity, and raw material characteristics. For the material supplier, the raw material characteristics become final product characteristics and the cycle begins again.

The key performance and final product characteristics are identified during the early phases of product development. This is often done by a direct input from the customers. For example, in the food industry, such information is obtained through concept tests, taste tests, and other consumer tests using representative consumers. These known key product characteristics are often control charted for feedback process control. We could also identify and control-chart key process characteristics that control product performance. For example, if we knew the key material characteristic that controls some key final product characteristic, then the material characteristic could be control-charted, making it possible to control the final product characteristic in a feed-forward manner. Identification of key process characteristics requires a good understanding of the process and of tools such as flow charts, cause-and-effect diagrams, regression analysis, and design of experiments.

In making the final selection of key characteristics, the following considerations apply:

1. Key final product characteristics are often used for process control. Efforts should be made to identify the controlling key process

characteristics so that the process can be cost effectively controlled at the earliest possible processing step, thus preventing the manufacture of poor quality products.

2. If the selected characteristics are highly correlated with each other and there is a causal relationship between them, then perhaps only one of the correlated characteristics needs to be monitored. It follows that if a selected characteristic is difficult to measure then another easy-to-measure but highly correlated characteristic may be used. For example, if volume is the selected characteristic but weight is easier to measure, then weight may be used so long as specific gravity is constant.
3. It is often possible to reduce the number of key characteristics on the basis of a good understanding of the process. For example, there may be several dimensions on a molded product that are key. However, it may be concluded that if any one dimension is in control, all other dimensions are bound to be in control. In such a case, the dimension with the tightest tolerance should be monitored using a control chart. Other dimensions need not be monitored as long as this understanding of the process holds.

Rational Subgroup. The logical selection of a subgroup depends upon the purpose of the control chart. Typically, a control chart is used to distinguish between common cause and special cause variability. The control limits are based upon within-subgroup variance and if the between-subgroup variability is large, it is identified by the control chart as being due to special causes. This suggests that subgroups need to be rationally or logically selected to only include common cause variability. In most instances, a small group of consecutively produced products is likely to be a rational subgroup because a special cause is less likely to occur in this short duration of time. This short-term variation is assumed to be due to common causes. The control chart, in effect, compares the differences between subgroups to the short-term variation and, as conditions change, signals special causes. However, the control chart may be used for a different purpose, for example, for product acceptance purposes. In this case, the subgroup may consist of a small random sample of all products produced over the sampling interval. Furthermore, the subgroup size may be selected such that the total number of products sampled over the production of a lot equals the number of products necessary to execute the acceptance sampling plan. In this manner, a control chart can be used in lieu of the acceptance sampling plan.

In general, in defining a rational subgroup, it is necessary to think about the sources of variation that should be included in the within-subgroup variance and the sources of variation that should be included in the between-subgroup variance. This may be seen as follows. Consider a production process that consists of three parallel machines as shown in Figure 4.17 (A.I.A.G.). Each machine has four stations, called A, B, C, and D. Every hour, a 12-part sample is collected. Table 4.10 shows an example of the data. Similar situations occur with multicavity molds, multihead filling machines, and where multiple lanes of products are produced.

In this case, there are three sources of variation: machine-to-machine variation captured by the differences in the rows, station-to-station variation captured by differences in columns, and hour-to-hour variation captured by the differences between the 12-part samples taken every hour. There are several ways to define subgroups (see Figure 4.17):

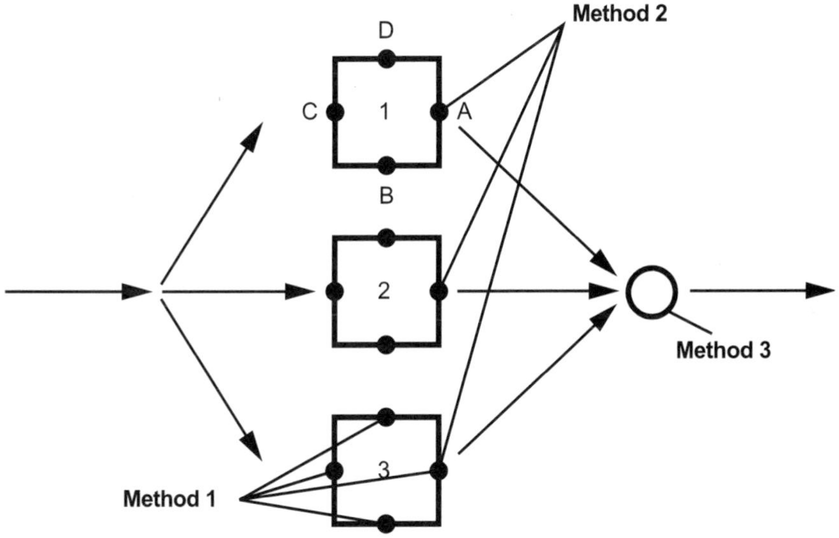

FIGURE 4.17
Rational subgroups.

TABLE 4.10 Data From the 12-Part Sample

	Station			
Machine	A	B	C	D
Machine 1	10	11	10	12
Machine 2	12	9	9	8
Machine 3	11	13	10	10

1. In method 1, each row is defined as a subgroup. A separate control chart would be needed for each machine. The within-subgroup variance will include station-to-station variability. Hour-to-hour variation will appear between subgroups and machine-to-machine differences will be obtained by comparing different control charts.
2. In method 2, each column is defined as a subgroup. There could be four separate control charts, one for each station. Machine-to-machine variability is captured within the subgroup and hour-to-hour variation appears between subgroups. Station-to-station differences can be obtained by comparing the different control charts.
3. In method 3, products are sampled from the combined output of all machines. There is a single control chart. Assuming mixing, the within-subgroup variability includes both the machine-to-machine and station-to-station variability. The hour-to-hour variation appears between subgroups.

Which method of subgrouping is rational? The answer depends upon the relative magnitude of the variability due to the different sources of variation and the specific questions the control chart is intended to answer. For example, if the objective is to determine special causes over time and if the station-to-station differences are small and machine-to-machine differences are big, then method 1 is appropriate. If both machine-to-machine and station-to-station differences are small, then method 3 is sufficient. The subject of determining the contribution due to each source of variation is a very important one and will be covered in Chapters 7 and 8.

As another example, consider the production of wood boards with a nominal thickness of one inch. In this case, the thickness variation may be partitioned into three components: variability within a board, variability between successive boards, and longer-term variability from hour to hour. Subgroups may be formed in two ways:

1. In the first approach, a subgroup consists of thickness measurements taken at the middle and the two ends of the board. The within-subgroup variability only includes the within-board variance. The between-subgroup variability includes both the board-to-board variability and the hour-to-hour variability. If the within-board variability is very small, the control limits will be narrow and every subgroup may be out of control.
2. In the second approach, a subgroup consists of three successive

boards with one thickness measurement taken at a random location per board. The within-subgroup variance includes both the within-board variability and the variability between successive boards. The hour-to-hour variation appears between subgroups.

In this case, the second approach is appropriate unless it is felt that the variation between successive boards is due to special causes that could be eliminated by the operator.

Rational subgroups should be defined by first understanding the causes of variation, their relative contribution to the total variability, and the purpose of the control chart. For real-time control applications, it is important that the subgroups be defined in such a way that when an out-of-control signal is produced, it can be acted upon by the operator.

Control Charts. Many control charts are described in Chapters 4 and 6. Each of these charts is optimal under certain assumptions regarding the process. So what is required is a selection of an appropriate chart based upon a good understanding of the process. As one example, $\bar{X} - S$ charts are good at detecting large shifts in the process but they are relatively slow at identifying small shifts. They are better suited for processes with moderate capability. They apply if the same product is being manufactured over long periods of time. So if the process meets these criteria, $\bar{X} - S$ charts may be selected. On the other hand, if it is important to identify small shifts rapidly, or the same process is used to manufacture multiple products, then other charts described in Chapter 6 may be used.

Control Limits. Control limits should be based upon at least 100 observations. Twenty-five subgroups of size 5 or 50 subgroups of size 3 are recommended because with this many observations, σ_{short} is estimated within 10% to 15% of the true value and satisfactory control limits are established. What if very little or no prior data are available to establish limits? In this case, a study may be conducted to collect necessary data if such data could be rapidly collected. Alternately, trial control limits may be established based upon limited data and updated when adequate data are available.

Subgroup Size. Explicit determination of subgroup size based upon the desired risk in capturing a specified shift in the mean is described in the risk-based control chart section of Chapter 6. Presently, the following points should be considered in selecting a subgroup size:

1. For $\bar{X}$ and R, and $\bar{X}$ and S control charts, subgroup sizes of three to five are often used. A subgroup size of five allows one to detect

a mean shift of 1.5 σ_{short} almost immediately. Lower subgroup sizes may be used if the cost of sampling and testing is high (e.g., destructive testing) or if the process is highly capable; namely, the process variability is very small such that the chance of producing out-of-specification product is very low. The converse is also true.

2. For X and mR charts, the subgroup size is one.
3. For np and p charts, the necessary subgroup sizes are very large, in the range of 50 to 100 or more. Chapter 6 describes the determination of these subgroup sizes.
4. For the c chart, the subgroup size is one, equal to the sample or area of opportunity. The considerations involved in determining the area of opportunity or the subgroup size for the u chart are briefly described in Chapter 6.

Sampling Interval. Selecting the appropriate sampling interval is as important a decision as selecting control limits and subgroup size. Sampling interval influences the speed with which process shifts are detected and also affects the cost of sampling. The following factors should be considered:

1. An important factor is the expected frequency of deliberate or unexpected process shifts. The smaller the interval between process shifts, the smaller the sampling interval.
2. If the process variability is very small so that the chance of producing an out-of-specification product is negligible, then such a process may be monitored infrequently. This is particularly so if the cost of sampling and testing is high.
3. The sampling interval can be event-driven. The sampling interval may be made to coincide with shift changes, operator changes, raw material lot changes, and so on.
4. It may be desirable to sample frequently until sufficiently large amounts of data are initially collected and sufficient process knowledge is gained to obtain a better estimate of the sampling interval.

Control Chart Redesign. The initial design of the control chart, namely, the selected control limits, subgroup size, and sampling interval, should be reassessed after a reasonable period of time has elapsed and sufficient data have been collected. Thereafter, it is a good practice to periodically review and, if necessary, update the control chart design.

Any redesign of the control chart or updating of control limits must be based upon objective evidence. For example, if the initially designed R chart shows several out-of-control ranges, then the initial estimate of $\bar{R}$ may be inaccurate and may have to be revised. If the initial estimate of the frequency of process changes is wrong, the sampling interval will have to be reassessed. Control limits may have to be revised either because they were initially based upon inadequate data or the process has been intentionally changed and the collected data signal that revised control limits are necessary.

Taking Corrective Actions. For a successful real-time implementation of control charts, out-of-control action procedures need to be developed prior to implementation. These procedures describe which control chart rules will be used and, in cases of rule violation, what specific investigatory, corrective, and other actions will be taken by the process operators. An out-of-control signal does not require the production to be stopped, but it should require an investigation of the possible reasons for the signal. If these special causes are found, appropriate permanent corrective actions should be taken to improve the process.

As an aid to identify special causes, it is a good practice to annotate the control charts with notes. Any significant event that may help interpret the chart should be noted. Examples include changes in supplier, material lot, operators, deliberate process changes, unexpected changes such as power outages, and so on. When special causes are identified, they should also be noted along with the corrective actions taken. These notes are also helpful in identifying what does not matter.

5

Process Capability

The concept of process capability is different from the concept of process stability. Process capability is the ability of the process to produce products that meet specifications. A process is said to be capable if essentially all the products to be produced are predicted to be within specifications on the basis of objective evidence regarding the performance of the process. Capability cannot be determined without knowing product specifications. On the other hand, a process is said to be stable if it is only influenced by common causes. Knowledge of product specifications is not necessary to judge process stability.

For a stable process, the characteristic of interest has a constant and predictable distribution over time. Therefore, meaningful predictions can be made regarding the probability of products being out of specification in the future. Depending upon this probability, a stable process may be classified as capable or not capable. If the process is unstable, the future distribution of the characteristic cannot be predicted with confidence and it is not possible to predict if the products produced in the future will be within specification or not. Therefore, stability is a prerequisite for defining capability. However, even for unstable processes, the following distinction may be made. If an unstable process is made stable by removing special causes, the resultant stable process will either turn out to be capable or incapable. Thus, an unstable process may be said to be unstable but potentially capable, or unstable and incapable. A process may, therefore, be characterized as stable and capable, stable and incapable, unstable but potentially capable, or unstable and incapable. Such a process characterization is very useful for identifying improvement actions.

This chapter deals with the quantification of process capability, for both stable and unstable processes, in terms of capability and performance indices. Methods to estimate these indices and their associated confidence intervals are described. The connection between a capability index and the tolerance interval is established. The meaning of and the rationale for the six sigma goal is explained. The use of capability and performance indices to set goals, to assess the current process, and to identify improvement actions is explained.

5.1 CAPABILITY AND PERFORMANCE INDICES

The capability indices measure what a process would be capable of if it were stable. The performance indices measure the current performance of the process regardless of whether it is stable or not. This section defines two capability indices, C_p and C_{pk}, and two performance indices, P_p and P_{pk}.

5.1.1 C_p Index

The C_p index makes the following four assumptions:

1. The specification is two-sided.
2. The process is perfectly centered in the middle of the specification.
3. The process is stable.
4. The process is normally distributed.

Figure 5.1 shows a stable process with a short-term standard deviation σ_{short}. The process is perfectly centered on target T in the middle of a two-sided specification. Since the process is stable, the short-term standard deviation and the total standard deviation over longer periods of time are equal. The $\pm\, 3\sigma_{\text{short}}$ distance around the mean μ is defined as the process width. So process width equals $6\sigma_{\text{short}}$. The C_p index is defined to be:

$$C_p = \frac{\text{Specification width}}{\text{Process width}} = \frac{USL - LSL}{6\sigma_{\text{short}}} \tag{5.1}$$

where
USL = Upper specification limit
LSL = Lower specification limit

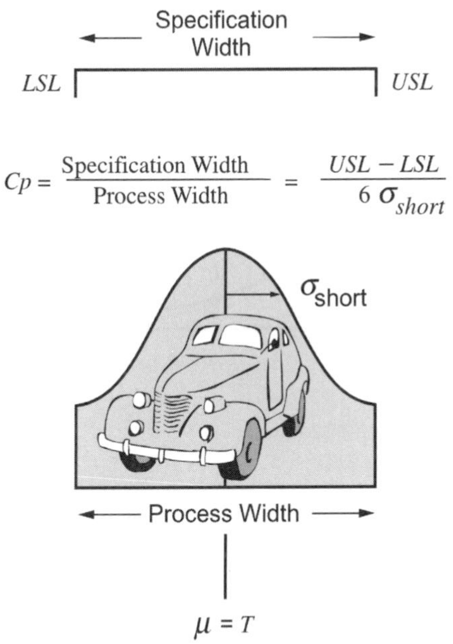

FIGURE 5.1 The car analogy.

If the process width and the specification width are equal, as shown in Figure 5.1, then the C_p index is 1.

If we think of the process in Figure 5.1 as a fancy car, and the specification as the garage we wish to park the car in, then a C_p index of 1 means that we will have a banged-up car. A higher C_p index will be desirable. What is the C_p index for a standard car on the roads in the United States? The C_p index in this case is the ratio of the width of the lane to the width of the car and is somewhere around 1.5. With this large a C_p index, most of the time, most of us have little difficulty staying in our lane.

The C_p index can take values from zero to infinity. Figure 5.2 shows the distributions of individual values for C_p indices of 1.0, 1.33, and 2.0. A higher C_p index means less nonconforming product. A C_p index of 2 means that we have a single car that we wish to park in a two-car garage, a job that perhaps could be done with eyes closed. The C_p index can be improved in two ways: by widening the specification and by reducing short-term variability.

5.1.2 C_{pk} Index

From a practical viewpoint, the C_p index makes too many unrealistic assumptions. For example, if the specification is one-sided, the C_p index cannot be computed. Also, the process is assumed to be perfectly cen-

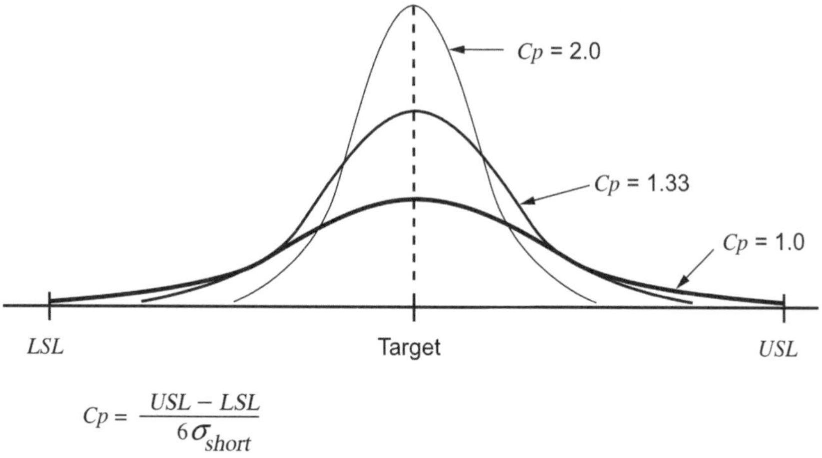

FIGURE 5.2 C_p index.

$$Cp = \frac{USL - LSL}{6\sigma_{short}}$$

tered in the middle of a two-sided specification. Such is rarely the case in practice. The C_{pk} index relaxes these two assumptions while retaining the following two assumptions:

1. The process is stable.

2. The process is normally distributed.

Figure 5.3 shows processes where the mean is off-target. The C_{pk} index is defined as

$$\frac{USL - Mean}{3\sigma_{short}} \quad \text{or} \quad \frac{Mean - LSL}{3\sigma_{short}} \tag{5.2}$$

whichever is smaller. If there is only one specification limit, then only the relevant of the above two calculations is performed.

The C_{pk} index can take values from $-\infty$ to $+\infty$. It is negative when the process mean is outside the specification and may be improved in three ways: by widening the specification, by reducing the short-term variability, and by changing the process mean. When the mean is perfectly centered in a two-sided specification, C_{pk} index becomes equal to the C_p index.

5.1.3 P_p Index

The P_p index measures the performance of the process without assuming it to be stable. It makes the following three assumptions:

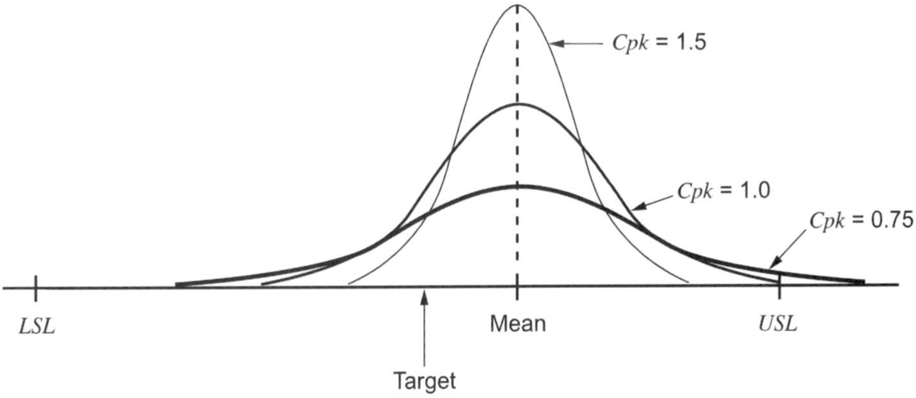

Cpk is the smaller of

$$Cpk = \frac{USL - Mean}{3\sigma_{short}}$$

$$Cpk = \frac{Mean - LSL}{3\sigma_{short}}$$

FIGURE 5.3
C_{pk} index.

1. The specification is two-sided.
2. The process is perfectly centered in the middle of the specification.
3. The process is normally distributed.

The P_p index is defined similarly to the C_p index:

$$P_p = \frac{USL - LSL}{6\sigma_{\text{total}}} \tag{5.3}$$

The only difference between C_p and P_p is that C_p uses σ_{short} and P_p uses σ_{total} to include both the common cause and the special cause variability. Since $\sigma_{\text{total}} \geq \sigma_{\text{short}}$, $C_p \geq P_p$. The P_p index can take values between zero and infinity. It can be improved in three ways: by widening the specification, by controlling the process to reduce special cause variability, and by reducing the short-term (common cause) variability.

5.1.4 P_{pk} Index

The P_{pk} index measures the current performance of the process without assuming a two-sided specification or assuming the process to be stable or centered. It only assumes that the process is normally distributed. The P_{pk} index is defined similarly to the C_{pk} index:

$$P_{pk} = \frac{USL - Mean}{3\sigma_{\text{total}}} \quad \text{or} \quad \frac{Mean - LSL}{3\sigma_{\text{total}}} \tag{5.4}$$

whichever is smaller. The only difference between C_{pk} and P_{pk} is that C_{pk} uses σ_{short} and P_{pk} uses σ_{total} to include both the common cause and the special cause variability. Since $\sigma_{\text{total}} \geq \sigma_{\text{short}}$, $C_{pk} \geq P_{pk}$. If there is only one specification limit, then only the relevant of the above two calculations is performed.

The P_{pk} index can take values from $-\infty$ to $+\infty$. It is negative when the process mean is outside specifications and may be improved in four ways: by widening the specification, by changing the process mean, by controlling the process to reduce special cause variability, and by reducing the short-term (common cause) variability.

5.1.5 Relationships between C_p, C_{pk}, P_p and P_{pk}

The four indices are computed under different assumptions regarding the process, regardless of whether the process actually satisfies the assumptions or not. Table 5.1 summarizes these assumptions. All assume a normally distributed process. This assumption of normality is not a requirement to compute the index; rather, it is used to interpret the index in terms of the implied fraction defective. The normality assumption is more likely to be true for stable processes compared to unstable processes. Hence, the performance indices are more susceptible to the impact of departures from normality.

The following relationships between C_p, C_{pk}, P_p, and P_{pk} hold for two-sided specifications. For one-sided specifications, only C_{pk} and P_{pk} can be computed and $C_{pk} > P_{pk}$ unless the process is perfectly stable, in which case $C_{pk} = P_{pk}$.

1. P_{pk} will always be the smallest and C_p will always be the largest of the four indices, with C_{pk} and P_p in between; i.e., $P_{pk} < C_{pk} < C_p$ and $P_{pk} < P_p < C_p$.

TABLE 5.1 Capability and Performance Index Assumptions

Assumptions	C_p	C_{pk}	P_p	P_{pk}
Two-sided specification	✓		✓	
Centered process	✓		✓	
Stable process	✓	✓		
Normal distribution	✓	✓	✓	✓

2. The four indices are related by the following equality:

$$P_p C_{pk} = C_p P_{pk} \tag{5.5}$$

If the values of three of the four indices are known, the remaining index can be calculated. For example,

$$P_p = \frac{C_p P_{pk}}{C_{pk}}$$

3. If process off-centering is more detrimental than process instability, $P_p > C_{pk}$. If process instability is more detrimental than off-centering, $P_p < C_{pk}$.

4. When the process is stable, $P_{pk} = C_{pk}$ and $P_p = C_p$. When the process is centered, $P_{pk} = P_p$ and $C_{pk} = C_p$. When the process is stable and centered, all indices are equal.

The computations and relationships between the four indices are graphically illustrated in Figure 5.4. The illustrated process is unstable and is not centered. Its current performance is measured by $P_{pk} = 1$. If the process could first be stabilized, the performance would improve to $C_{pk} = 2$. If it could then be centered, the performance would further improve to $C_p = 4$. Alternately, if the process could first be centered, the performance would improve to $P_p = 2$. Then, if it could be stabilized, the performance would further improve to $C_p = 4$. In this case, it has turned out that $C_{pk} = P_p$, implying that off-centering and instability are equally important for this process.

5.2 ESTIMATING CAPABILITY AND PERFORMANCE INDICES

This section describes ways of estimating the capability and performance indices and their associated confidence intervals. The sample sizes necessary to estimate the indices to a desired degree of precision and the connection between these indices and tolerance intervals are also discussed.

5.2.1 Point Estimates for Capability and Performance Indices

If the process mean μ, the short-term standard deviation σ_{short}, and the long-term standard deviation σ_{total} are known, all four indices can be calculated. So the question really is how to estimate μ, σ_{short}, and σ_{total}. We now consider the estimation of μ, σ_{short}, and σ_{total} for subgrouped data and individual data, assuming the data to be serially uncorrelated.

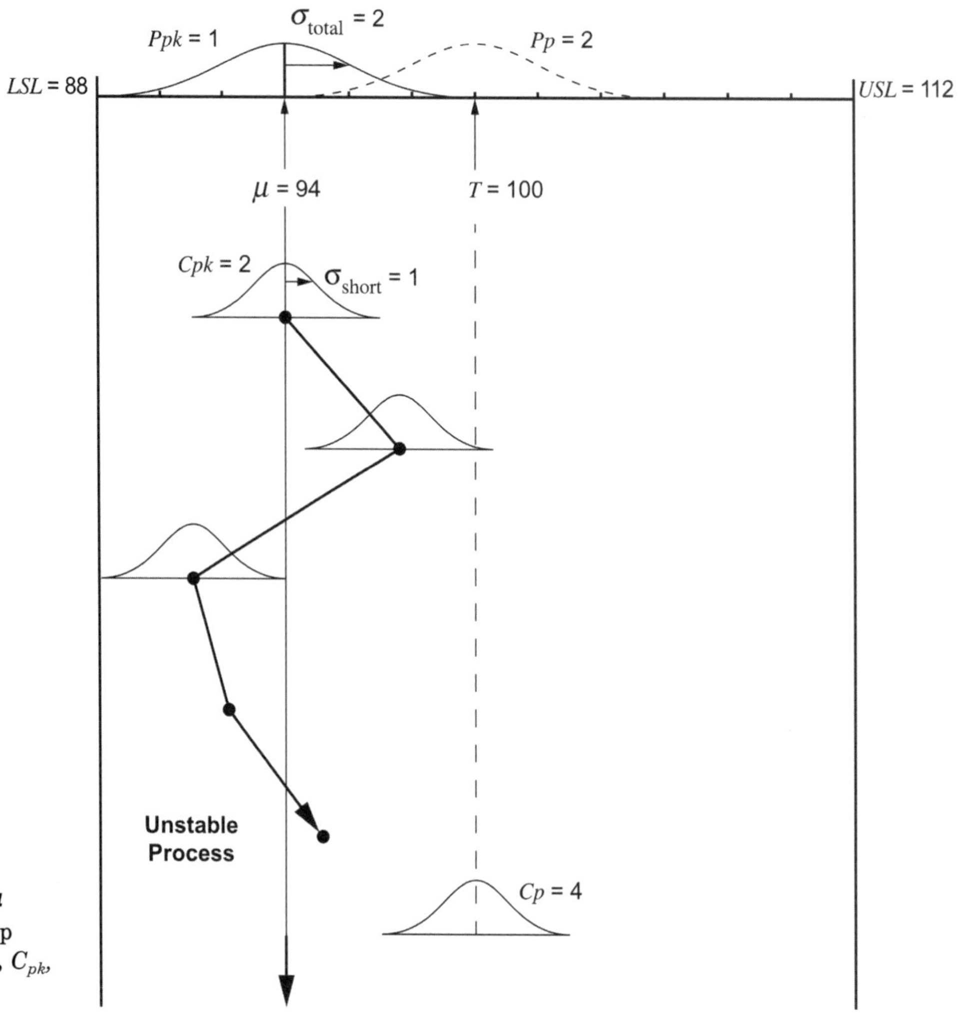

FIGURE 5.4
Relationship between C_p, C_{pk}, P_p, and P_{pk}.

Subgrouped Data. Here data are available as k subgroups of size n. The $\overline{X}-R$ and $\overline{X}-S$ charts are examples of such data. For subgrouped data:

1. $\mu \approx \bar{\bar{x}}$.
2. $\sigma_{\text{short}} \approx \overline{R}/d_2$, $\bar{s}/c_4$, or s_w based upon $k(n-1)$ degrees of freedom, where s_w is the pooled within-subgroup standard deviation, as explained in Section 4.3.1.
3. $\sigma_{\text{total}} \approx s_{\text{total}} = \sqrt{[\Sigma(x_i - \bar{\bar{x}})^2/(nk-1)]}$ based upon $(nk-1)$ degrees of freedom, namely, σ_{total} is estimated by the standard deviation of

all collected data. An alternate procedure for estimating σ_{total} is to use variance components analysis to estimate the within subgroup (s_w) and between-subgroup (s_b) standard deviations, as explained in Chapter 7. Then $\sigma_{\text{total}} \approx \sqrt{s_w^2 + s_b^2}$.

Individual Data. Here data are available as n individual values collected over time. The $X - mR$ chart, short-term capability studies, and process validation provide examples of such data. From individual data,

1. $\mu \approx \bar{x}$.
2. $\sigma_{\text{short}} \approx m\bar{R}/1.128$ based upon $(n - 1)$ degrees of freedom.
3. $\sigma_{\text{total}} \approx s_{\text{total}} = \sqrt{[\Sigma(x_i - \bar{x})^2/(n - 1)]}$ based upon $(n - 1)$ degrees of freedom.

Product Weight Example (continued). Section 4.3.1 described an example of an $\bar{X} - R$ chart dealing with the weight of a product. The declared weight was 250 grams, which becomes the single-sided lower specification limit. Analysis in Section 4.3.1 resulted in the following estimates of μ, σ_{short}, and σ_{total}:

$$\mu \approx \bar{\bar{x}} = 253.66$$

$$\sigma_{\text{short}} \approx \frac{\bar{R}}{d_2} = 1.074$$

$$\sigma_{\text{total}} \approx s_{\text{total}} = 1.098$$

In this case, only C_{pk} and P_{pk} can be calculated because the specification is single-sided:

$$C_{pk} = \frac{253.66 - 250}{3(1.074)} = 1.14$$

$$P_{pk} = \frac{253.66 - 250}{3(1.098)} = 1.11$$

Note that C_{pk} and P_{pk} are essentially equal because the process is essentially stable.

5.2.2 Confidence Intervals for Capability and Performance Indices

There can be much uncertainty associated with the estimated capability and performance indices. The uncertainty in the C_p and P_p indices results from the fact that σ_{short} is not known precisely. In estimating the C_{pk} and P_{pk} indices, there is added uncertainty since the process mean is also not precisely known. These uncertainties are quantified by computing a confidence interval for the capability and performance indices. The confidence intervals vary as a function of the degrees of freedom available to estimate the index and are wider for the C_{pk} and P_{pk} indices.

Exact confidence intervals for C_p and P_p indices can be obtained using the confidence interval for σ discussed in Chapter 2. Since C_{pk} and P_{pk} involve two sources of uncertainty, namely μ and σ, more complicated procedures are required. Equation (5.6) provides approximate $100(1 - \alpha)\%$ two-sided confidence intervals as shown by Bissell in 1990 (How reliable is your capability index? *Applied Statistics, 39,* 3, 331–340). The approximation improves with sample size.

$$\text{For the } C_p \text{ index: } C_p \pm Z_{\alpha/2} C_p \sqrt{\frac{1}{2\nu}}$$

$$\text{For the } P_p \text{ index: } P_p \pm Z_{\alpha/2} P_p \sqrt{\frac{1}{2\nu}}$$

$$\text{For the } C_{pk} \text{ index: } C_{pk} \pm Z_{\alpha/2} C_{pk} \sqrt{\frac{1}{9NC_{pk}^2} + \frac{1}{2\nu}} \qquad (5.6)$$

$$\text{For the } P_{pk} \text{ index: } P_{pk} \pm Z_{\alpha/2} P_{pk} \sqrt{\frac{1}{9NP_{pk}^2} + \frac{1}{2\nu}}$$

where ν = degrees of freedom to estimate σ_{short} (for C_p and C_{pk}) or σ_{total} (for P_p and P_{pk}) and N is the total number of observations. As an example, for k subgroups of size n, $N = nk$ and $\nu = k(n - 1)$ to estimate σ_{short} and $(nk - 1)$ to estimate σ_{total}.

For 95% confidence, $Z_{\alpha/2} = 2$. Based upon Equation 5.6, Table 5.2 shows the approximate confidence intervals for any value of C_p or P_p and for values of C_{pk} or P_{pk} close to 1. Similar computations can be made for other estimated values of C_{pk} and P_{pk}. For most realistic values of C_{pk} and P_{pk}, the results in Table 5.2 are satisfactory.

TABLE 5.2 Approximate Confidence Intervals for C_p, P_p, C_{pk}, and P_{pk}

Degrees of Freedom (ν)	95% Confidence Interval as % of Estimated Value (for C_p or P_p)	95% Confidence Interval as % of Estimated Value (for C_{pk} or P_{pk})
10	± 44%	± 50%
50	± 20%	± 22%
100	± 14%	± 16%
200	± 10%	± 11%
500	± 6%	± 7%

The conclusion is that we need more than 100 to 200 observations to get a reasonable estimate of capability and performance indices. Even then, the uncertainty will be on the order of ± 10%.

In many practical applications, we are interested in a lower bound on capability and performance indices. This implies a one-sided confidence interval, which is obtained by replacing $Z_{\alpha/2}$ by Z_α. The 100(1 – α)% one-sided lower bounds for the capability and performance indices are:

$$\text{For the } C_p \text{ index: } C_p - Z_\alpha C_p\sqrt{\frac{1}{2\nu}}$$

$$\text{For the } P_p \text{ index: } P_p - Z_\alpha P_p\sqrt{\frac{1}{2\nu}} \tag{5.7}$$

$$\text{For the } C_{pk} \text{ index: } C_{pk} - Z_\alpha C_{pk}\sqrt{\frac{1}{9NC_{pk}^2} + \frac{1}{2\nu}}$$

$$\text{For the } P_{pk} \text{ index: } P_{pk} - Z_\alpha P_{pk}\sqrt{\frac{1}{9NP_{pk}^2} + \frac{1}{2\nu}}$$

As an illustration, for the product weight example, there were 22 subgroups of size 5. Hence ν = 22 (5 – 1) = 88 and N = 110. The C_{pk} index was estimated to be 1.14. Therefore,

1. The two-sided 95% confidence interval for C_{pk} is

$$1.14 \pm (1.96)(1.14)\sqrt{\frac{1}{9(110)(1.14)^2} + \frac{1}{2(88)}} = 1.14 \pm 0.18 = 0.96 \text{ to } 1.32$$

   We are 95% sure that the true C_{pk} is between 0.96 to 1.32.

2. The 95% lower bound on C_{pk} is

$$1.14 - (1.64)(1.14)\sqrt{\frac{1}{9(110)(1.14)^2} + \frac{1}{2(88)}} = 1.14 - 0.15 = 0.99$$

   We are 95% sure that the true C_{pk} is greater than 0.99.

5.2.3 Connection with Tolerance Intervals

Recall from Chapter 2 that a tolerance interval can be constructed to contain 100 (1 – p)% of the population with 100 (1 – α)% confidence. To validate a process, we may collect n observations over a relatively short period of time and construct the tolerance interval $\bar{x} \pm ks$, where the constant k depends upon α, p, and n. If the tolerance interval is within specification limits, validation passes. For example, it is a common practice in some industries to construct a 95/95 tolerance interval. This means that we are 95% sure that 95% of the population is within the constructed tolerance interval. If this tolerance interval is within specification limits, the process is said to be validated.

In the context of process validation, there is a connection between the tolerance interval and the lower bound on the C_{pk} index or the lower bound on the P_{pk} index if the process is unstable.

Two-Sided Specifications. Consider the case in which the two-sided tolerance interval is inside the specification interval. The limiting case will be when one tolerance limit exactly matches the corresponding specification limit or when the tolerance interval exactly matches the specification interval. In both these cases, we are $100(1 - \alpha)\%$ sure that no more than $100p\%$ of the product is outside the specification. This means that the distance between the estimated process mean and the nearest specification limit must be at least $\sigma Z_{p/2}$. Hence, in this case, we are $100(1 - \alpha)\%$ sure that

$$C_{pk} \geq \frac{Z_{p/2}}{3} \tag{5.8a}$$

One-Sided specification. In this case, a one-sided tolerance interval is constructed. Consequently, $Z_{p/2}$ is replaced by Z_p. If validation passes, we are $100\ (1 - \alpha)\%$ sure that

$$C_{pk} \geq \frac{Z_p}{3} \tag{5.8b}$$

If the validation acceptance criterion is to assume that the process is validated with $100(1 - \alpha)\%$ confidence provided that $100\ (1 - p)\%$ of the population is enclosed inside the specification interval, then, Table 5.3 shows the correspondence between the validation acceptance criterion

TABLE 5.3 Validation Acceptance Criterion and Implied Minimum C_{pk}

% Population Enclosed = $100\ (1 - p)\%$	Implied Minimum C_{pk}	
	Two-Sided Specification	One-Sided Specification
90%	0.55	0.43
95%	0.67	0.55
99%	0.86	0.78
99.7%	1.00	0.93
99.9%	1.09	1.03

and the implied minimum C_{pk} for all values of α and selected values of 100 $(1 - p)\%$. From Table 5.3, a 95/95 tolerance interval is equivalent to a minimum C_{pk} of 0.67 for a two-sided specification.

For purposes of process validation, calculating the lower bound on C_{pk} is better than using tolerance intervals. Instead of simply concluding that validation either passes or fails, the lower bound on C_{pk} allows for an assessment of the goodness of the process on a continuous scale. The maximum fraction defective can be predicted. If long-term data are available, by computing all four indices, the consequences of process stability and centering can be examined. The process can be judged against known benchmark C_{pk} values and improvement strategies can be formulated as discussed in Section 5.4.

5.3 SIX SIGMA GOAL

We want the process to be stable. For a stable process, $C_p = P_p$ and $C_{pk} = P_{pk}$. What should the targets be for the C_p and C_{pk} indices? Clearly, the larger the value of C_{pk} the better, but are there some specific values that may be defined as targets to strive for? A C_{pk} of one means that better than 99.73% of the individual characteristic values are within specification or no more than three individual values out of 1000 are expected to be outside specifications. Is this acceptable? The answer depends upon the following considerations:

1. The consequence of a characteristic being outside specifications. For a critical characteristic involving safety, the risk of 3 in 1000 is not acceptable. C_{pk} values much higher than one are desirable. On the other hand, if the consequence is a minor degradation in performance, a C_{pk} of one may well be reasonable.

2. The number of key product characteristics that control the total performance of the product. The C_{pk} index is usually defined for a single characteristic. There may be dozens or even hundreds of key characteristics that affect product performance. A food product such as frozen pizza has about a dozen characteristics of interest such as weights of various ingradients, physical properties of crust, etc. A complex system may consist of many components, each of which may have several characteristics of interest. Each of these characteristics needs to be within its specification for the entire system to function properly. For 10 and 100 such independent characteristics, each with a C_{pk} of one, the probabilities that the system will perform satisfactorily are

$$(0.9973)^{10} = 97.3\%$$

and

$$(0.9973)^{100} = 76.3\%$$

These probabilities are too small to be acceptable and a C_{pk} of greater than one would be desirable. For example, if the C_{pk} for each characteristic is 1.33, then the probabilities that the system will function correctly are

$$(0.999937)^{10} = 99.93\%$$

and

$$(0.999937)^{100} = 99.4\%$$

3. The closeness of the estimated C_{pk} to the population C_{pk}. There are wide confidence limits associated with the estimated C_{pk} value. Therefore, the estimated C_{pk} needs to be greater than one to have sufficient confidence that the real C_{pk} exceeds one. Also, the computations assume a normal distribution for individual values. Departures from normality may cause the true C_{pk} values to be lower than estimated.

4. The assumption is that the process is perfectly stable and continues to be so. Such is never the case. Furthermore, as discussed in Chapter 6, small changes in the process mean or variability are hard to detect using practical sample sizes for control charts. Therefore, in designing the product and the process, provision must be made for some undetected and uncorrected process shifts during manufacturing. The process needs to be designed to a C_{pk} greater than one in order to achieve a C_{pk} of one in practice, i.e., in order for the P_{pk} to be one during manufacturing.

Table 5.4 integrates the above considerations. The design C_p is the capability that the product and process are designed to deliver if there are no shifts during manufacturing and the process stays perfectly centered. This capability can also be measured in terms of the sigma level of the process, computed as the number of standard deviations from the process mean to the nearest specification. This design capability degrades during manufacturing. Mean shifts smaller than $\pm 1.5\sigma_{\text{short}}$ are difficult to detect and correct using usual subgroup sizes for control charts. The manufacturing C_{pk} column shows the C_{pk} realized in manufacturing assuming $\pm 1.5\sigma$ shift in mean. The next column indicates the level of nonconformance in parts per million (PPM) corresponding to the manufacturing C_{pk} and assuming normal distribution of individual values. In practice, the distribution may not be normal and this may increase the level of nonconformance. The percent conforming column

TABLE 5.4 Setting Capability Goals

Design Capability					Number of Characteristics for System Acceptance Probability of:	
Sigma Level	C_p	Manufacturing C_{pk}	Nonconforming PPM	% Conforming	99%	99.9%
3 σ	1.00	0.50	66800	93.32	0	0
4 σ	1.33	0.83	6210	99.379	1	0
5 σ	1.67	1.17	233	99.9767	43	4
5.5 σ	1.83	1.33	63	99.9937	159	15
6 σ	2.00	1.50	3.4	99.99966	2955	294

shows the percentage of the population expected to be within specification for a single characteristic. The last two columns show the permissible maximum number of key characteristics for the probability of system acceptance to be 99% and 99.9%. For example, for a four-sigma process, $C_p = 1.33$, manufacturing $C_{pk} = 0.83$, and the product may have only one key characteristic for the product failure rate to be less than 1%. The selected sigma level and the C_{pk} targets should be based upon the acceptable product failure rate, the number of key characteristics that control product performance, and the ability to control production processes. Table 5.4 is helpful in this regard. Many companies are striving for a six-sigma goal, which is equivalent to the requirement that design $C_p \geq 2$ and manufacturing $C_{pk} \geq 1.5$.

5.4 PLANNING FOR IMPROVEMENT

Planning for improvement means deciding whether the process needs to be improved and, if so, how? A strategic plan answers three questions: where are we? where do we want to go? and how are we going to get there? Similarly, process improvement plan needs to answer three questions: what is the current performance of our process? what should the process capability targets be? and what improvement actions are necessary?

The C_p, C_{pk}, P_p, and P_{pk} indices permit an assessment of the process in terms of stability, centering, and capability. Figure 5.5 displays the four indices for a variety of situations. C_p, C_{pk}, and P_{pk} are shown by dots. The three indices will always maintain the relationship $C_p \geq C_{pk} \geq P_{pk}$. The P_p index is shown by a dash. The key to reading the display is shown in the right-hand portion of Figure 5.5. This figure can be used as an at-a-glance display of the performance of all key characteristics asso-

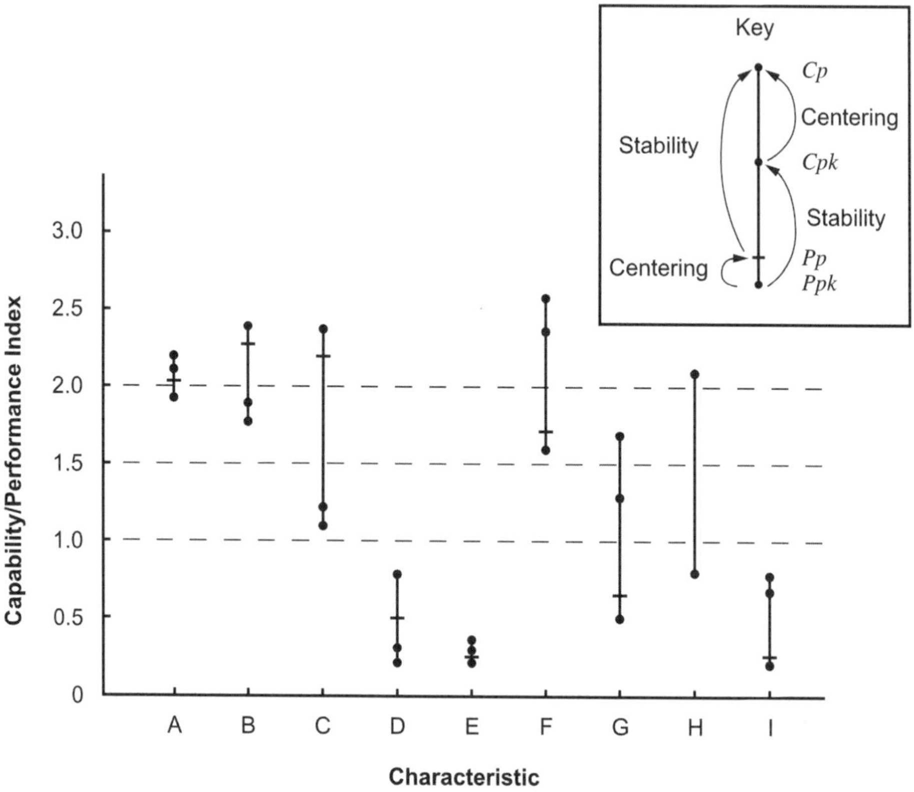

FIGURE 5.5 Assessing process performance.

ciated with a particular product. Such a display is useful not only for planning improvements but also for tracking and communicating results.

Based upon the values of the process capability and performance indices, processes may be grouped into the following six categories obtained by classifying the process as stable or unstable, and then as capable, potentially capable, or incapable. The capability targets should correspond to the goals set by the company. The following discussion assumes the capability targets corresponding to the six-sigma goal, namely, $C_p \geq 2$ and $P_{pk} \geq 1.5$.

1. **Stable and Capable.** These are stable processes whose current performance, as measured by the P_{pk} index, meets the target. The signature of such a process is shown in Figure 5.5, characteristic A, which shows a stable, centered, and capable process. If the process is not well centered, the signature may be as for characteristic B. The current performance meets the target but could be improved by centering the process. Stable and capable processes

do not require significant improvement effort. Such processes should be monitored using a control chart to ensure that the process does not degrade and they should be centered if necessary.

2. **Stable and Potentially Capable.** These are stable processes whose current performance as measured by P_{pk} does not meet the target; but P_{pk} will become satisfactory if the process is better centered. In other words, although P_{pk} does not meet the target, P_p does. The signature of such a process is shown in Figure 5.5, characteristic C. Such processes need to be control charted and better centered.

3. **Stable and Incapable.** These are stable processes that, even if centered properly, will not produce acceptable performance. For such processes, the C_p index is too low to be acceptable. The signature of such processes is shown in Figure 5.5, characteristics D and E. To expect a control chart to significantly improve such processes is unrealistic. The process needs to be changed to reduce short-term variability or a different process needs to be used.

4. **Unstable but Capable.** These are unstable processes that are currently capable. The P_{pk} index meets the desired target. Such processes may or may not be centered correctly. Figure 5.5, characteristic F shows one signature of such a process. Even though the current levels of instability and noncentering are satisfactory, such a process will benefit from being control charted, stabilized, and centered. If the process is made stable, it will provide greater assurance of continued good performance in the future.

5. **Unstable but Potentially Capable.** These are unstable processes whose current performance is unsatisfactory but can be made satisfactory by improving stability and centering. Figure 5.5, characteristic G and H show signatures of such processes. For characteristic G, P_{pk} = 0.5, P_p = 0.65, C_{pk} = 1.3, and C_p = 1.7. This means that the current performance of the process is unsatisfactory, but if the process could be stabilized and centered, it would become a nearly capable process. For characteristic H, there are only two dots, meaning that the specification is one-sided and only P_{pk} (lower dot) and C_{pk} (upper dot) can be calculated. Again, the current performance is not satisfactory, but if the process is made stable, the process will become very capable. A control chart is extremely useful for such processes. It is necessary is to monitor the process using a control chart, find and remove special causes as they arise, and adjust the mean as necessary.

6. **Unstable and Incapable.** These are unstable processes whose current performance is unsatisfactory. Although their performance may be improved by stabilizing and centering the process, the improved process will still not be satisfactory. Figure 5.5, characteristic I is a signature of such a process. The process is unstable, since P_{pk} and C_{pk} differ. The process happens to be well centered since C_p and C_{pk} are close together. However, even if the process is made stable, it would continue to be incapable since C_p is not satisfactory. By controlling and centering such processes, some improvement in performance is possible. Again, control charts are useful for this purpose. However, in addition, common cause variability will have to be reduced. Key causes of variability will have to be identified using tools such as variance components, measurement systems analysis, design of experiments, and so on. Or, a completely new process will have to be used.

6

Other Useful Charts

A control chart is a way of monitoring, controlling, and improving a process; i.e., a way of managing a process. This process management involves risks—the risk of false alarm and the risk of not detecting a process shift when it has occurred. It also entails costs, such as the cost of off-target products, the cost of sampling, and the cost of corrective action. At times, it involves special circumstances not considered by the traditional variable and attribute control charts discussed in Chapter 4. This chapter considers five additional control charts: risk-based charts, modified limit charts, charts to detect small shifts, short-run charts, and charts for nonnormal distributions. These charts are useful in many practical applications. Risk-based charts explicitly manage the two risks of making wrong decisions. Modified limit charts are useful when it becomes uneconomical to adjust the process every time it goes out of control. Such is the case when the process capability is high and the cost of adjustment is also high. Charts to detect small shifts are useful when it is important to rapidly detect small but sustained shifts, as would be the case when process capability is low. Short-run charts deal with the situation in which the same process is used to produce multiple products, each for a short period of time. In this case, keeping a separate chart for each product becomes ineffective and inefficient. Charts for nonnormal distributions apply when the distribution of the plotted point departs significantly from a normal distribution.

6.1 RISK-BASED CONTROL CHARTS

In making decisions with a control chart, there are two risks of making a wrong decision, α and β risks:

> α risk is the probability of concluding that a process is out of control when it is not.
>
> β risk is the probability of concluding that a process is in control when it is not.

α risk leads to false alarms and wasted efforts to detect a process shift when none exists. β risk implies inability to detect process shifts when they have occurred, causing larger amounts of off-target product to be produced. These α and β risks can be controlled by a proper selection of control limits and subgroup size.

6.1.1 Control Limits, Subgroup Size, and Risks

The effect of changing control limits and subgroup size on α and β risks is now illustrated using a "single point outside the control limits" as the out-of-control rule. Figure 6.1 (a) shows the α and β risks for an X chart (subgroup size n = 1) with 3 sigma limits. The α risk is 0.135% + 0.135%

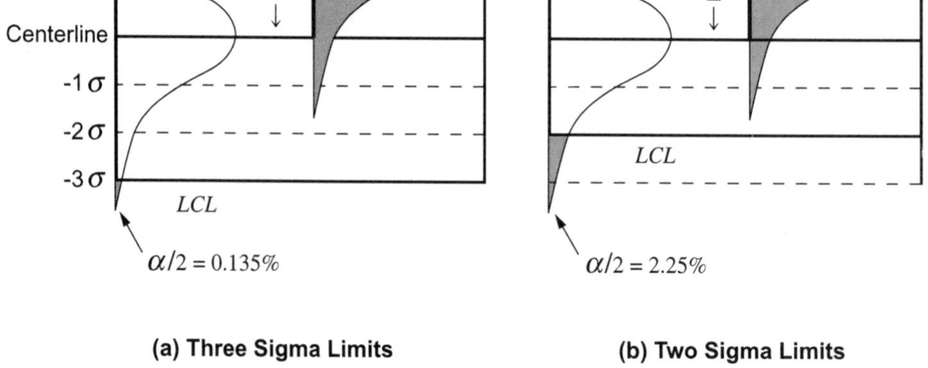

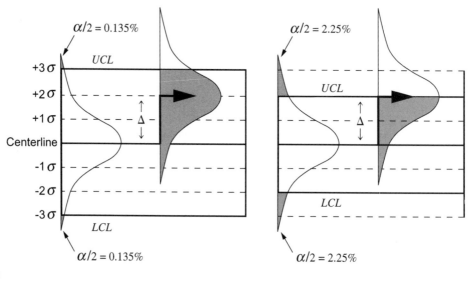

FIGURE 6.1 Effect of changing control limits on α and β risks.

= 0.27%. This is the risk of false alarm; namely, the probability that a plotted point will fall outside the control limits purely by chance and the wrong conclusion will be drawn that the process is out of control. If the process mean shifts by an amount Δ, shown equal to 2σ as an illustration in Figure 6.1 (a), then the process is clearly out of control. However, the probability that the plotted point immediately after the shift will fall inside the control limits is 84%. This is the β risk. In this case, there is an 84% chance of not detecting a sporadic shift as big as 2σ. If the special cause produces a sustained shift in the mean of magnitude $\Delta = 2\sigma$, then the probability of not detecting it in k subsequent subgroups is $(0.84)^k$. Thus, for k equal to 2, 4, 6, 8, and 10, the β risk becomes progressively smaller, taking values of 70%, 50%, 35%, 25% and 17%, respectively. At some point in time after the sustained shift has occurred, it is detected.

If 2 sigma control limits are used, then from Figure 6.1(b), $\alpha = 4.5\%$, and for the same shift $\Delta = 2\sigma$, the β risk is 50%. Thus, narrower control limits increase the α risk and reduce the β risk; wider control limits do the opposite.

Figure 6.2 shows the effect of changing subgroup size. As subgroup size increases from one to four, the standard deviation of $\overline{X}$ reduces by a factor of two and the 3 sigma control limits tighten and become half as wide as they were before. Figure 6.2 (b) shows these new control limits. The α risk remains unchanged because the control limits in Figure 6.2

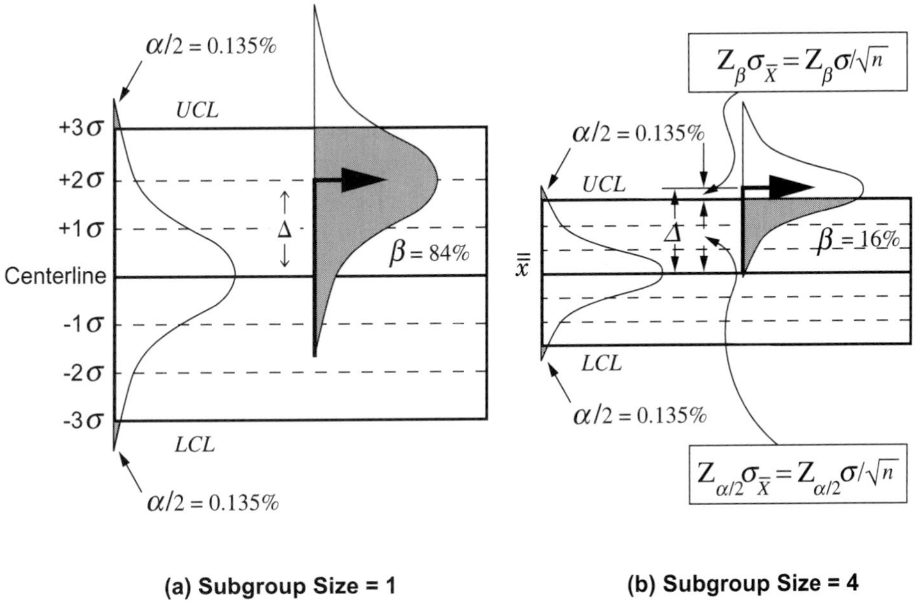

FIGURE 6.2 Effect of changing subgroup size on β risk.

(b) are still 3 sigma limits, but for the same shift Δ, the β risk decreases from 84% to 16%. Thus, for a fixed α risk, increasing subgroup size reduces the β risk. By properly selecting control limits and subgroup size, any desired α and β risks can be attained.

6.1.2 Risk-Based $\overline{X}$ Chart

For an $\overline{X}$ chart, the control limits and subgroup size can be determined to meet any specified α and β risks based upon a "single point outside control limits" as the out-of-control rule. With reference to Figure 6.2 (b),

$$\text{Control limits} = \bar{\bar{x}} \pm Z_{\alpha/2}\frac{\sigma}{\sqrt{n}} \tag{6.1}$$

To determine the subgroup size n, again referring to Figure 6.2 (b),

$$UCL = \bar{\bar{x}} + Z_{\alpha/2}\frac{\sigma}{\sqrt{n}} = (\bar{\bar{x}} + \Delta) - Z_{\beta}\frac{\sigma}{\sqrt{n}}$$

$$\text{Hence, } n = \left(\frac{Z_{\alpha/2} + Z_{\beta}}{d}\right)^2 \tag{6.2}$$

where $d = \Delta/\sigma$. For the traditional 3 sigma limits, $Z_{\alpha/2} = 3$ and the control limits and subgroup size are:

$$\text{Control limits} = \bar{\bar{x}} \pm 3\frac{\sigma}{\sqrt{n}} \tag{6.3}$$

$$n = \left(\frac{3 + Z_{\beta}}{d}\right)^2 \tag{6.4}$$

where $d = \Delta/\sigma$. σ is the short-term standard deviation of individual values and may be obtained from historical data. Table 6.1 shows the approximate values of subgroup size n for various values of d and β for 3 sigma limit $\overline{X}$ charts. As an example, for $d = 2$ and $\beta = 25\%$, the subgroup size $n = 4$. Recall that β is the probability of not detecting a shift in the first subgroup after the shift.

TABLE 6.1 Subgroup Size n ($\alpha = 0.27\%$)

d	$\beta = 50\%$	$\beta = 25\%$	$\beta = 10\%$	$\beta = 5\%$
3.0	1	2	2	3
2.5	2	2	3	4
2.0	3	4	5	6
1.5	4	6	8	10
1.0	9	14	18	22

The table shows that sporadic shifts less than 1.5σ ($d < 1.5$) cannot usually be immediately detected using the typical subgroup sizes of 3 to 5. With a subgroup size of 5, shifts greater than 1.5σ can be detected with reasonable β risk. Much higher subgroup sizes are necessary to immediately detect smaller shifts.

What about the ability of an $\overline{X}$ chart to detect sustained shifts? The probability that a shift will be detected on the kth subgroup following the shift is $\beta^{k-1}(1-\beta)$. Therefore, the expected number of subgroups to detect a sustained shift is

$$\sum_{k=1}^{\infty} k\beta^{k-1}(1-\beta) = \frac{1}{1-\beta} \tag{6.5}$$

From Table 6.1, for $n = 4$, the β risk of not detecting a shift of magnitude 1.5σ is approximately 50%. This means that the probability of detecting such a shift in the first subgroup after the shift is 50%, in the second subgroup it is 25%, in the third subgroup it is 12.5%, and so on. The average or expected number of subgroups with a sustained shift of this magnitude is $1/(1-\beta) = 2$. Hence, it should be recognized that whereas a shift may have been detected in the kth subgroup, it may have occurred not just in the interval between the $(k-1)$th and kth subgroups, but much prior to that. The investigative efforts need to take this into account.

A similar analysis shows the R chart to be very ineffective in detecting changes in variability. If σ doubles, which is a very large change in variability, a subgroup of size 5 has a β risk of 60%, or only a 40% chance of detecting the shift immediately after it has occurred. For a 50% increase in σ, the β risk is approximately 85% (Duncan).

Product Weight Example (continued). For the product weight example in Chapter 4, $\sigma = \overline{R}/d_2 = 1.074$. For a mean shift $\Delta = 1.5\sigma$, $d = \Delta/\sigma = 1.5$. The subgroup size was 5. Therefore, from Equation (6.4),

$$5 = \left(\frac{3 + Z_\beta}{1.5}\right)^2$$

which gives $Z_\beta = 0.354$. The corresponding β risk is 37%. This means that a sporadic shift of 1.5σ will go undetected 37% of the time in the first subgroup after the shift. If the shift is sustained, then the probabilities that the shift will go undetected for 2, 3, and 4 sampling periods are 14%, 5%, and 2% respectively. If it is desirable to detect a shift of 1.5σ in one sampling period with a β risk of 10%, then from Table 6.1, the necessary subgroup size is 8.

***X* Chart.** Since the subgroup size is fixed equal to one, the β risk is uncontrolled and is generally very large for a chart of individual values.

When $n = 1$, from Equation (6.4), $Z_\beta = d - 3$. For $d = \Delta/\sigma = 3.0, 2.5, 2.0, 1.5$, and 1.0, the respective β risks are 50%, 70%, 84%, 93%, and 98%. The X chart has a very limited ability to detect shifts rapidly.

Nested Subgroup. Consider the situation in which a subgroup consists of k samples with n measurements per sample. Such a subgroup has a nested structure (see Chapter 7) with two variance components σ_w^2 (variance within a sample) and σ_b^2 (variance between samples). These variance components can be estimated using methods described in Chapter 7. Given the values of σ_b^2 and σ_w^2, values of k and n may be determined to achieve the desired β risk. Since $\overline{X}$ is obtained by averaging n measurements from each of the k samples,

$$\sigma_{\overline{X}}^2 = \frac{\sigma_b^2}{k} + \frac{\sigma_w^2}{nk}$$

From Figure 6.2 (b),

$$UCL = \overline{\overline{x}} + 3\sigma_{\overline{X}} = \overline{\overline{x}} + \Delta - Z_\beta \sigma_{\overline{X}}$$

Hence,

$$\sigma_{\overline{X}} = \frac{\Delta}{3 + Z_\beta}$$

and

$$\frac{\sigma_b^2}{k} + \frac{\sigma_w^2}{nk} = \left(\frac{\Delta}{3 + Z_\beta}\right)^2$$

For specified values of Δ and β, there will be multiple pairs of values for k and n that will satisfy the above equation. From these, the final selection of k and n can be made on the basis of cost considerations.

6.1.3 Risk-Based Attribute Charts

For $\alpha = 0.27\%$, the sample size for p and u charts is approximately determined using Equation (6.4). For specified values of Δ and β,

$$n = \left(\frac{3 + Z_\beta}{d}\right)^2 \tag{6.6}$$

where $d = \Delta/\sqrt{p(1-p)}$ for the p chart and $d = \Delta/\sqrt{u}$ for the u chart. In these formulae, p denotes fraction defective and u denotes the average number of defects per product.

For attribute charts, very large sample sizes are required to achieve meaningfully small β risk. As one example, if the average fraction defective is 5% ($p = 0.05$) and we want to detect a shift of 2% ($\Delta = 0.02$) with a β risk of 20% ($\beta = 0.2$), then $Z_\beta = 0.84$, $d = 0.092$, and from Equation

(6.6), $n = 1740$! If a subgroup size of 100 is used, the β risk corresponding to $\Delta = 0.02$ can be calculated from Equation (6.6) and is approximately 98%. Attribute control charts with practical sample sizes are not able to capture meaningful shifts rapidly.

6.2 MODIFIED CONTROL LIMIT $\overline{X}$ CHART

The usual Shewhart 3 sigma control limit charts are designed to distinguish between common and special causes of variation. When the C_{pk} index is high and the cost of finding and correcting special causes is very large, the 3 sigma control limits may become uneconomical. Modified control limit charts are designed to reduce the cost of corrective action while ensuring that no out-of-specification product is produced. This is achieved by letting the process drift just high enough and just low enough before taking a corrective action while keeping out of trouble with specifications. Such charts are also known as acceptance control charts (Montgomery).

This situation is exemplified in Figure 6.3 for a chart of individual measurements. The specification limits are 1 ± 0.03 and the standard deviation of individual values is $\sigma = 0.002$. The Shewhart control limits

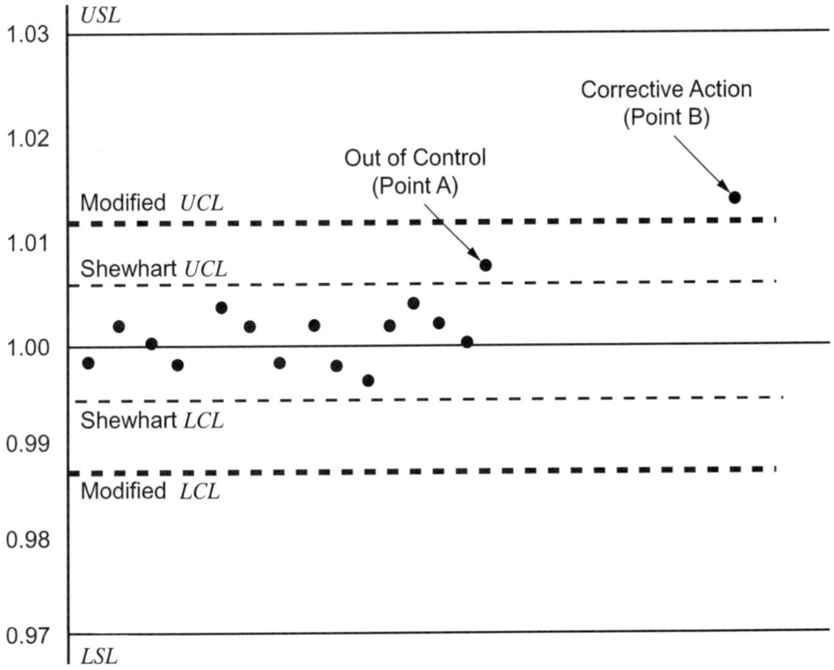

FIGURE 6.3
Concept of modified control limits.

are 1 ± 0.006 and the C_{pk} index, assuming the process to be centered, is 5.0 (= 0.03/3σ). Point A provides the first out-of-control signal. However, in this case, even though the process is out of control, the probability of producing an out-of-specification product is negligible. If the cost of corrective action is high, it may be uneconomical to correct the process. In such cases, modified control limit charts with wider control limits may be used and corrective actions taken only when a point falls outside the modified control limits, as in the case of Point B.

Thus, modified control limits have a completely different function than the Shewhart control limits. Shewhart control limits, along with the out-of-control rules, identify when the process is out of control. Modified control limits are approximate economic action limits intended to signal the need for corrective action. Typically, the process is out of control much before modified control limits provide a corrective action signal.

6.2.1 Chart Design

The question of determining economic action limits for the $\overline{X}$ chart should explicitly consider the cost of sampling, the cost of identifying and taking corrective actions, and the cost of off-target products. Modified control limit charts provide approximate economic action limits under the following assumptions:

1. The cost of identifying and correcting special causes is much higher than the cost of off-target products, so that the primary reason to seek wider limits exists.
2. The range chart is in control and the within-subgroup standard deviation may be assumed to be constant.
3. There are a large number of observations (>200) available to estimate the within-subgroup standard deviation. Therefore, σ is assumed to be known, say, equal to $\overline{R}/d_2$.
4. The product characteristic is essentially equally acceptable as long as it is anywhere within the specification limits. Thus, the process mean may be allowed to drift from μ_L to μ_U, as shown in Figure 6.4 (a), as long as δ, the probability of out-of-specification product when μ is at μ_L or μ_U is acceptably small.
5. Even when the above assumptions are satisfied, modified control limit charts cannot be designed in all cases. The need for the modified limits to be wider than the 3 sigma limits places restrictions on the permissible values of C_{pk} and the sample size n, as explained later.

With reference to Figure 6.4 (a), μ_U and μ_L may be set as follows:

$$\mu_U = USL - Z_\delta \sigma$$

$$\mu_L = LSL + Z_\delta \sigma$$

Even if the true process mean drifts to μ_U or μ_L, the probability of out-of-specification product is no more than δ, which can be set to some acceptably small level.

The true process mean is unknown and is estimated by $\overline{X}$. Figure 6.4 (b) shows the distribution of $\overline{X}$ for $\mu = \mu_L$ and $\mu = \mu_U$. The UCL is drawn in such a way that if the observed $\bar{x} < UCL$, the probability of $\mu \geq \mu_U$ is less than β. The lower control limit is drawn such that if the observed $\bar{x} > LCL$, then probability of $\mu \leq \mu_L$ is less than β. Thus, so long as $\bar{x}$ is within LCL to UCL, we are at least $100(1 - \beta)\%$ sure that $\mu_L \leq \mu \leq \mu_U$ and, hence, the probability of out-of-specification product is less than δ. With reference to Figure 6.4 (b)

$$\text{Modified } UCL = USL - \left(Z_\delta + \frac{Z_\beta}{\sqrt{n}}\right)\sigma$$

$$\text{Modified } LCL = LSL + \left(Z_\delta + \frac{Z_\beta}{\sqrt{n}}\right)\sigma \tag{6.7}$$

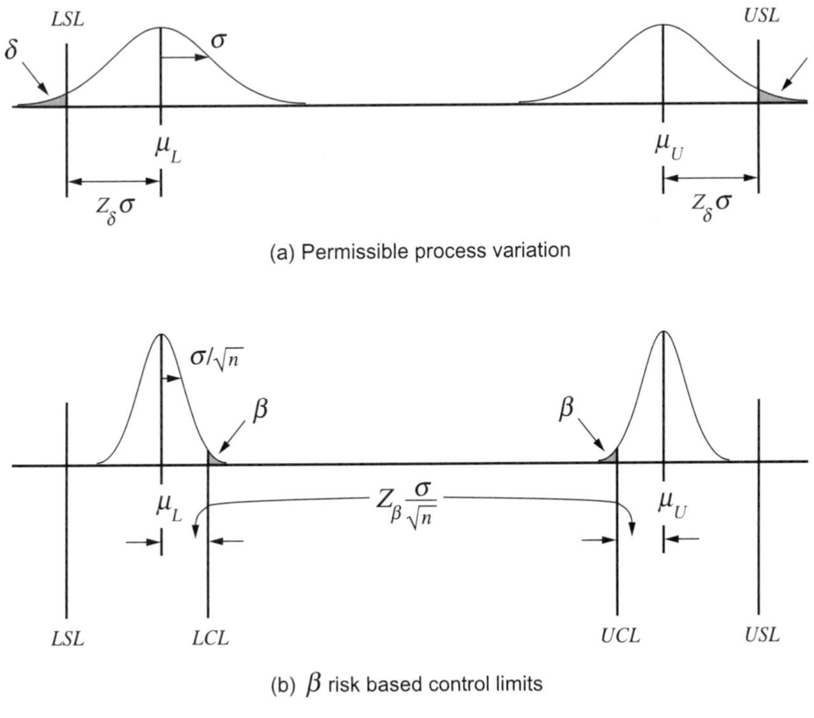

FIGURE 6.4
Modified control limits.

If both δ and β are 0.0013, then $Z_\delta = Z_\beta = 3$ and

$$\begin{aligned} \text{Modified } UCL &= USL - 3\sigma\left(1 + \frac{1}{\sqrt{n}}\right) \\ \text{Modified } LCL &= LSL + 3\sigma\left(1 + \frac{1}{\sqrt{n}}\right) \end{aligned} \tag{6.8}$$

6.2.2 Required Minimum C_{pk}

For the modified limits to be wider than the usual Shewhart limits, C_{pk} must exceed a certain lower bound. Let μ_0 be the centerline, and if USL is closer to μ_0 then

$$C_{pk} = \frac{USL - \mu_0}{3\sigma}$$

The distance from the modified UCL to μ_0 must be larger than the distance between Shewhart UCL and μ_0.

$$\text{Modified } UCL - \mu_0 \geq \text{Shewhart } UCL - \mu_0$$

$$USL - \left(Z_\delta + \frac{Z_\beta}{\sqrt{n}}\right)\sigma - \mu_0 \geq \mu_0 + \frac{3\sigma}{\sqrt{n}} - \mu_0$$

Which gives

$$C_{pk} \geq \frac{1}{\sqrt{n}} + \frac{\left(Z_\delta + \frac{Z_\beta}{\sqrt{n}}\right)}{3} \tag{6.9}$$

For $\delta = \beta = 0.0013$, $Z_\delta = Z_\beta = 3$ and

$$C_{pk} \geq 1 + \frac{2}{\sqrt{n}} \tag{6.10}$$

Equations (6.9) and (6.10) describe the minimum required C_{pk} to implement modified control limit charts. For example, for $n = 1$, C_{pk} must exceed 3. For the usual values of n from 3 to 6, C_{pk} needs to be approximately 2 or more.

Example. The steps in constructing a modified limit chart are now illustrated with an example.

Step 1. Ensure that the basic assumptions are satisfied; namely, that the cost of correction is high compared to the cost of producing off-target products, the range chart is in control, product anywhere within specification is essentially equally satisfactory, and σ can be rather precisely estimated.

Step 2. Calculate the C_{pk} index. For this example, let us assume that the specification is 1.0 ± 0.03, the mean is centered, and $\sigma = \overline{R}/d_2 = 0.002$ so that $C_{pk} = C_p = 5.0$.

Step 3. Select the values of δ and β. Presently $\delta = \beta = 0.0013$ and $Z_\delta = Z_\beta = 3$.

Step 4. Compute the minimum subgroup size to ensure that the C_{pk} index is greater than or equal to the minimum required C_{pk} given by Equation (6.9). Presently $C_{pk} = 5.0 \geq 1 + (2/\sqrt{n})$ for all n. Therefore, no restrictions are placed on the required subgroup size. As a further illustration, if $C_{pk} \leq 1.0$, then a modified control limit chart cannot be designed for any subgroup size. If $C_{pk} = 2.0$, then $2.0 \geq 1 + (2/\sqrt{n})$ requires n to be greater than or equal to 4.

Step 5. Compute the modified control limits using Equation (6.7) based upon the desired subgroup size n. If $n = 4$, then

$$\text{Modified } UCL = USL - 3\sigma\left(1 + \frac{1}{\sqrt{n}}\right) = 1.021$$

$$\text{Modified } LCL = LSL + 3\sigma\left(1 + \frac{1}{\sqrt{n}}\right) = 0.979$$

Note that the Shewhart control limits are 1 ± 0.006 so that the computed modified limits are much wider.

Step 6. Select an appropriate sampling interval based upon the usual considerations. Also ensure that the probability of the plotted point going from within-modified control limits to outside-specification limits in one sampling interval is negligible.

Step 7. Implement the modified control limit chart. If points fall outside the modified control limits, immediate corrective action needs to be taken. If a point falls outside the specification limit, the products produced since the previous sample need to be 100% screened.

6.3 MOVING AVERAGE CONTROL CHART

The $\overline{X}$ chart with usual subgroup sizes cannot easily detect small shifts in the mean. A number of other charts, such as the moving average (*MA*) chart, exponentially weighted moving average (*EWMA*) chart, and cumulative sum (*CUSUM*) chart are more effective in detecting small shifts (Montgomery). This section considers the moving average chart.

Let $\bar{x}_1, \bar{x}_2 \ldots \bar{x}_t \ldots$ denote the subgroup means, with a constant subgroup size n. The moving average of span w at time t is defined as

$$M_t = \frac{\bar{x}_t + \bar{x}_{t-1} + \ldots + \bar{x}_{t-w+1}}{w} \tag{6.11}$$

Since $\bar{X}$ is normally distributed with mean μ and variance σ^2/n, it follows that M_t is normally distributed with mean μ and variance σ^2/nw. The grand average $\bar{\bar{x}}$ estimates μ, which gives the control limits for the *MA* chart as

$$\text{Control limits for } MA \text{ chart} = \bar{\bar{x}} \pm \frac{3\sigma}{\sqrt{nw}}$$

For $n > 1$, $3\sigma/\sqrt{n}$ is estimated by $A_2\bar{R}$ and for $n = 1$, it is estimated by $2.66\ m\bar{R}$. Hence,

$$\text{Control limits for } MA \text{ chart} = \bar{\bar{x}} \pm \frac{A_2\bar{R}}{\sqrt{w}} \qquad (\text{for } n > 1)$$

$$\text{Control limits for } MA \text{ chart} = \bar{\bar{x}} \pm \frac{2.66m\bar{R}}{\sqrt{w}} \qquad (\text{for } n = 1) \tag{6.12}$$

Essentially, by taking an average of w subgroups, the control limits become narrower by a factor of $\sqrt{w}$ compared to the corresponding $\bar{X}$ or X chart. The narrower limits make it easier to detect smaller sustained shifts. Smaller sustained shifts are better detected by larger w. Since the $\bar{X}$ chart is good at detecting large sporadic shifts and the *MA* chart is good at detecting small sustained shifts, both charts may be used simultaneously to improve the overall effectiveness in detecting shifts.

Product Weight Example (continued). For the product weight example in Chapter 4, the 22 subgroup averages are shown in Table 6.2. Recall from Chapter 4 that the subgroup size $n = 5$, $\bar{R} = 2.5$, and $\bar{\bar{x}} = 3.66$. For $n = 5$, $A_2 = 0.577$. From Equation (6.12) the control limits for the *MA* chart with span $w = 4$ are:

$$3.66 \pm \frac{0.577(2.5)}{\sqrt{4}} = 3.66 \pm 0.72$$

These control limits apply for $t \geq w$. For $t < w$, the control limits are given by

$$\bar{\bar{x}} \pm \frac{3\sigma}{\sqrt{nt}} = \bar{\bar{x}} \pm \frac{A_2\bar{R}}{\sqrt{t}}$$

In particular, the control limit for $t = 1$ is the same as for the $\bar{X}$ chart. The computations are in Table 6.2 and the *MA* chart is shown in Figure 6.5. Since the successive M_t values are correlated, it seems appropriate to only use the first out-of-control rule.

TABLE 6.2 Data and Calculations for the *MA* Chart

t	$\overline{X}$	M_t	LCL	UCL
1	3.30	3.30	2.22	5.10
2	4.30	3.80	2.64	4.68
3	3.22	3.60	2.83	4.49
4	3.94	3.69	2.94	4.38
5	4.16	3.91	2.94	4.38
6	3.86	3.79	2.94	4.38
7	3.88	3.96	2.94	4.38
8	1.60	3.38	2.94	4.38
9	3.46	3.20	2.94	4.38
10	4.18	3.28	2.94	4.38
11	3.38	3.16	2.94	4.38
12	3.50	3.63	2.94	4.38
13	3.94	3.75	2.94	4.38
14	3.16	3.50	2.94	4.38
15	4.44	3.76	2.94	4.38
16	3.80	3.84	2.94	4.38
17	3.36	3.69	2.94	4.38
18	4.30	3.98	2.94	4.38
19	4.12	3.90	2.94	4.38
20	3.67	3.86	2.94	4.38
21	3.50	3.90	2.94	4.38
22	3.50	3.70	2.94	4.38

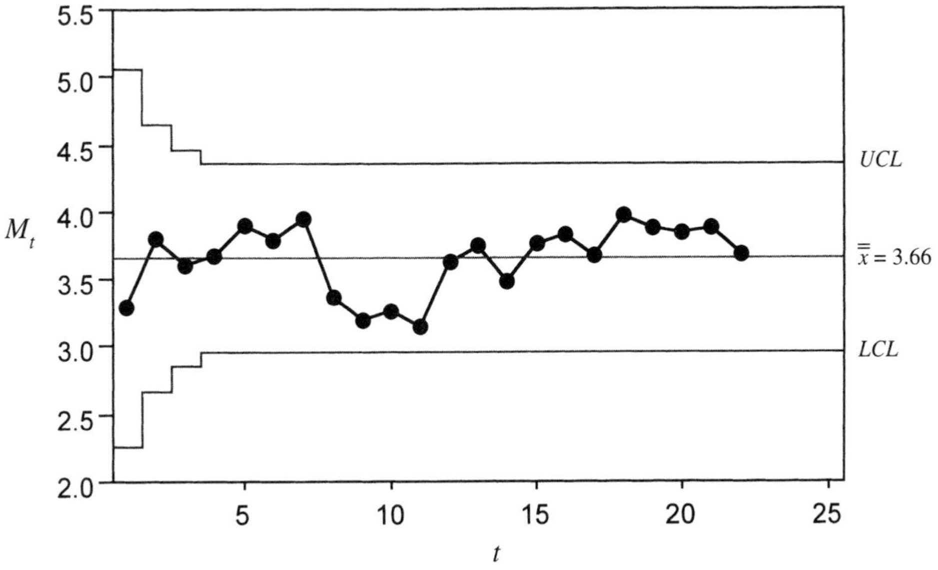

FIGURE 6.5 Moving average chart for product weight data.

6.4 SHORT-RUN CONTROL CHARTS

Traditional control charts assume a long-run production of the same product. It is possible to collect 20 to 25 subgroups, establish appropriate control limits, and monitor and improve the process thereafter. On the other hand, a short run is a situation in which a particular product is manufactured only for a short period of time. Such a situation occurs frequently. Multiple products are often produced using the same manufacturing process, each perhaps periodically over a short time duration. The just-in-time system demands short production runs. With short production runs and changing product streams, it may appear that control chart techniques will not apply. But such is not the case (Wheeler).

This section describes the difficulties that arise when conventional control charts are used in a short-run situation. The basic principle of converting a short-run situation to the long-run situation is then explained. This is followed by a description of short-run variable and attribute control charts.

Table 6.3 shows the data from a short-run shop. The same machine is used to produce products A and B. Due to just-in-time considerations, the production alternates between these two products. The characteristic being measured is the cutoff length of the product. The target length for Product A is 20 and for Product B it is 30.

In general, the data could be the individual measurements for each item, individual measurements taken at a predetermined sampling interval, or even subgroup averages or attribute measurements. In all these cases, traditional control chart methods may be used to create separate control charts for each product by separating the data by product.

TABLE 6.3 Data From a Short-Run Shop

Subgroup	Product	x	Subgroup	Product	x
1	A	19	13	B	31
2	A	20	14	B	31
3	B	32	15	B	33
4	B	29	16	A	23
5	B	27	17	A	21
6	A	20	18	A	20
7	A	18	19	B	29
8	B	29	20	B	29
9	B	31	21	A	17
10	A	21	22	B	31
11	A	22	23	A	20
12	A	21	24	B	32

| x | 19 | 20 | 20 | 18 | 21 | 22 | 21 | 23 | 21 | 20 | 17 | 20 |
| mR | | 1 | 0 | 2 | 3 | 1 | 1 | 2 | 2 | 1 | 3 | 3 |

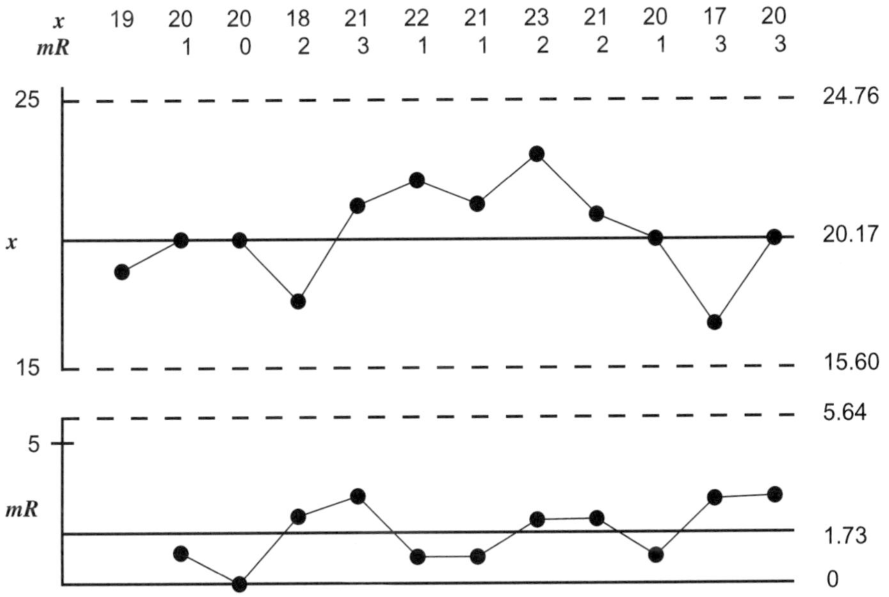

FIGURE 6.6
X and mR charts for product A.

Figures 6.6 and 6.7 show such individual and moving-range charts for product A and product B. The process appears to be in control. As production continues, additional data on products A and B may be separately plotted on these product control charts for A and B. As sufficient data become available, the control limits may be revised as appropriate.

This individual product control chart approach has the advantage of viewing each product independently by using traditional control chart methods. However, the approach has many disadvantages. It leads to a proliferation of control charts per process. The generated out-of-control products may have been shipped a long time ago and not be available for further analysis. A longer time period is required to establish meaningful control limits for each product. The approach does not apply to the manufacture of few of a kind products. Most importantly, this approach fragments the continuous running record of the process, making it difficult to visualize the long-term performance of the process.

Thus, it would be beneficial if data from multiple products could be charted on the same control chart. This cannot be achieved by simplistically plotting the raw data (e.g., x values from Table 6.3) for multiple products on the same chart because each product may have a different mean and standard deviation and, even under statistical control, the plotted statistic X will not have a fixed distribution over time. The basic idea behind short-run control charts is to select a statistic to plot that, under the assumption of statistical control, has a fixed distribution over

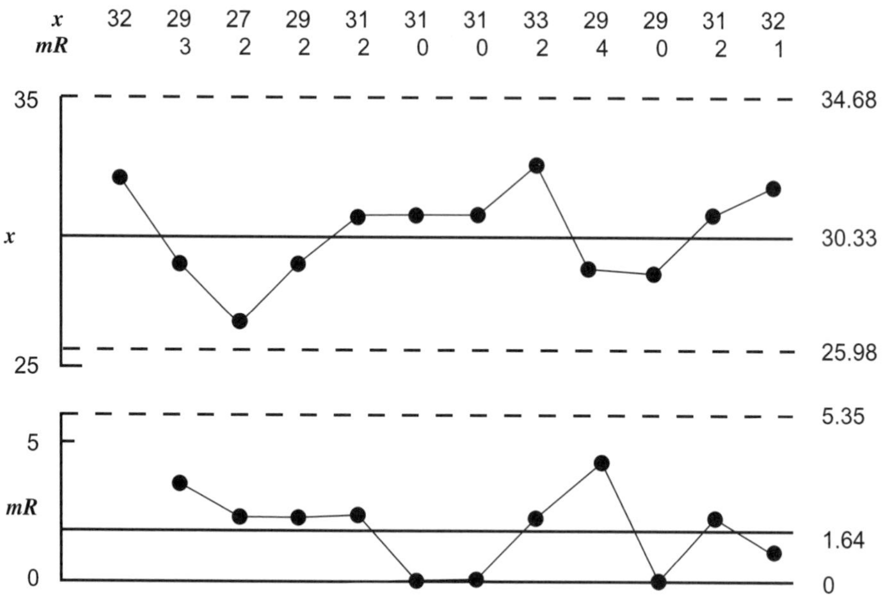

FIGURE 6.7 X and mR charts for product B.

time regardless of the product. In this manner, the short-run situation is effectively converted into a long-run situation and the process performance is continuously monitored and improved.

In general, if X denotes the individual values, averages, or counts and has a mean μ and standard deviation σ, then the statistic

$$Z = \frac{(X - \mu)}{\sigma}$$

has a distribution with zero mean and unit standard deviation. Therefore, even when μ and σ are changing from product to product, under statistical control, Z has a fixed distribution over time regardless of the product. The Z values can be plotted for all products on the same chart.

6.4.1 Short Run Individual and Moving Range Charts

Consider the situation in which subgroup size $n = 1$ and variable data are being collected for each product. Two cases arise. Products may differ only in terms of their target values or they may also differ in terms of their variability. These lead to $\Delta X - mR$ and $Z - W$ charts, respectively.

ΔX and *mR* Charts. If products differ only in terms of target values and the variability is constant from product to product, then standardization with respect to σ is not necessary. One may simply plot the difference Δx given by

$$\Delta x = x - T_i$$

where T_i is the target value or the historical average value for the ith product. The mR values are computed as successive differences between the Δx values. Figure 6.8 shows this computation of Δx and mR values for the data in Table 6.3.

The centerline of the ΔX chart is taken to be zero. To the extent that T_i differs from μ_i, ΔX will have a nonzero mean equal to $(\mu_i - T_i)$. This will manifest itself as a shift and produce an out-of-control signal on the ΔX chart, which will prove useful for bringing the process mean on target. The standard deviation of ΔX is equal to σ, which is estimated as $m\overline{R}/d_2$. With $n = 2$, $d_2 = 1.128$. Therefore, for the ΔX chart,

$$Control\ Limits_{\Delta X} = 0 \pm 2.66\, m\overline{R} \tag{6.14}$$

Since the variance is constant from product to product, the mR chart is designed traditionally with

$$\begin{aligned} Centerline_{mR} &= m\overline{R} \\ UCL_{mR} &= 3.268\, m\overline{R} \\ LCL_{mR} &= 0 \end{aligned} \tag{6.15}$$

The short-run ΔX and mR charts are shown in Figure 6.8 for the data in Table 6.3. The ΔX chart shows the process to be out of control because

Product	A	A	B	B	B	A	A	B	B	A	A	A	B	B	B	A	A	A	B	B	A	B	A	B
x	19	20	32	29	27	20	18	29	31	21	22	21	31	31	33	23	21	20	29	29	17	31	20	32
Target	20	20	30	30	30	20	20	30	30	20	20	20	30	30	30	20	20	20	30	30	20	30	20	30
Δx	-1	0	2	-1	-3	0	-2	-1	1	1	2	1	1	1	3	3	1	0	-1	-1	-3	1	0	2
mR		1	2	3	2	3	2	1	2	0	1	1	0	0	2	0	2	1	1	0	2	4	0	2

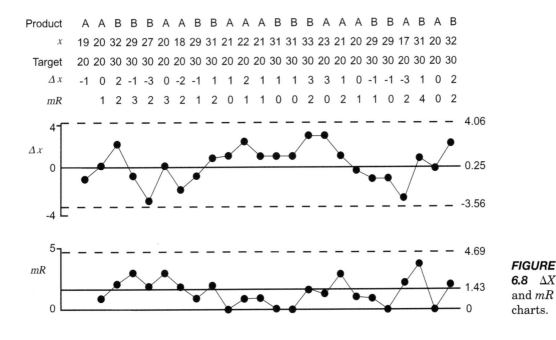

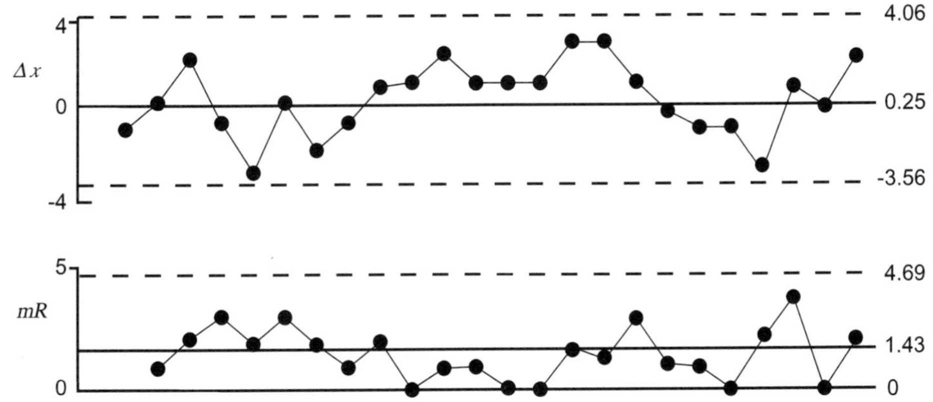

FIGURE 6.8 ΔX and mR charts.

there are nine points in a row above the centerline. The product control charts did not produce this signal because product control charts segment the long-run record of the process.

Z and W Charts. In addition to the target values, if the variability also changes from product to product, then a Z chart is used, where

$$Z = \frac{X - T_i}{\sigma_i}$$

where T_i is the target value or the average for the ith product and σ_i is estimated as $m\overline{R}_i/d_2$. For $n = 2$, $d_2 = 1.128$ and $m\overline{R}_i$ is the average moving range obtained from the Δx values for the ith product. Hence,

$$Z = \frac{1.128(X - T_i)}{m\overline{R}_i}$$

The centerline for the Z chart is taken to be zero. Since $\sigma_Z = 1$, the Z chart is designed with

$$\textit{Control Limits}_Z = 0 \pm 3 \tag{6.16}$$

Corresponding to each value of X, the Z value is computed and plotted on the Z chart. If μ_i differs from T_i, the control chart produces a signal indicating that the mean is not on target, so that appropriate corrective action may be taken.

The variability is monitored by using the W chart, where the W values are absolute differences between the successive Z values. The W statistic has a distribution with mean d_2 and standard deviation d_3. For $n = 2$, $d_2 = 1.128$ and $d_3 = 0.8525$. Therefore, for the W chart

$$\textit{Centerline}_w = 1.128$$

$$UCL_w = 1.128 + 3(0.8525) = 3.686 \tag{6.17}$$

$$LCL_w = 0$$

***CV* Constant.** A special case occurs when each product has a different mean and standard deviation but in a manner that keeps the coefficient of variation $CV(=\sigma_i/\mu_i)$ constant. In this case, we can control chart X/T_i, which will have a mean of 1 and a standard deviation equal to CV. Hence, for the X/T_i chart,

$$\textit{Control Limits}_{X/T_i} = 1 \pm 3\,CV \tag{6.18}$$

A chart of moving ranges can be constructed by taking the absolute differences between the successive x/T_i values. Such a moving range will have a mean equal to 1.128 CV and a standard deviation equal to 0.8525 CV. Hence, for this moving range chart,

$$
\begin{aligned}
Centerline &= 1.128\ CV \\
UCL &= 3.686\ CV \\
LCL &= 0
\end{aligned}
\tag{6.19}
$$

The selection of the type of short-run individual and moving range charts depends upon whether the variability is constant from product to product. This could be assessed using methods in Chapter 3. If variability changes from product to product, Z and W charts should be used, the CV constant being a special case. Otherwise, ΔX and mR charts should be used.

6.4.2 Short-Run Average and Range Charts

Given the scarcity of data in short-run situations, the individual and moving range charts will generally be more applicable. However, if data are naturally available in subgroups, short-run average and range charts may be used. The rationale for short-run average and range charts is similar to that discussed above.

$\Delta\overline{X}$ and R Charts. In this case, products differ only in terms of their target values. Subgroups of size n are taken at the desired sampling interval and the values of $\bar{x}$ and R are determined for each subgroup. Since the variance is assumed constant from product to product, the R chart is designed conventionally with

$$
\begin{aligned}
Centerline_R &= \overline{R} \\
UCL_R &= D_4\overline{R} \\
LCL_R &= D_3\overline{R}
\end{aligned}
\tag{6.20}
$$

The target values differ from product to product and the $\bar{x}$ values are standardized to obtain $\Delta\bar{x}$:

$$\Delta\bar{x} = \bar{x} - T_i$$

where T_i is the target value or the historical average for the ith product.

The centerline for the $\Delta\overline{X}$ chart is taken to be zero. $\Delta\overline{X}$ has a distribution with a standard deviation equal to $\sigma/\sqrt{n}$. σ is estimated as $\overline{R}/d_2$, where $\overline{R}$ is the average range for all products. Therefore, for the $\Delta\overline{X}$ chart,

$$Control\ Limits_{\Delta\overline{X}} = 0 \pm A_2\overline{R} \tag{6.21}$$

$\overline{Z}$ and $\overline{W}$ Charts. In addition to the target values, if the variability also changes from product to product, then it is necessary to use the $\overline{Z}$ chart. For the ith product, the plotted values of $\overline{Z}$ are obtained from

$$\overline{Z} = \frac{(\bar{x} - T_i)}{\sigma_i/\sqrt{n}} = \frac{d_2\sqrt{n}(\bar{x} - T_i)}{\overline{R}_i}$$

The centerline for the $\overline{Z}$ chart is taken to be zero and since the standard deviation of $\overline{Z}$ is one,

$$Control\ Limits_{\overline{Z}} = 0 \pm 3 \tag{6.22}$$

Variability is monitored using the $\overline{W}$ chart. The within-subgroup ranges R are converted into $\overline{W}$ values using

$$\overline{W} = \frac{R}{\sigma_i} = \frac{d_2 R}{\overline{R}_i}$$

$\overline{W}$ has a mean equal to d_2 and a standard deviation equal to d_3, where d_2 and d_3 are based upon the subgroup size. For the $\overline{W}$ chart,

$$Control\ Limits_{\overline{W}} = d_2 \pm 3d_3 \tag{6.23}$$

***CV* Constant.** A special case occurs when each product has a different mean and standard deviation but in a manner as to keep the coefficient of variation $CV(=\sigma_i/\mu_i)$ constant. In this case, we can control chart $\overline{X}/T_i$, which will have a mean equal to 1 and standard deviation equal to $CV/\sqrt{n}$. Hence, for the $\overline{X}/T_i$ chart,

$$\text{Control limits} = 1 \pm \frac{3\,CV}{\sqrt{n}} \tag{6.24}$$

The within-subgroup variability can be monitored by plotting R/T_i. For this standardized range chart,

$$\text{Control limits} = (d_2 \pm 3d_3)CV \tag{6.25}$$

The selection of the type of short-run $\overline{X}$ and R charts depends upon whether the variability from product to product can be considered to be constant or not, which can be assessed using methods in Chapter 3. If products have different variabilities, $\overline{Z}$ and $\overline{W}$ charts should be used, the CV constant being a special case. Otherwise, $\Delta\overline{X}$ and R charts are preferred. The short-run variable charts are summarized in Table 6.4.

6.4.3 Short Run Attribute Charts

The Z statistic can be directly applied to obtain short-run attribute charts. Table 6.5 summarizes the short-run statistics for all attribute charts.

6.5 CHARTS FOR NONNORMAL DISTRIBUTIONS

There are situations in which the characteristic of interest does not have a normal distribution. Microbiological counts and particulate counts are

TABLE 6.4 Short-Run Variable Charts

Chart	Centerline	UCL	LCL	Plotted Point
ΔX	0	$2.66\ m\overline{R}$	$-2.66\ m\overline{R}$	$\Delta x = x - T_i$ T_i = Target for ith product
mR	$m\overline{R}$	$3.268\ m\overline{R}$	0	mR = Absolute difference between successive Δx values
Z	0	3	-3	$Z = 1.128\,(x - T_i)/m\overline{R}_i$
W	1.128	3.686	0	W = Absolute difference between successive Z values
X/T_i	1	$1{+}3\ CV$	$1 - 3\ CV$	x/T_i
mR	$1.128\ CV$	$3.686\ CV$	0	mR = Absolute difference between successive x/T_i values
$\Delta\overline{X}$	0	$A_2\overline{R}$	$-A_2\overline{R}$	$\Delta\overline{x} = \overline{x} - T_i$
R	$\overline{R}$	$D_4\overline{R}$	$D_3\overline{R}$	R = Within-subgroup range
$\overline{Z}$	0	3	-3	$\overline{Z} = d_2\sqrt{n}(\overline{x} - T_i)/\overline{R}_i$
$\overline{W}$	d_2	$d_2 + 3d_3$	$d_2 - 3d_3$	$\overline{W} = d_2R/\overline{R}_i$
$\overline{X}/T_i$	1	$1 + 3\ CV/\sqrt{n}$	$1 - 3\ CV/\sqrt{n}$	$\overline{x}/T_i$
R/T_i	d_2CV	$(d_2 + 3d_3)CV$	$(d_2 - 3d_3)CV$	R/T_i

Note: A_2, D_3, D_4, d_2, and d_3 are determined based upon the subgroup size from Appendix G.

not normally distributed. Whenever the characteristic of interest can take values that are orders of magnitude apart, the characteristic is unlikely to have a normal distribution. Length of time between occurrences of events is usually not normally distributed. Thus, the length of time from entering a restaurant to being served, called the waiting time, may be exponentially distributed. Particle diameter often does not have a

TABLE 6.5 Short Run Attribute Charts

Chart	Centerline	UCL	LCL	Plotted Point (Z)
Short-run np	0	3	-3	$\dfrac{np - n\overline{p}}{\sqrt{n\overline{p}(1-\overline{p})}}$
Short-run p	0	3	-3	$\dfrac{p - \overline{p}}{\sqrt{\overline{p}(1-\overline{p})/n}}$
Short-run c	0	3	-3	$\dfrac{c - \overline{c}}{\sqrt{\overline{c}}}$
Short-run u	0	3	-3	$\dfrac{u - \overline{u}}{\sqrt{\overline{u}/n}}$

normal distribution. Normal distribution takes values from $-\infty$ to $+\infty$. Whenever a measured characteristic can only take positive values, and in particular when the values are close to zero, the distribution may not be normal.

For nonnormal distributions, the ±3 sigma limits do not enclose 99.73% of the population. The α risk changes. If the distribution is not symmetric, the run rules for out of control change. This lack of normality is not a significant issue for an $\overline{X}$ chart because the central limit theorem provides some assurance that regardless of the distribution of X, the distribution of $\overline{X}$ will be close to normal. Nonnormality can have significant impact on the chart of individual values.

Two approaches may be taken to design an X chart when the distribution of X is not normal. One approach is to first identify the appropriate distribution for X, based upon a combination of theory and data. Distributions such as exponential, lognormal, and gamma are of particular interest in this regard. The parameters of the distribution can be estimated from the data. A centerline that corresponds to the mean (or median) of the fitted distribution and control limits that correspond to an α risk of 0.027 can then be determined, assuming the data to have been collected under stability.

This approach can be easily illustrated for a uniform distribution. If $X(a \leq X \leq b)$ has a uniform distribution, then $\mu_X = (a + b)/2$ and $\sigma_X = (b - a)/\sqrt{12}$. For $a = 10$ and $b = 16$, the $\mu \pm 3\sigma$ control limits would have been 13 ± 5.2, considerably wider than the actual range of X. For an α risk of 0.27%, the actual control limits are $a + \delta$ to $b - \delta$, where $\delta = 0.0135\ (b - a)$. For $a = 10$ and $b = 16$, $\delta = 0.008$ and the control limits are 13 ± 2.92, which are very different from the 3σ limits.

The second approach is to transform the distribution of X into a normal distribution by finding an appropriate transformation $Y = f(X)$. A control chart for the transformed Y values can then be designed by traditional methods. Whenever the transformation makes physical sense, this would be the preferred approach.

Lognormal Distribution. A number of characteristics of interest are likely to have a lognormal distribution. Particle diameters are often lognormally distributed. One may expect microbiological and particulate counts to have a Poisson distribution, but the Poisson distribution has the restriction that the variance is equal to the mean. This restriction can become constraining and a lognormal distribution, with independent mean and variance, may provide a better fit. Lognormal is also a likely distribution when the data take values close to zero.

A lognormal distribution is a positively skewed distribution. A very useful property is that if X has a lognormal distribution, then $Y = \log_e X$ has a normal distribution. Therefore, it is very easy to transform the X

data to a normal distribution by taking $\log_e X$ and then use conventional control charts for $\log_e X$.

This transformation approach breaks down if the X data contain many zero values because $\log 0 = -\infty$. In this case, an upper control limit may be obtained as follows by fitting the lognormal distribution. If X has a lognormal distribution with mean μ_X and standard deviation σ_X, then $Y = \log_e X$ has a normal distribution with the following mean and standard deviation (Johnson and Kotz):

$$\mu_Y = \log_e(\mu_X) - \frac{1}{2}\log_e\left(\frac{\sigma_X^2}{\mu_X^2} + 1\right)$$
$$\sigma_Y = \sqrt{\log_e\left(\frac{\sigma_X^2}{\mu_X^2} + 1\right)} \tag{6.26a}$$

The upper control limit for Y is $\mu_Y + 3\,\sigma_Y$. For X, the upper control limit becomes

$$UCL_X = e^{\mu_Y + 3\sigma_Y} \tag{6.26b}$$

Since μ_X and σ_X can be estimated even when X data contain zeros, the upper control limit can be estimated from Equation (6.26).

For example, if $\mu_X = 1$ and $\sigma_X = 3$, then

$$\mu_Y = \log_e(1) - \frac{1}{2}\log_e(10) = -1.151$$
$$\sigma_Y = \sqrt{\log_e(10)} = 1.517$$
$$UCL_X = e^{-1.151+3(1.517)} = 30$$

7 Variance Components Analysis

To reduce variability, it is important to understand the extent of variability, the causes of variability, and the contribution of each cause to the total variability. Such an understanding of process variability is indispensable to ensure that the variance reduction efforts are correctly targeted. For example, if the total variability could be partitioned into that due to the procured material, the production process, and the test method, the improvement efforts could be correctly focused on the largest source of variability.

Toward this objective, the histogram provides useful information on the total variability. A control chart detects the presence of special causes and thereby helps identify special causes of variability. Variance transmission analysis and variance components analysis are two methods used to assess the contribution of each cause to the total variability. The first requires an equation relating the process output to the key input factors. Variance transmission analysis can then be used to estimate the variance contribution due to variability in each input factor. This approach will be briefly considered in the next chapter. On the other hand, variance components analysis allows the total variability to be partitioned into variability due to each cause solely on the basis of data. Such data are often readily available. This underutilized technique of variance components analysis is the subject of this chapter. The practical utility of variance components analysis, particularly in a manufacturing environment, lies in the fact that, whereas planned experiments are often difficult to conduct on a full-scale manufacturing process, variability in input and output factors naturally occurs during manufacturing and thus provides a way to identify key causes of variability and their contribution. This is a way of listening to Murphy speak.

This chapter begins by introducing the idea of variance components, first for an $\overline{X}$ chart and then for a one-way classification with a fixed factor. These two simple applications of variance components will prove very useful in practice. The topic of structured data collection or structured capability studies is considered next. The differences between fixed and random factors, nested and crossed classifications, along with the mathematics of variance components are explained. By matching data collection to the likely major causes of variability, structured studies permit data to be collected and analyzed in a manner suitable to understand not only the total variability but also the variability due to each major cause of variation.

7.1 $\overline{X}$ CHART (RANDOM FACTOR)

Fruit Tray Example. Table 7.1 shows a portion of the data collected on the dry weight, in grams, of a molded-fiber fruit tray. The data shown are for a single mold cavity. Weights of three successive trays were measured, so the subgroup size $n = 3$. The sampling interval was one hour.

Table 7.1 also shows various computations necessary to construct $\overline{X} - R$ charts, which are shown in Figure 7.1. The range chart is in control,

TABLE 7.1 Fruit Tray Weight Data

Subgroup	Weights (x)			$\bar{x}$	R	s^2
1	79.6	78.5	79.1	79.07	1.1	0.30
2	77.5	78.1	79.0	78.20	1.5	0.57
3	76.6	77.5	76.2	76.77	1.3	0.44
4	78.0	78.5	77.7	78.07	0.8	0.16
5	78.9	79.5	79.7	79.37	0.8	0.17
6	80.3	78.4	79.1	79.27	1.9	0.92
7	80.1	80.2	79.5	79.93	0.7	0.14
8	77.2	78.5	78.1	77.93	1.3	0.44
9	80.4	81.2	80.6	80.73	0.8	0.17
10	78.7	80.1	78.5	79.10	1.6	0.76
				$\bar{\bar{x}} = 78.84$	$\bar{R} = 1.18$	$s^2_{\text{pooled}} = 0.41$

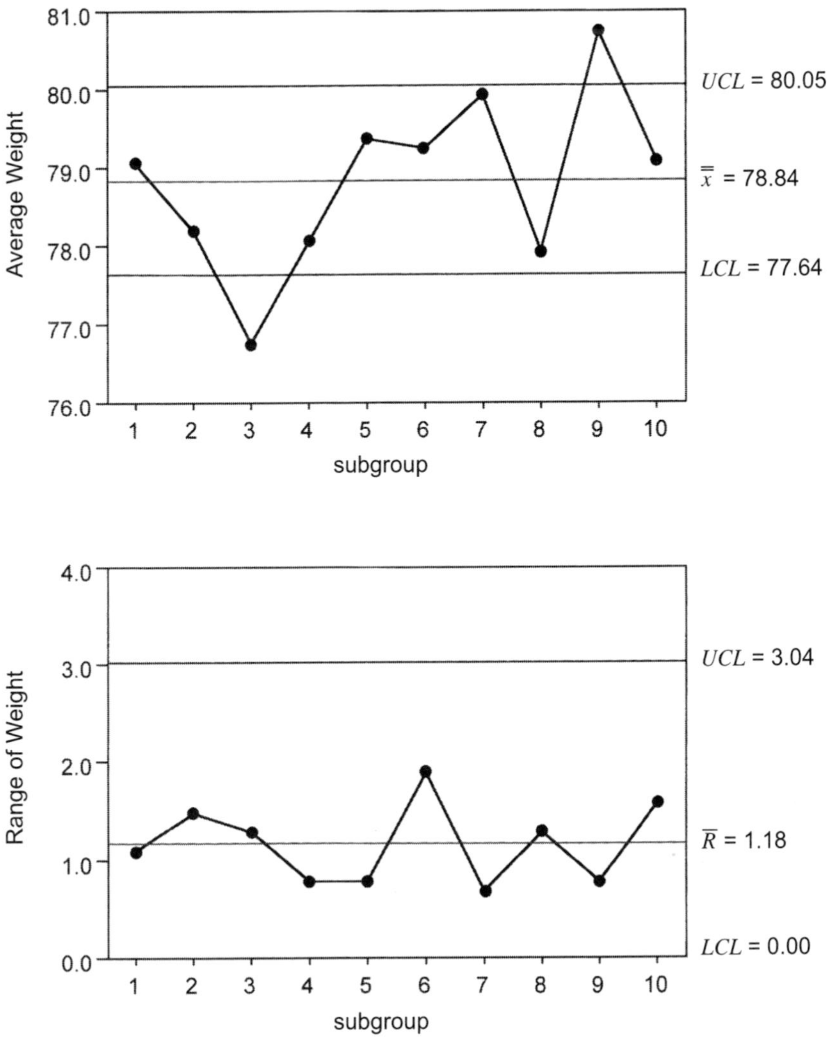

FIGURE 7.1 $\overline{X}$ and R charts for fruit tray data.

meaning that the within-subgroup variability is constant. The $\overline{X}$ chart shows the process mean to be unstable.

7.1.1 Nested Structure

The data collection scheme for an $\overline{X}$ chart has the nested structure shown in Figure 7.2. The figure shows the 10 subgroups and the three products per subgroup. The individual observations and subgroup averages are also shown. Note that even though the products in each sub-

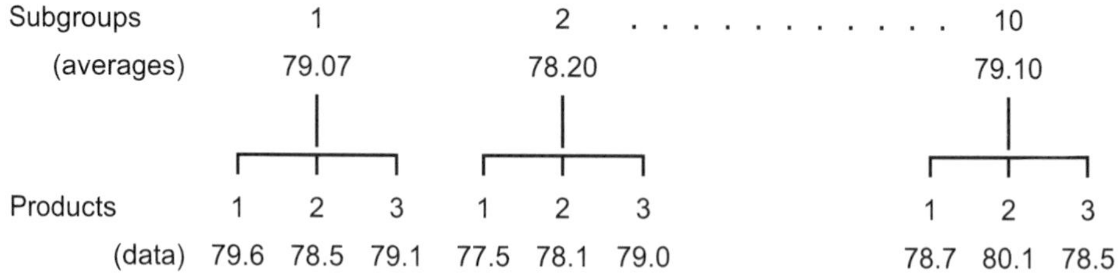

FIGURE 7.2 Nested structure for $\overline{X}$ chart.

group are labeled 1, 2, and 3 for ease of understanding, each subgroup consists of three different products. The product labeled 1 in subgroup 1 is not the same as product labeled 1 in subgroup 2. Both subgroups and products are random factors. The 10 subgroups are a random sample from the infinitely many possible subgroups. Similarly, the three products are a random sample from the population of products produced at that time. Products are said to be nested inside the subgroups, providing the structure with a tree-diagram appearance.

The total variance of fruit tray weight can be partitioned into two components of variance called the between-subgroup and within-subgroup variance components:

$$\sigma_t^2 = \text{total variance}$$

$$\sigma_b^2 = \text{variance between subgroups}$$

$$\sigma_w^2 = \text{product-to-product variance within a subgroup}$$

This partitioning of variance is graphically shown in Figure 7.3. Subgroup means come from a distribution with mean μ and standard deviation σ_b. For the ith subgroup, individual products come from a distribution with mean μ_i and standard deviation σ_w. A randomly selected product from a randomly selected subgroup has a distribution with mean μ and a variance given by

$$\sigma_t^2 = \sigma_b^2 + \sigma_w^2 \qquad (7.1)$$

Estimating the between- and within-variance components is very useful. σ_w represents the short-term (common cause) variability of the process and σ_b represents the special cause variation. If σ_b is large, controlling the process helps reduce variance. If σ_b is small, merely using a control chart to control the process is not enough; the process needs to be changed to reduce variance. The factors influencing σ_w and σ_b are often different. For example, measurement variability influences σ_w, whereas differences in batches of raw material influence σ_b. Hence, knowing the variance components helps focus attention on the right causes.

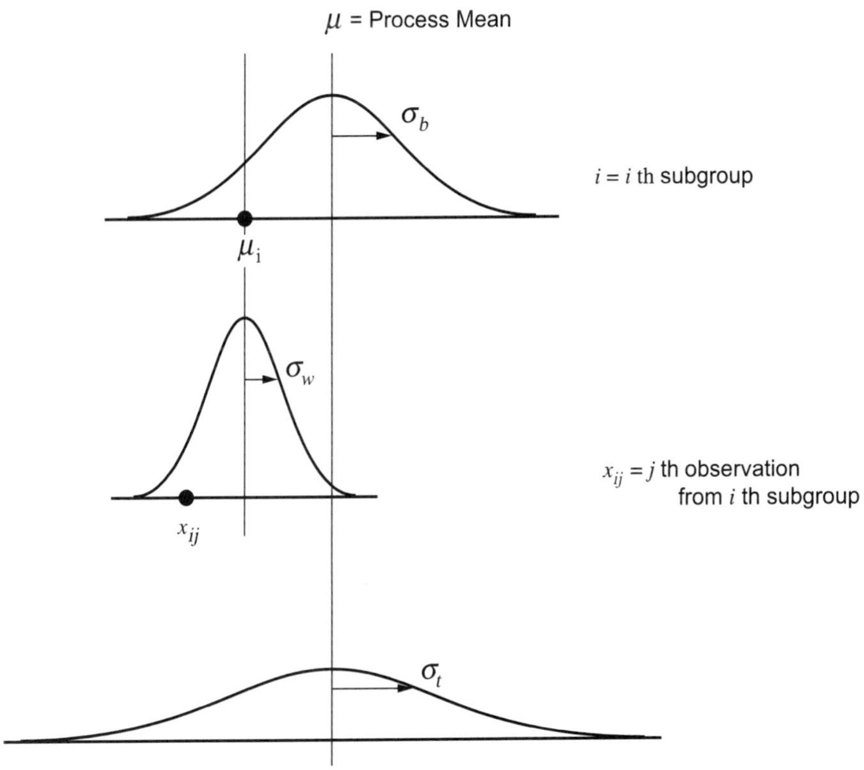

FIGURE 7.3 Variance components for $\overline{X}$ chart.

7.1.2 Estimating Variance Components

We now consider several ways of estimating variance components for the $\overline{X}$ chart and also for the X chart.

Using Variance of $\overline{X}$. With reference to Figure 7.2, the three observations within subgroup 1 are different from each other because of product-to-product variability within a subgroup, namely σ_w^2. Hence, one estimate of σ_w^2 is obtained by computing the variance between these three observations. From Table 7.1, for subgroup 1, an estimate of within-subgroup variance is 0.30. Similar estimates can be obtained from each subgroup and are shown in the last column of Table 7.1. Since the subgroup size is constant, a pooled estimate of σ_w^2, denoted by s_w^2, is obtained by taking the average of the 10 individual estimates:

$$s_w^2 = s_{\text{pooled}}^2 = 0.41$$

Again with reference to Figure 7.2, there are two reasons why subgroup averages differ from each other. One reason is that the true (not just the observed) subgroup means are different, leading to the between-subgroup standard deviation σ_b. The second reason is that the product-to-product variance within a subgroup contributes σ_w^2/n to the variance of subgroup averages. The variance computed from the 10 subgroup averages 79.07, 78.20, . . . , 79.10 in Table 7.1 is 1.27, and the subgroup size n is 3. Therefore,

$$1.27 = s_b^2 + \frac{s_w^2}{3}$$

Hence, $s_b^2 = 1.14$ which is an estimate of $\sigma_b{}^2$. The total variance σ_t^2 is estimated by

$$s_t^2 = s_w^2 + s_b^2 = 1.55$$

The conclusion is that approximately 25% of the total variance is due to short-term variability and 75% of the total variance is due to factors that influence between-subgroup variance. Identifying and removing special causes should significantly reduce variability in this case.

Using Analysis of Variance. This is a very general method to compute variance components. The method is complicated and may well be one reason why variance components are not as frequently used in industry as they should be. This approach is discussed in Section 7.3.

Using $\bar{x}$ and R Data Only. Situations arise in which the primary data are not available; only the $\bar{x}$ and R values are available for each subgroup. In this case, the variance components can be determined as follows:

$$s_w = \frac{\overline{R}}{d_2} \tag{7.2}$$

where $\overline{R}$ is the average within-subgroup range and d_2 may be obtained from Appendix G corresponding to subgroup size n. If $V(\bar{x})$ represents the variance computed from the $\bar{x}$ values (equal to 1.27 in this case), then for a fixed subgroup size n,

$$V(\bar{x}) = s_b^2 + \frac{s_w^2}{n} \tag{7.3}$$

from which s_b can be computed.

Using X – mR Chart. Frequently, the subgroup size is equal to one leading to $X - mR$ charts. In this case, σ_w represents the short-term (between successive observations) standard deviation and is estimated as

$$s_w = \frac{m\overline{R}}{1.128} \tag{7.4}$$

If $V(x)$ represents the variance computed from all the x values, then

$$V(x) = s_b^2 + s_w^2 \tag{7.5}$$

from which s_b can be computed.

7.2 ONE-WAY CLASSIFICATION (FIXED FACTOR)

Consider an experiment involving k fryers, or k different levels of a single factor, or k formulations. These are the only fryers, or levels, or formulations of interest to us. Suppose n observations have been made at each level giving a total of kn observations. For example, one may obtain 12 finished fry moistures from each of three fryers. Here $k = 3$ and $n = 12$ for a total of 36 observations. The data are shown in Table 7.2.

The total variance of moisture can be partitioned into two variance components:

σ_w^2 = product-to-product variance within a fryer

σ_b^2 = variance between fryers

TABLE 7.2 Finished Fry Moisture Data

	Moisture		
Number	Fryer 1	Fryer 2	Fryer 3
1	63.79	65.74	63.70
2	62.13	64.99	64.07
3	64.20	65.21	63.72
4	65.07	65.76	63.14
5	63.99	63.99	64.12
6	64.24	65.02	63.21
7	62.34	64.25	63.91
8	64.95	64.52	64.36
9	65.37	65.76	63.76
10	64.73	66.62	64.86
11	64.88	64.72	63.78
12	64.53	65.74	64.11
Average	64.185	65.193	63.895
Variance	1.045	0.5784	0.2195
$\bar{\bar{x}} = 64.42$	Pooled within-fryer variance = s^2_{pooled} = 0.6143		

This situation is very similar to the $\overline{X}$ chart discussed earlier. The $\overline{X}$ chart could also be described as a one-way classification. The only difference is that whereas subgroup is a random factor, fryer is a fixed factor. The three fryers are not a random sample from a universe of fryers, but are the only fryers to be used on the production line. In this sense, they represent the entire population of fryers of interest, which makes fryer a fixed factor.

Figure 7.4 (a) graphically represents the situation with fryers considered fixed. Each fryer has a fixed mean μ_i and the moisture observations from the ith fryer are distributed around μ_i with standard deviation σ_w. When the fryers are considered to be a fixed factor, the between fryer variance is

$$\text{(Fryers considered fixed) } \sigma_b^2 = \frac{\Sigma(\mu_i - \mu)^2}{k} \tag{7.6}$$

where $\mu = \Sigma\mu_i/k$. The denominator in Equation 7.6 is k and not $k - 1$ because the entire population is included.

On the other hand, as shown in Figure 7.4 (b), if fryers are considered to be a random factor, then the three fryers represent a random sample from a distribution with an unknown mean μ and variance σ_b^2, and

$$\text{(Fryers considered random) } s_b^2 = \frac{\Sigma(\mu_i - \overline{\mu})^2}{(k - 1)} \tag{7.7}$$

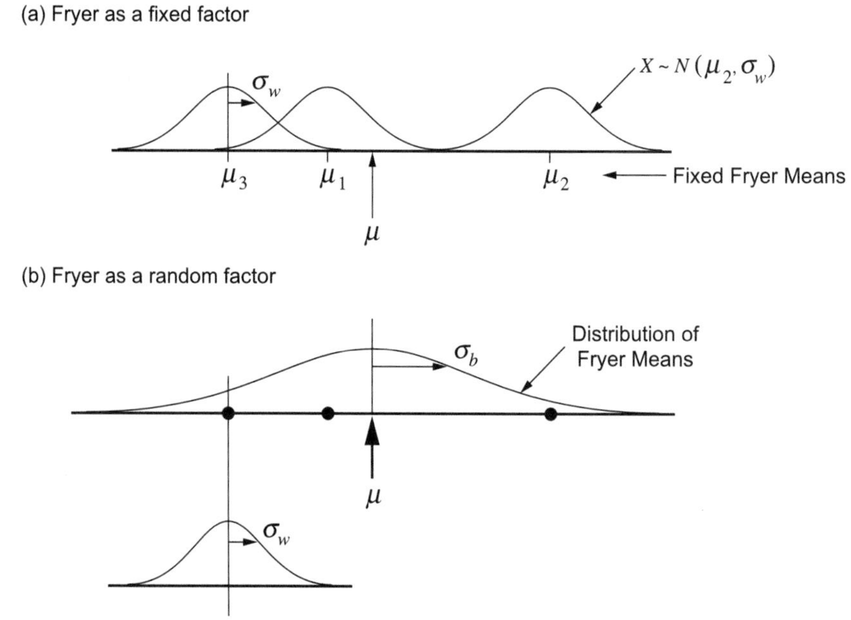

FIGURE 7.4
Fryer as a fixed and as a random factor.

where $\overline{\mu} = \Sigma\mu_i/k$ estimates μ. The denominator in Equation 7.7 is $(k - 1)$ because μ_i are a sample and do not constitute the entire population.

From Equations 7.6 and 7.7,

$$\sigma_b^2 \text{ (Fryer considered fixed)} = \frac{k-1}{k}\sigma_b^2 \text{ (Fryers considered random)} \quad (7.8)$$

Equation 7.8 allows us to estimate variance components for the fryers considered fixed case using the variance of $\overline{X}$ method described earlier, which applies for the fryers considered random case. From the data in Table 7.2, a pooled estimate of within-fryer variance is

$$s_w^2 = s_{\text{pooled}}^2 = 0.6143.$$

The variance computed from the three fryer averages is 0.46439. If fryers are considered random, then from Equation 7.3,

$$0.46439 = s_b^2 \text{ (Fryers considered random)} + \frac{0.6143}{12}$$

Hence,

$$s_b^2 \text{ (Fryers considered random)} = 0.4132$$

and from Equation 7.8,

$$s_b^2 \text{ (Fryers considered fixed)} = \frac{2}{3}(0.4132) = 0.275$$

For the fryers-considered-fixed case, the total moisture variance is estimated as

$$s_t^2 = s_w^2 + s_b^2 = 0.6143 + 0.275 = 0.89$$

The finished fry moisture specification was 65 ± 2. Based upon this specification and the results of the variance components analysis, the following conclusions may be drawn:

1. From Table 7.2, the overall mean moisture is 64.42, which is slightly off-target, the target being 65.
2. The standard deviation of moisture is $\sqrt{0.89} = 0.94$.
3. The current $P_{pk} = (64.42 - 63.00)/3(0.94) = 0.5$. Note that this P_{pk} index cannot be directly related to out-of-specification product by the usual methods since the distribution of moisture is likely to be a trimodal distribution rather than the normal distribution.
4. If the mean could be perfectly centered at 65, P_{pk} would improve to 0.7.

5. Approximately 30% of the total moisture variance is due to differences between fryers; the remaining 70% is due to product-to-product differences within a fryer.
6. If the differences between fryers could be eliminated, perhaps by adjusting the temperature, then the total moisture standard deviation would reduce to $\sqrt{0.6143} = 0.78$ and the P_{pk} index would improve to 0.85.
7. If the within-subgroup variance could be reduced by 60%, it would become approximately 0.25. The new moisture standard deviation would be 0.5 and the P_{pk} index would become 1.33.
8. Further structured studies and variance components analyses will be needed to understand the key sources of variability that contribute to s_w. How much of this variability is due to variation in the batches of french fries? How much is due to frying time and temperature variation? How big is the measurement error? Answers to such questions will dictate the improvement efforts. In this manner, variance components analysis becomes a tool to plan process improvement efforts.

7.3 STRUCTURED STUDIES AND VARIANCE COMPONENTS

In order to determine the total process variability and how the various sources of variability contribute to the total variability, it is necessary to collect and analyze data in a structured manner. This section deals with the design of structured data collection schemes, which may be called structured studies or structured process capability studies, and the technique of variance components analysis. Together, they permit the desired decomposition of total process variability into variability due to the constituent causes of variation. The section begins by describing the difference between fixed and random factors. Then the difference between nested and crossed classifications is explained. Finally, many practically useful structured studies and the associated mathematics of variance decomposition are described.

7.3.1 Fixed and Random Factors

Whether a factor is fixed or random depends upon the way the levels of the factor are selected and the inferences to be made from the analysis. A factor is fixed if the levels of the factor are systematically chosen to be of particular interest and the inferences are limited to those levels. In this sense, the entire population of levels is included in the experiment. As an example of a fixed factor, suppose a manufacturing plant has three similar machines to produce a product. If we are interested in determining the differences between the three machines, then machine is a fixed factor. If there are five formulations of particular interest that an experimenter wishes to evaluate and these represent the entire set of formulations regarding which the experimenter wishes to make inferences, then formulation is a fixed factor.

A factor is random if the selected levels of the factor represent random drawings from a much larger (infinite) population. Inferences are to be made regarding the infinite population, not just the levels represented in the experiment. Raw material lots are an example of a random factor. A few lots of raw material may be included in the study but the interest centers not on the specific lots but on the population of lots to be used in future production.

Sometimes, it may be difficult to decide whether a given factor is fixed or random. The main distinction lies in whether the levels of the factor can be considered to be random samples from a large population or represent the entire population of interest. Consider a factor such as ambient humidity. Let us assume that this environmental factor has a large influence on the quality of the product produced. In order to investigate this effect, two widely different levels of humidity (say 20% and 70%) are chosen and appropriate experiments are conducted in humidity chambers. Is humidity a random factor or a fixed factor? The answer is that during production, humidity will vary throughout the year according to some distribution. So during actual production, humidity is a random factor and we are ultimately interested in predicting the effect of this humidity distribution on product quality. However, in the experiment itself, two levels of humidity were specifically chosen (not randomly chosen) to represent a wide range of humidity. Therefore, in the experiment, humidity is a fixed factor.

What will be the effect of misclassifying humidity? If we arbitrarily call the low level of humidity in the experiment –1 and the high level +1, then the distribution of humidity in the experiment is shown in Figure 7.5 (a). This is called a Bernoulli distribution with 50% of the data at –1 and 50% of the data at +1. Since –1 and +1 represent the entire population of humidity in the experiment, the mean is zero and the variance of humidity in the experiment is

$$V(\text{Humidity considered fixed}) = \frac{(-1-0)^2 + (1-0)^2}{2} = 1$$

We divide by 2 (i.e., n, not $n - 1$) because μ is known to be zero in this case. Alternately, if –1 and 1 are considered to be random samples from

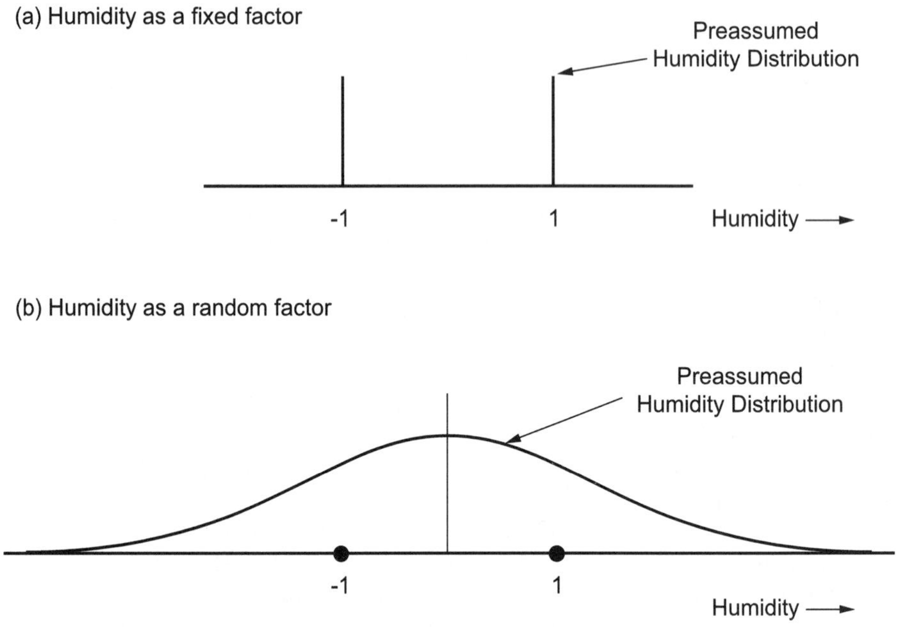

FIGURE 7.5
Humidity as a fixed and as a random factor.

a population with a continuous distribution as shown in Figure 7.5 (b), then the presumed population variance of humidity is

$$V(\text{Humidity considered random}) = \frac{(-1-0)^2 + (1-0)^2}{2-1} = 2$$

We divide by $(n - 1)$ in this case because μ is unknown and is estimated to be zero. Depending upon whether humidity is considered to be a fixed or a random factor, its variance contribution will be computed assuming variance of humidity to be 1 or 2. Larger variance of humidity implies greater contribution of humidity to product quality variation. The effect of misclassifying humidity on variance components analysis will be very large in this case.

7.3.2 Nested and Crossed Factors

Figure 7.6 (a) shows a nested structure involving lots and products within a lot. Products are said to be nested inside lots. There are a lots. From each lot b products are selected for evaluation. Even though the products for each lot are labeled 1, 2, . . . b, product 1 for lot 1 is a different product than product 1 for lot 2. In this sense, the b levels of product are different for each lot. If all product 1 results are averaged, this average does not have a practically different meaning compared to the average of, say, products labeled 2.

(a) Nested Structure

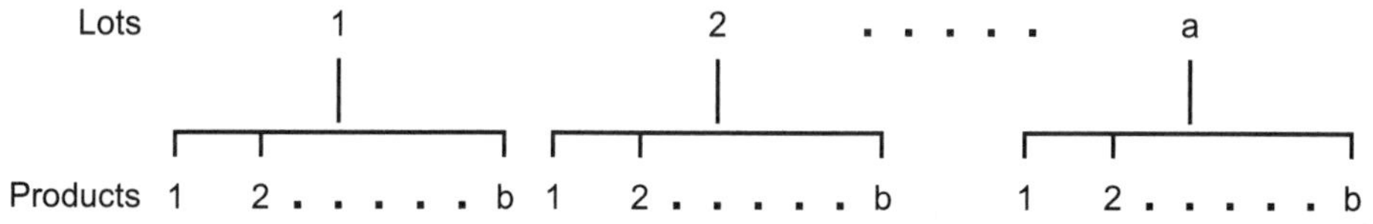

(b) Crossed Structure

Lanes	1	2	Times	b
1	X	X		X
2	X	X		X
. . . .				
a	X	X		X

FIGURE 7.6 Nested and crossed structures.

A nested structure or classification is a hierarchical structure and looks like a tree diagram. Levels of the nested factor are all different for each level of the prior factor. A nested structure may have multiple levels of nesting so that factor B may be nested inside A, C may be nested inside B, and so on. The factors themselves may be random or fixed. Lots and products are typically random factors. Instead, there may be a machines and b products per machine. Here machines is a fixed factor and products are random. If there were three similar machines with four stations per machine, both machines and stations would be considered as fixed factors. A nested structure allows total variability to be partitioned into variability due to each factor.

Figure 7.6 (b) shows a two-factor crossed structure. The two factors are lanes and times. There are multiple lanes of products being simultaneously produced as with multihead filling machines. There are a lanes and b times. What distinguishes a crossed structure from a nested structure is the fact that time 1 for lane 1 is exactly the same time as time 1 for lane 2. Similarly, lane 1 for time 1 is exactly the same lane as lane 1 for all levels of time. In a crossed structure, data cells are formed by combining each level of one factor with each level of every other factor. The resultant structure looks like a matrix. In this case, the average result for time 1 has a practically meaningful and different interpretation than the average result for time 2.

A crossed structure may have more than two factors. The factors may be fixed or random. In the above example, lanes is a fixed factor and time is random. A crossed structure allows the total variability to be partitioned into variability due to each factor and variability due to interactions between factors. In the above example, lanes and times are said to interact if the difference between lanes changes with time. For a nested structure, if B is nested inside A, then A and B cannot interact.

7.3.3 One-Way Classification

We now begin the discussion of some practically useful structured studies and the use of analysis of variance as a method to calculate variance components (Sahai and Ageel; also see Duncan). Consider an experiment with a levels of a single factor A. Suppose n replicated observations are taken at each level of A. The collected data may be shown as in Figure 7.7 (a) or it could equally well be shown as the nested structure in Figure 7.7 (b). Factor A may be a fixed or random factor. Replicated observations are always treated as random.

Such a simple structure has many practical applications. For the $\overline{X}$ chart in Section 7.3.1, A is a random factor called subgroups and replicates are observations within a subgroup. For the fryer example in Section 7.3.2, A is a fixed factor called fryers and several replicated observations were taken per fryer. A could be a fixed factor called shifts, or machines, and replicates could be products made by each shift or machine. A could be a fixed factor called formulations and observations could be the sensory data collected on each formulation. A may be operators and replicates may be the repeated observations taken by each operator of the same product, and so on.

The analysis of variance model for this structured study is

$$x_{ij} = \mu + \alpha_i + e_{ij}(i = 1, \ldots, a, j = 1, \ldots, n)$$

where μ is the true grand mean, α_i is the change in μ due to the ith level of factor A, and e_{ij} is the random error associated with the jth observation at the ith level of A. The following assumptions are made:

1. The errors e_{ij} are randomly distributed with mean zero and variance σ_e^2.

(a) One-way Classification

		Factor A			
		1	2	i	a
Observations	1	x_{11}	x_{21}	x_{i1}	x_{a1}
	2	x_{12}	x_{22}	x_{i2}	x_{a2}
	j	x_{1j}	x_{2j}	x_{ij}	x_{aj}
	n	x_{1n}	x_{2n}	x_{in}	x_{an}
		$\bar{x}_1$		$\bar{x}_i$	$\bar{x}_a$

Grand Mean = $\bar{\bar{x}}$

(b) Nested Structure

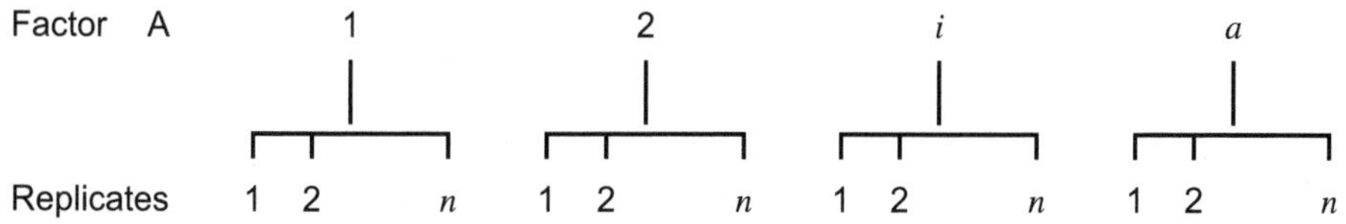

FIGURE 7.7 One-way classification.

2. When A is random, α_i's are assumed to be randomly distributed with mean zero and variance σ_A^2 estimated as $\Sigma\alpha_i^2/(a - 1)$.
3. When A is fixed, $\Sigma\alpha_i = 0$ and α_i have zero mean and variance σ_A^2 equal to $\Sigma\alpha_i^2/a$.

The variance of individual observations σ_X^2 can be partitioned into two variance components σ_A^2 and σ_e^2 such that

$$\sigma_X^2 = \sigma_A^2 + \sigma_e^2$$

The analysis of variance procedure is now briefly explained.

Sum of Squares. With reference to Figure 7.7 (a), we have the following identity:

$$(x_{ij} - \bar{\bar{x}}) = (x_{ij} - \bar{x}_i) + (\bar{x}_i - \bar{\bar{x}})$$

By squaring both sides and summing over i and j (note that the cross product term vanishes),

$$\sum_{ij}(x_{ij} - \bar{\bar{x}})^2 = \sum_{ij}(x_{ij} - \bar{x}_i)^2 + n\sum_i(\bar{x}_i - \bar{\bar{x}})^2$$

or

$$SS_T = SS_e + SS_A$$

where
SS_T = Total sum of squares
SS_e = Sum of squares due to error
SS_A = Sum of squares due to A

Mean Square and Expected Mean Square. SS_e may be viewed as follows. For the ith level of factor A,

$$\frac{\sum_j (x_{ij} - \bar{x}_i)^2}{(n - 1)} = s_e^2$$

and this estimate of σ_e^2 is based upon $(n - 1)$ degrees of freedom. Since the are a levels of factor A, a pooled estimate of σ_e^2 is obtained as

$$\frac{\sum_{ij}(x_{ij} - \bar{x}_i)^2}{a(n - 1)} = \frac{SS_e}{a(n - 1)} = MS_e = s_e^2$$

based upon $a(n - 1)$ degrees of freedom. Error sum of squares SS_e divided by error degrees of freedom $a(n - 1)$ is known as the error mean square MS_e, whose expected value is σ_e^2.

SS_A is determined from a levels of factor A and has $(a - 1)$ degrees of freedom. It may be viewed as follows by first considering factor A to be a random factor. In this case, the $\bar{x}_i$ values are a random sample from an infinite population and the variance computed from the $\bar{x}_i$ values is

$$\frac{\sum_i(\bar{x}_i - \bar{\bar{x}})^2}{(a - 1)} = \frac{SS_A}{n(a - 1)}$$

Since there are n observations averaged together to obtain $\bar{x}_i$, this computed variance estimates $\sigma_e^2/n + \sigma_A^2$, i.e.,

$$\frac{SS_A}{n(a - 1)} = \frac{s_e^2}{n} + \frac{\Sigma\alpha_i^2}{(a - 1)}$$

or

$$\frac{SS_A}{(a - 1)} = MS_A = s_e^2 + n\frac{\Sigma\alpha_i^2}{(a - 1)}$$

where MS_A denotes the mean square due to factor A.

When A is random, $\Sigma\alpha_i^2/(a-1)$ estimates σ_A^2, hence MS_A estimates $\sigma_e^2 + n\sigma_A{}^2$. When A is fixed,

$$\sigma_A^2 = \frac{\Sigma\alpha_i^2}{a}$$

hence MS_A estimates $\sigma_e^2 + [an/(a-1)]\sigma_A^2$. These results are summarized in Table 7.3. As a becomes large, the expected mean squares for the random and fixed cases converge.

Applications. The actual applications of this method to estimate variance components are straightforward. We compute mean squares from the data, equate them to their expectations, and solve the resultant equations. A number of software packages provide mean square computations as a part of the analysis of variance table.

As one example, for the fruit tray weight data in Table 7.1, subgroup corresponds to random factor A and weights within subgroup correspond to error in Table 7.3. The subgroup size n is 3. From the data, the mean square values can be computed and lead to the following two equations:

$$MS_A = 3.83 = \sigma_e^2 + 3\sigma_A^2$$

$$MS_e = 0.41 = \sigma_e^2$$

The two equations can be simultaneously solved to obtain the estimated within-subgroup variance of 0.41 and between-subgroup variance of 1.14, exactly the same as computed before.

As another example, consider the finished fry moisture data in Table 7.2. Fryer is a fixed factor. There were 3 fryers, so $a = 3$. There were 12 observations per fryer, so $n = 12$. We have the following two equations:

$$MS_A = 5.57 = \sigma_e^2 + 18\sigma_A^2$$

$$MS_e = 0.61 = \sigma_e^2$$

Hence, the estimated within-fryer variance is 0.61 and the estimated between-fryer variance is 0.2755, the same as computed before.

TABLE 7.3 One Way Classification. Model: $x_{ij} = \mu + \alpha_i + e_{ij}$

		Expected Mean Square	
Source	Mean Square	A Fixed	A Random
A $i = 1, 2 \ldots a$	$\frac{n}{(a-1)}\sum_i (\bar{x}_i - \bar{\bar{x}})^2$	$\sigma_e^2 + \frac{an}{(a-1)}\sigma_A^2$	$\sigma_e^2 + n\sigma_A^2$
Error $j = 1, 2 \ldots n$	$\frac{1}{a(n-1)}\sum_{i,j} (x_{ij} - \bar{x}_i)^2$	σ_e^2	σ_e^2

7.3.4 Nested Classification

The one-way classification of the last section is a nested classification where factor A can be a random or fixed factor and error may be replicate observations or any other random factor. In a similar fashion, one may have a two-factor nested, a three-factor nested, or a higher-order nested structure. Figure 7.8 shows a three-factor nested structure, the factors being A, B, and C, with e denoting replicated observations (or a fourth nested random factor). Factors A, B, and C, could be fixed or random; e is always random.

There are a very large number of applications of such nested structures in industry. Here is one example:

A = Raw material lot (random)

B = Production batches within a raw material lot (random)

C = Products within a production batch (random)

e = Duplicate measurements on the product (random)

Another example may involve shifts (fixed), products within a shift (random), and duplicate measurements per product (random). Yet another may involve machines (fixed), stations (fixed), products (random), and duplicate measurements (random).

For a three-factor nested structure, the analysis of variance model is

$$x_{ijkl} = \mu + \alpha_i + \beta_{j(i)} + \gamma_{k(ij)} + e_{l(ijk)}$$

where α_i are effects due to A, $\beta_{j(i)}$ are effects due to B, and by putting i in brackets, the nomenclature indicates that B is nested within A and $\gamma_{k(ij)}$

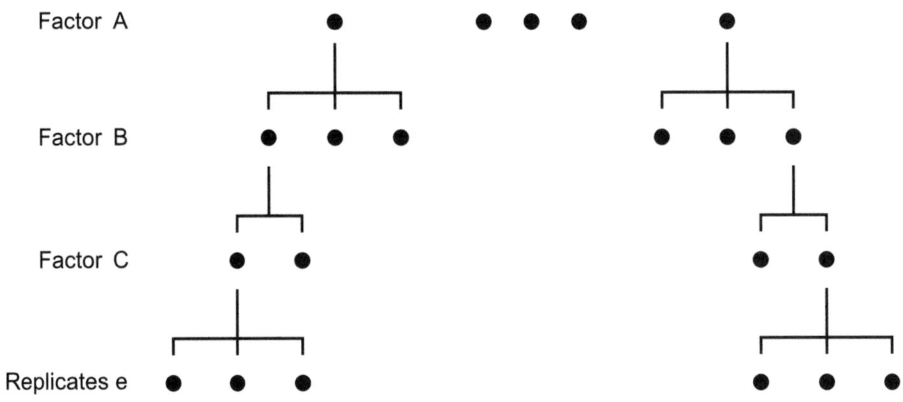

FIGURE 7.8 Three-factor nested structure with replicates.

are effects due to factor C. Tables 7.4 and 7.5 summarize the mean squares and the expectation of mean squares for two-factor nested and three-factor nested structures. In practical applications of these tables, the mean squares are computed from the data and equated to the appropriate expected value to obtain the variance components. Several applications are considered in Chapter 8.

7.3.5 Crossed Classification

Two crossed structures, without and with interaction, are considered here. Figure 7.6 (b) shows a two-factor crossed structure with a single observation per cell. The two factors may both be fixed or random, or one factor may be fixed with the other factor being random. The mean squares and the expected mean squares are shown in Table 7.6. The analysis of variance model is

$$x_{ij} = \mu + \alpha_i + \beta_j + e_{ij}$$

where α_i and β_j are effects due to factors A and B, respectively. The total variance is partitioned into three components as follows:

$$\sigma_X^2 = \sigma_A^2 + \sigma_B^2 + \sigma_e^2$$

Figure 7.9 shows the two-factor crossed structure with replicated measurements in each cell. The replicates are nested inside the two factors. In this sense, this structure involves both crossed and nested classifications. Once again, the two factors may both be fixed or random, or one

Factor A	Factor B			
	1	2	. . . j . . .	b
1	XX	XX		XX
2	XX	XX		XX
i				
a	XX	XX		XX

FIGURE 7.9 Two-factor crossed structure with replicates.

TABLE 7.4 Two-Factor Nested Structure.* Model: $x_{ijk} = \mu + \alpha_i + \beta_{j(i)} + e_{k(ij)}$

		Expected Mean Square			
Source	Mean Square	A and B Fixed	A and B Random	A Fixed, B Random	A Random, B Fixed
A $i = 1, 2 \ldots a$	$\frac{bn}{(a-1)}\sum_i (\bar{x}_i - \bar{\bar{x}})^2$	$\sigma_e^2 + \frac{abn}{(a-1)}\sigma_A^2$	$\sigma_e^2 + n\sigma_B^2 + bn\sigma_A^2$	$\sigma_e^2 + n\sigma_B^2 + \frac{abn}{(a-1)}\sigma_A^2$	$\sigma_e^2 + bn\sigma_A^2$
B $j = 1, 2 \ldots b$	$\frac{n}{a(b-1)}\sum_{i,j} (\bar{x}_{ij} - \bar{\bar{x}}_i)^2$	$\sigma_e^2 + \frac{bn}{(b-1)}\sigma_B^2$	$\sigma_e^2 + n\sigma_B^2$	$\sigma_e^2 + n\sigma_B^2$	$\sigma_e^2 + \frac{bn}{(b-1)}\sigma_B^2$
Error $k = 1, 2 \ldots n$	$\frac{1}{ab(n-1)}\sum_{i,j,k} (x_{ijk} - \bar{x}_{ij})^2$	σ_e^2	σ_e^2	σ_e^2	σ_e^2

*Error could be replicates or third nested random factor.

TABLE 7.5 Three-Factor Nested Structure.* Model: $x_{ijkl} = \mu + \alpha_i + \beta_{j(i)} + \gamma_{k(ij)} + e_{l(ijk)}$

		Expected Mean Square		
Source	Mean Square	A, B, C Fixed	A, B, C Random	A and B Fixed, C Random
A $i = 1, 2 \ldots a$	$\frac{bcn}{(a-1)} \sum (\bar{x}_i - \bar{\bar{x}})^2$	$\sigma_e^2 + \frac{abcn}{(a-1)} \sigma_A^2$	$\sigma_e^2 + n\sigma_C^2 + cn\sigma_B^2 + bcn\sigma_A^2$	$\sigma_e^2 + n\sigma_C^2 + \frac{abcn}{(a-1)} \sigma_A^2$
B $j = 1, 2 \ldots b$	$\frac{cn}{a(b-1)} \sum_{ij} (\bar{x}_{ij} - \bar{x}_i)^2$	$\sigma_e^2 + \frac{bcn}{(b-1)} \sigma_B^2$	$\sigma_e^2 + n\sigma_C^2 + cn\sigma_B^2$	$\sigma_e^2 + n\sigma_C^2 + \frac{bcn}{(b-1)} \sigma_B^2$
C $k = 1, 2 \ldots c$	$\frac{n}{ab(c-1)} \sum_{ijk} (\bar{x}_{ijk} - \bar{x}_{ij})^2$	$\sigma_e^2 + \frac{cn}{(c-1)} \sigma_C^2$	$\sigma_e^2 + n\sigma_C^2$	$\sigma_e^2 + n\sigma_C^2$
Error $l = 1, 2 \ldots n$	$\frac{1}{abc(n-1)} \sum_{ijkl} (x_{ijkl} - \bar{x}_{ijk})^2$	σ_e^2	σ_e^2	σ_e^2

*Error could be replicates or fourth nested random factor.

TABLE 7.6 Two Factors Crossed Structure without Interaction.* Model: $x_{ij} = \mu + \alpha_i + \beta_j + e_{ij}$

		Expected Mean Square		
Source	Mean Square	*A* and *B* Fixed	*A* and *B* Random	*A* Fixed, *B* Random
A $i = 1, 2 \ldots a$	$\frac{b}{(a-1)}\sum_i(\bar{x}_i - \bar{\bar{x}})^2$	$\sigma_e^2 + \frac{ab}{(a-1)}\sigma_A^2$	$\sigma_e^2 + b\sigma_A^2$	$\sigma_e^2 + \frac{ab}{(a-1)}\sigma_A^2$
B $j = 1, 2 \ldots b$	$\frac{a}{(b-1)}\sum_j(\bar{x}_j - \bar{\bar{x}})^2$	$\sigma_e^2 + \frac{ab}{(b-1)}\sigma_B^2$	$\sigma_e^2 + a\sigma_B^2$	$\sigma_e^2 + a\sigma_B^2$
Error	$\frac{1}{(a-1)(b-1)}\sum_{ij}(x_{ij} - \bar{x}_i - \bar{x}_j + \bar{\bar{x}})^2$	σ_e^2	σ_e^2	σ_e^2

*Error is the interaction term.

TABLE 7.7 Two Factors Crossed Structure with Interaction and Replicates. Model: $x_{ijk} = \mu + \alpha_i + \beta_j + (\alpha\beta)_{ij} + e_{ijk}$

		Expected Mean Square		
Source	Mean Square	A and B Fixed	A and B Random	A Fixed, B Random
A $i = 1, 2 \ldots a$	$\frac{bn}{(a-1)}\sum_i (\bar{x}_i - \bar{\bar{x}})^2$	$\sigma_e^2 + \frac{abn}{(a-1)}\sigma_A^2$	$\sigma_e^2 + n\sigma_{AB}^2 + bn\sigma_A^2$	$\sigma_e^2 + n\sigma_{AB}^2 + \frac{abn}{(a-1)}\sigma_A^2$
B $j = 1, 2 \ldots b$	$\frac{an}{(b-1)}\sum_j (\bar{x}_j - \bar{\bar{x}})^2$	$\sigma_e^2 + \frac{abn}{(b-1)}\sigma_B^2$	$\sigma_e^2 + n\sigma_{AB}^2 + an\sigma_B^2$	$\sigma_e^2 + an\sigma_B^2$
AB	$\frac{1}{(a-1)(b-1)}\sum_{ij} (\bar{x}_{ij} - \bar{x}_i - \bar{x}_j + \bar{\bar{x}})^2$	$\sigma_e^2 + \frac{abn}{(a-1)(b-1)}\sigma_{AB}^2$	$\sigma_e^2 + n\sigma_{AB}^2$	$\sigma_e^2 + n\sigma_{AB}^2$
Error	$\frac{1}{ab(n-1)}\sum_{ijk} (x_{ijk} - \bar{x}_{ij})^2$	σ_e^2	σ_e^2	σ_e^2

may be fixed and the other random. The mean squares and the expected mean squares are shown in Table 7.7. The model is

$$x_{ijk} = \mu + \alpha_i + \beta_j + (\alpha\beta)_{ij} + e_{ijk}$$

where α_i and β_j are effects due to factors A and B and $(\alpha\beta)_{ij}$ is the effect of the interaction between A and B. The AB interaction may be explained as follows. If the differences due to the various levels of B are the same at each level of A, then A and B do not interact or they act independently. Thus, if an interaction exists, it means that the effect of various levels of B depends upon the level of A. Table 7.7 helps partition the total variance into four components as follows:

$$\sigma_X^2 = \sigma_A^2 + \sigma_B^2 + \sigma_{AB}^2 + \sigma_e^2$$

There are several practical applications of a crossed structure. Many manufacturing operations produce multiple lanes of products simultaneously, as in the case of multihead filling machines, multicavity molds, and the multilane frozen pizza manufacturing process. In all such cases, lanes and times form a two-factor crossed structure, lanes being a fixed factor and times being a random factor. Other examples include machines (fixed), operators (fixed), and replicated measurements taken at each machine–operator combination. A crossed structure also forms the typical data collection scheme for conducting a measurement system variability study. These applications are considered in Chapters 8 and 9.

8

Quality Planning with Variance Components

This chapter has three purposes. The first is to illustrate the many applications of variance components analysis. Four applications are considered: (1) an application of the nested structure to the typical manufacturing scenario involving raw material variability, manufacturing variability, and measurement variability; (2) an application of the crossed structure to the multilane manufacturing process; (3) an application involving factorial designs; and (4) an application to set variance component targets in R&D to meet downstream manufacturing specifications.

The second purpose of this chapter is to illustrate how the knowledge of variance components and the ideas of process capability and economic loss can be coupled to plan future quality improvements and to make improvement decisions on an economic basis. The classical and the quadratic loss functions are introduced as ways to translate variability into its economic consequence. "What if" analysis is used as a planning and decision making tool.

Finally, two additional tools are introduced in this chapter. The first is called the multi-vari chart, which is a simple graphical approach to viewing variance components. The second is variance transmission analysis, which is an equation-based approach to compute variance components.

8.1 TYPICAL MANUFACTURING APPLICATION

This example concerns the production of a product for which the final product moisture is an important characteristic of interest. The total process is shown in Figure 8.1. It consists of feeding the incoming batches of raw material into a continuous manufacturing process that ultimately produces the product. At certain intervals, product is periodically withdrawn from the production line and multiple moisture tests are conducted to determine the moisture content. The target value for moisture is 10%, the specification being from 9% to 11%. To assess whether the process is satisfactory and, if not, how to improve it, the following specific questions need to be answered.

1. What is the extent of total variability in moisture? Given the specification, is the mean moisture on target? Is the variability in moisture acceptable? The answers to these questions help determine the extent of resources to be allocated for improvement purposes.
2. There are three causes of moisture variability: (a) moisture variability caused by raw material batch-to-batch variation, called supplier variability; (b) product-to-product moisture variability due to the manufacturing process, called manufacturing variability; and (c) test-to-test variability due to the measurement system, called test variability. Since the moisture measurement test is a destructive test, this test variability necessarily includes the sample-to-sample variability within a product. The specific question is: how much of the total variability is due to the supplier, the manufacturing process, and the test method? The answer to this question helps determine where to focus the improvement effort.

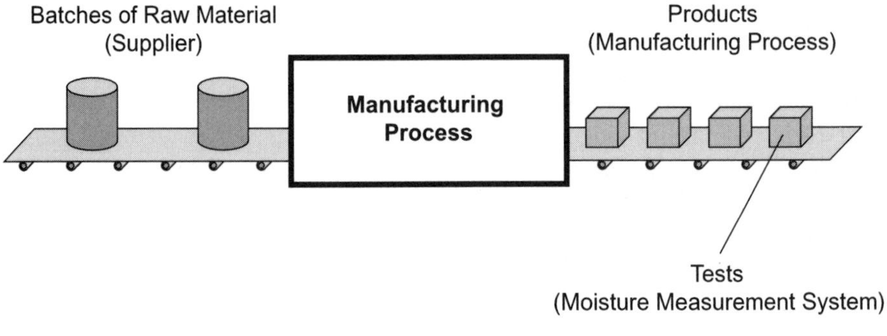

FIGURE 8.1
Manufacturing process.

3. The final question has to do with setting improvement targets and allocating resources, namely, developing a plan for improvement. If manufacturing variability is the largest source of variation, how much reduction in manufacturing variability is necessary to achieve the desired quality target? From a return on investment viewpoint, how much money can be spent to accomplish this objective? The answers to these questions establish a practical process improvement plan.

Structured Capability Study. In order to partition the total variability into that due to the supplier, the manufacturing process, and the test method, it is necessary to collect data in a structured manner. Simply measuring 30–50 values of moisture from consecutively produced products will not do. The selected nested study is shown in Figure 8.2 and involves three random factors, namely, batches, products, and tests. Moisture measurements were taken over 15 batches of raw material. Two products were selected from each batch and two moisture measurements were made on samples taken from each product. Note that this structured study differs from a designed experiment. A designed experiment is also structured but involves active, purposeful changes to the factors being investigated. A structured capability study merely involves passive observations of the process made in a structured manner.

This structured study with 15 raw material batches, two products per batch, and two tests per product leads to 60 moisture measurements shown in Table 8.1. A look at the data suggests that the variability in moisture is large compared to the specification limits. There are some moisture measurements outside the specification limits of 9% to 11%. This large variability suggests the need to improve the process.

How much data should be collected in such a capability study? The larger the amount of data, the better. Whenever possible there should be at least 30 degrees of freedom to estimate each variance component. Data of this type are often already collected in a manufacturing environment and it is merely necessary to properly analyze the existing data.

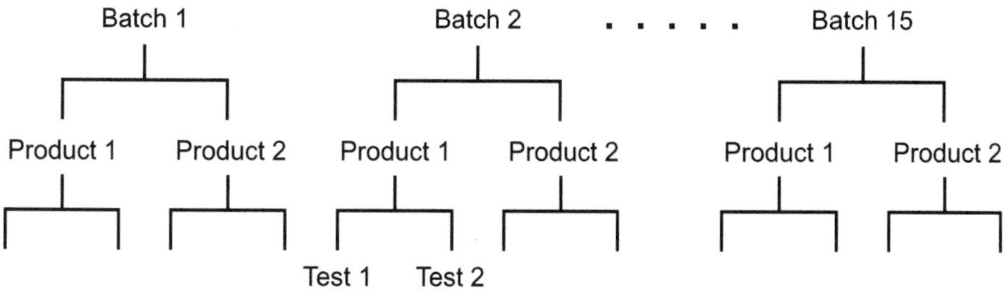

FIGURE 8.2 Nested capability study.

TABLE 8.1 Moisture Data for the Nested Study

Batches	Products	Tests	% Moisture	Batches	Products	Tests	% Moisture
1	1	1	11.6	8	2	1	10.5
1	1	2	11.1	8	2	2	10.8
1	2	1	10.6	9	1	1	10.3
1	2	2	10.8	9	1	2	10.4
2	1	1	10.2	9	2	1	10.7
2	1	2	10.4	9	2	2	10.7
2	2	1	10.1	10	1	1	8.9
2	2	2	10.2	10	1	2	9.2
3	1	1	10.5	10	2	1	10.3
3	1	2	10.3	10	2	2	10.0
3	2	1	9.0	11	1	1	10.1
3	2	2	9.2	11	1	2	9.9
4	1	1	11.2	11	2	1	10.0
4	1	2	11.4	11	2	2	10.3
4	2	1	10.3	12	1	1	10.5
4	2	2	10.6	12	1	2	10.6
5	1	1	9.5	12	2	1	10.7
5	1	2	9.7	12	2	2	10.8
5	2	1	9.3	13	1	1	9.2
5	2	2	9.1	13	1	2	9.3
6	1	1	10.9	13	2	1	10.2
6	1	2	10.8	13	2	2	10.1
6	2	1	10.2	14	1	1	9.9
6	2	2	10.0	14	1	2	10.0
7	1	1	9.5	14	2	1	10.1
7	1	2	9.6	14	2	2	10.4
7	2	1	10.5	15	1	1	11.5
7	2	2	10.4	15	1	2	11.3
8	1	1	11.0	15	2	1	10.2
8	1	2	10.8	15	2	2	10.4

Histogram. Figure 8.3 shows the histogram of the 60 moisture measurements. One purpose of the histogram is to assess if the mean is on target and if the variability is large compared to specifications. The overall mean is 10.27%, slightly above the target value of 10%. Whether the process mean can be considered to be on target or not may be assessed by constructing a confidence interval for the mean. Approximately 3.3% of the data are below the lower specification limit and 10% of the data are above the upper specification limit. The histogram provides an initial answer to the question, "Is there a major problem here?" The answer in this case is "yes."

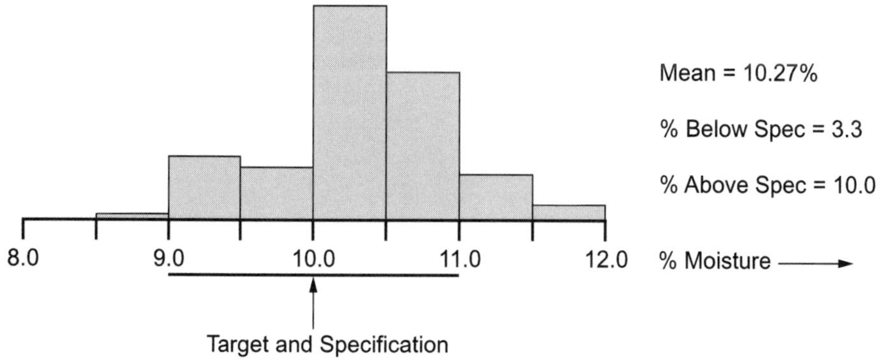

FIGURE 8.3 Moisture histogram.

A second purpose of the histogram is to identify outliers. Outliers should be suspected if one or more cells of the histogram are far removed from the main body. If an outlier is suspected, attempts should be made to identify the cause of the outlier and appropriate corrective actions should be taken. If the outlier is due to incorrect data recording, it should be replaced by the correct value or the value of a duplicate test. There appear to be no outliers here.

A third usual purpose of the histogram is to identify departures from the normal distribution. With structured capability studies, nonnormal distributions can occur because multiple potentially major sources of variability are included in the study. Variance components analysis helps pinpoint the causes if the causes are among the identified sources of variability. Otherwise, additional causes of variability should be suspected and investigated.

Multi-Vari Chart. This is a revealing graphical procedure (Duncan) used to analyze variation when multiple factors influence variability. As shown in Figure 8.4, application of the procedure in this case is to plot the range of test-to-test variation as a vertical line with the structure of the study shown along the horizontal axis. Moisture varies from approximately 9% to 11.6%. Compared to this variation, the test-to-test variability is very small. Both product-to-product variability and raw material batch-to-batch variability appear to be large. The graphical procedure serves the purpose of identifying large causes of variability. The observed pattern should correspond to the causes of variation which form the basis for the structured capability study. Otherwise, factors other than those in the study should be suspected, as would be the case if the chart showed a sudden large shift or time trend. A multi-vari chart does not provide a numerical measure of variance contribution due to each source of variability; variance components analysis does.

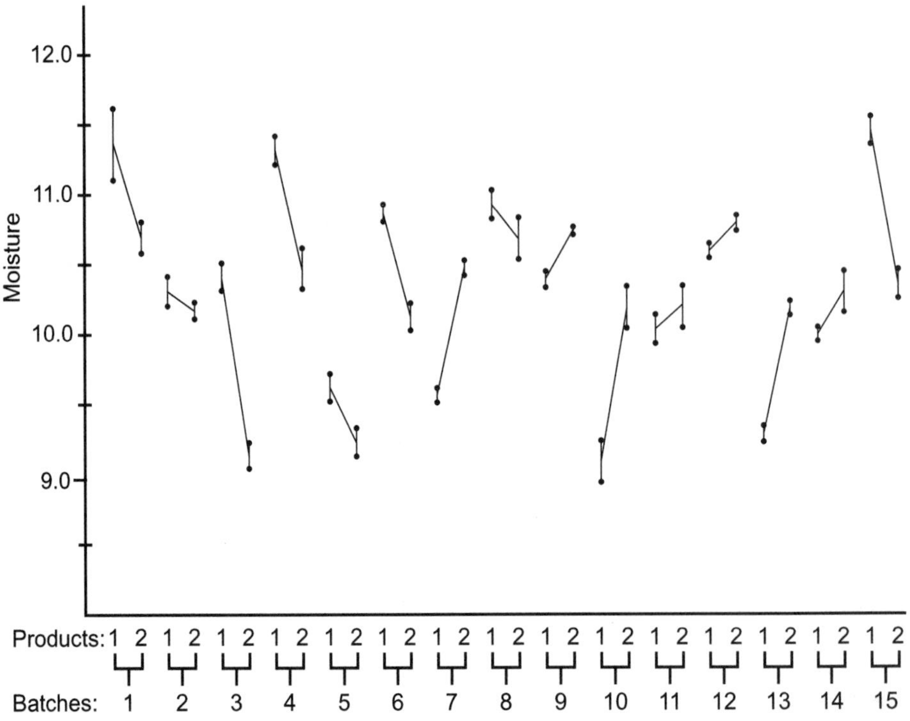

FIGURE 8.4 Multi-vari chart (moisture data).

Variance Components Analysis. This structured study is designed to estimate three variance components:

σ_B^2 = Moisture variability due to batches of raw material (supplier)

σ_P^2 = Product-to-product moisture variability within a batch of raw material (manufacturing process)

σ_e^2 = Test-to-test variability within a product (measurement method)

The concept of variance components analysis is graphically illustrated in Figure 8.5. μ is the overall mean moisture. The average moisture for the ith batch may be α_i away from mean μ. For this batch, the average moisture for a specific product may be $\beta_{j(i)}$ away from the batch mean. For this batch and product, the specific individual measurement may be $e_{k(ij)}$ away from the product mean. Therefore, the model for the individual moisture measurement x_{ijk} is

$$x_{ijk} = \mu + \alpha_i + \beta_{j(i)} + e_{k(ij)}$$

and

$$\sigma_t^2 = \sigma_B^2 + \sigma_P^2 + \sigma_e^2 \tag{8.1}$$

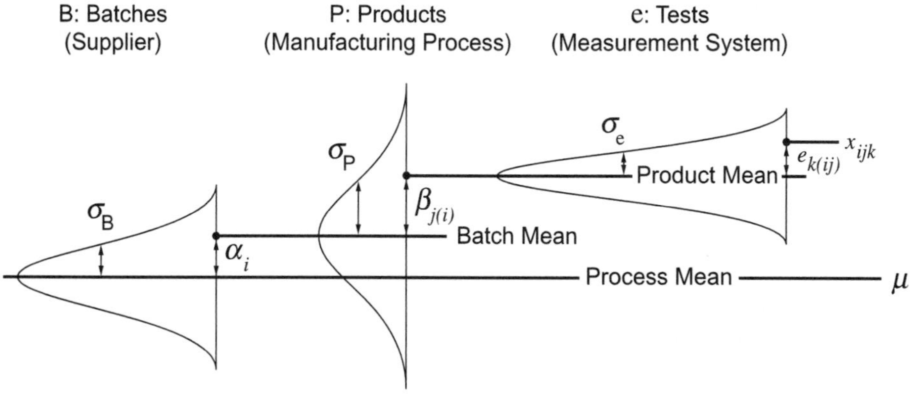

FIGURE 8.5 Graphical illustration of variance components.

where σ_t^2 represents the total long-run variance of moisture measurements. Thus, the total variance is partitioned into three variance components.

This variance decomposition may be accomplished as follows. Figure 8.6 shows the structured capability study and a partial set of observed data. For the first product of the first batch, there are two test results, 11.6 and 11.1. These results are different from each other only due to test variability (including the sample-to-sample variability within a product). An estimate of this test variance, obtained from this one pair of results, is 0.125. This estimate has one degree of freedom. There are 30 such pairs of test results. Pooling the 30 test variance estimates gives s_e^2 = 0.023 with 30 degrees of freedom.

Similarly, for the first batch there are two product average moistures, 11.35 and 11.70, obtained by averaging the corresponding two test results. The variance between these two averages is 0.06125. These two product averages differ from each other not only due to product-to-product differences, but also due to test variability. Since two test results are averaged, 0.06125 estimates $\sigma_P^2 + \sigma_e^2/2$. There are 15 such pairs of aver-

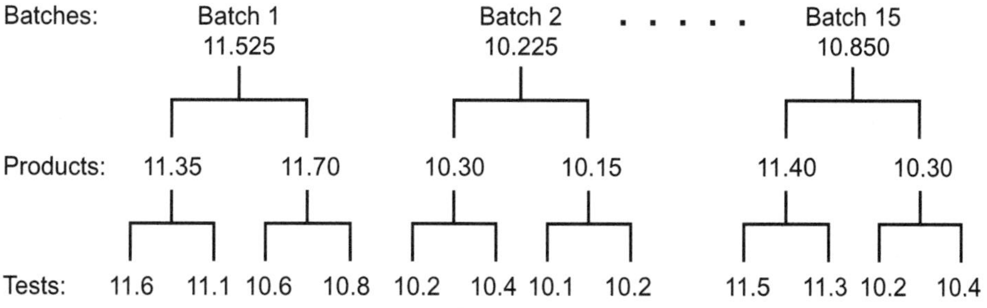

FIGURE 8.6 Structured study and data.

ages and the pooled estimate of the variance of product averages is 0.2653. Hence,

$$s_P^2 + \frac{s_e^2}{2} = 0.2653$$

Substituting for s_e^2, we get $s_P^2 = 0.254$ with 15 degrees of freedom.

Finally, there are 15 batch averages. The overall mean is 10.27. These batch averages differ from each other because of batch-to-batch, product-to-product, and test-to-test differences. Since each batch average consists of two products and four tests, the variance derived from the 15 batch averages estimates $\sigma_B^2 + \sigma_P^2/2 + \sigma_e^2/4$, i.e.

$$s_B^2 + \frac{s_P^2}{2} + \frac{s_e^2}{4} = \frac{(11.525 - 10.27)^2 + \ldots\ldots + (10.850 - 10.27)^2}{14} = 0.266$$

Hence, $s_B^2 = 0.133$ with 14 degrees of freedom.

A second approach to estimate variance components is to recognize that this structured study is a two-factor nested study with replicates, where both batches and products are random factors. Table 7.4 gives the expected mean squares for this structure. With respect to Table 7.4, the mean squares can be computed from the data and equated to their expectations as follows:

$$0.023 = s_e^2$$

$$0.530 = s_e^2 + 2s_B^2$$

$$1.063 = s_e^2 + 2s_P^2 + 4s_B^2$$

The variance components are: $s_e^2 = 0.023$, $s_P^2 = 0.254$, and $s_B^2 = 0.133$. These are summarized in Table 8.2. The last column of Table 8.2 shows the percent variance contribution of each source to the total moisture variance. Approximately, the supplier is responsible for 32% of the variance, the manufacturing process is responsible for 62% of the variance,

TABLE 8.2 Variance Components Table

Source	Degrees of Freedom	Variance	Standard Deviation	Percent Contribution
Batches (supplier)	14	0.133	0.365	32.4%
Products (manufacturing process)	15	0.254	0.504	62.0%
Test (measurement system)	30	0.023	0.152	5.6%
Total	59	0.410	0.640	100.0%

and the test error is small, accounting for about 6% of the variance. Any improvement efforts need to be focused on the larger variance components. We will return to this question of process improvements in Section 8.3.

8.2 ECONOMIC LOSS FUNCTIONS

We now turn to the question of how to relate variability to its economic consequence because then we could make improvement decisions on an economic basis. Variability leads to economic losses. As ammunition dispersion increases, the probability of hitting a target reduces and the cost of defeating a target increases. As various components of a system deviate from target, the system performance degrades and the customer costs increase due to reduced reliability, availability, and product life. As food product characteristics deviate from target, eating quality and shelf life decrease, causing market share to drop. A weakness of the process capability index as a measure of quality is that it is not immediately possible to comprehend the index in economic terms. If the P_{pk} index improves from 0.7 to 0.9, the economic significance of this improvement is not immediately obvious. If quality and quality improvement could be measured in economic terms, then in addition to the capability index, economic loss due to poor quality will serve as a guide to making quality-planning decisions.

This section deals with economic loss functions that permit the impact of process noncentering and variability to be measured in economic terms and in a timely manner. Two types of loss functions, called the classical loss function and the quadratic loss function, are considered as measures of quality.

8.2.1 Classical Loss Function

The classical loss function assumes that if the characteristic value is within specifications, then the product is acceptable and there is no economic loss. However, if the characteristic value is outside the specification, then the product is unacceptable. It may have to be scrapped, reworked, or sold at a lower price. The resultant economic loss function is shown in Figure 8.7, where the value of the characteristic X is plotted on the horizontal axis and the scrap and rework costs are plotted on the vertical axis. A similar rationale applies to a one-sided specification, in which case the loss function will also be one-sided.

The economic loss due to process variability and noncentering can be computed by first calculating the fraction of product above and below

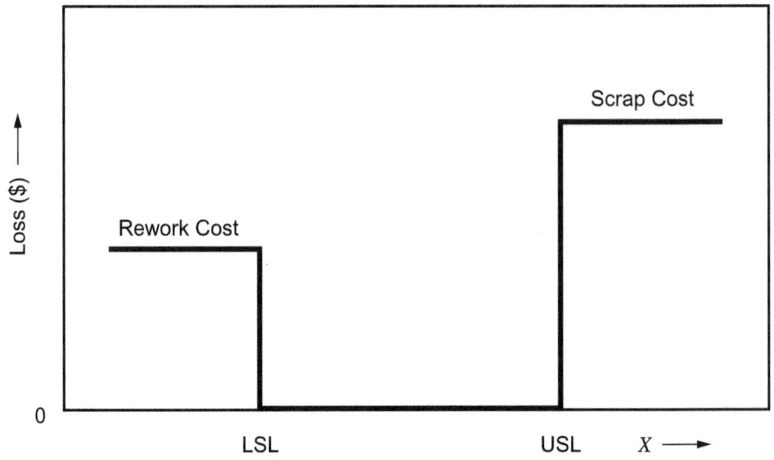

FIGURE 8.7 Classical loss function.

specification and then multiplying by the appropriate scrap and rework costs. Such computations can be easily done based upon C_p and C_{pk} indices for stable processes and P_p and P_{pk} indices for unstable processes assuming normality.

Classical Loss and C_p Index. As an example, consider the case of a two-sided specification with the process mean being at the specification midpoint. The product will be scrapped if it is outside specification and the scrap cost is $100 per unit. It is not possible to rework the product. The fraction of product outside the specification limits can be easily computed from the C_p index, knowing that the Z values at the specification limits are $\pm 3C_p$. Figure 8.8 shows the per-unit economic loss for various values of the C_p index.

Cp	Fraction Outside Specification	Per Unit Economic Loss ($)
0.4	0.2302	23.02
0.6	0.0718	7.18
0.8	0.0164	1.64
1.0	0.0026	0.26
1.2	0.0002	0.02
1.33	0	0

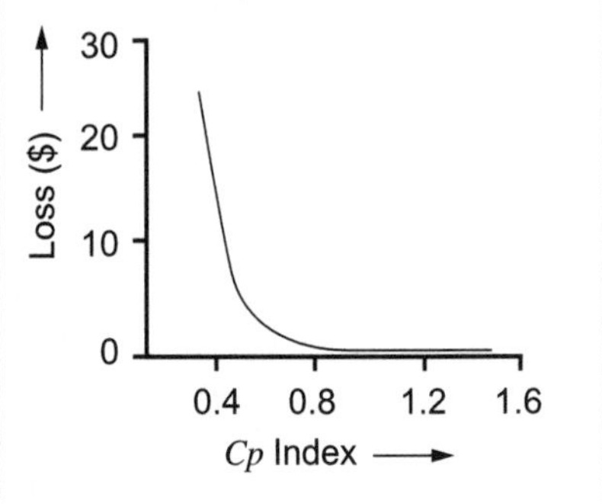

FIGURE 8.8 Economic loss as a function of *Cp* index (classical loss function).

From Figure 8.8, the following points should be noted with respect to the C_p index and the classical loss function:

1. The classical loss function permits the process variability to be converted to economic loss, thereby measuring the benefits of improvement in economic terms. For example, if the C_p index improves from 0.6 to 0.8, the loss per unit of product reduces from \$7.18 to \$1.64.
2. As the C_p index increases, loss continuously reduces so that the economic loss is a monotonic function of the C_p index. Furthermore, it appears from Figure 8.8 that there is not much incentive to improve the value of C_p beyond 1.0 since the expected reduction in economic loss is very small.
3. If the loss function is asymmetrical, namely, the rework cost is much lower than the scrap cost, then the process mean will have to be targeted at a value different from the specification mid-point such that the total loss is minimal.

Classical Loss and C_{pk} Index. Similar loss computations can be easily made for the C_{pk} index. For one-sided specifications, economic loss decreases as the C_{pk} index increases. However, for two-sided specifications, C_{pk} and economic loss do not have a one-to-one monotonic relationship, even for the symmetric classical loss function. This may be seen as follows. Let the specification limits be 70 to 80. Process A has mean = 80, σ = 2, and, therefore, C_{pk} = 0. Given a scrap cost of \$100, the per-unit loss is \$50. Process B has mean = 75, σ = 8, and, therefore, C_{pk} = 0.21. The percent out-of-specification product is 53.2% and the per-unit loss is \$53.2. In this case, a higher C_{pk} has resulted in a greater loss! Therefore, whenever possible, it is best to measure quality directly in terms of economic loss.

The classical loss function is simple to understand and use and allows quality to be measured in economic terms. However, its basic premise is flawed and can be improved upon. Classical loss function assumes that products within specification are equally good from the customer's viewpoint. This is not true. If the specifications are correctly set, products near the target value in the middle of a two-sided specification are much better than products that are just inside the specification. Similarly, there is usually no discernible difference between a product just inside the specification and another product just outside the specification. From a customer viewpoint, product quality continuously degrades as the characteristic value deviates from a well-set target. The classical loss function does not reflect this viewpoint and instead focuses on meeting specifications. What is necessary is a loss function that focuses on meeting the target.

8.2.2 Quadratic Loss Function

The quadratic loss function (Taguchi) presumes that the quality loss is zero if the characteristic X is on target T and the loss increases on both sides of the target in proportion to the square of the deviation of the quality characteristic from target. The quadratic loss function is expressed as

$$\text{Loss} = L(X) = k(X - T)^2 \tag{8.2}$$

k is called the quality loss coefficient. The quadratic loss function is graphically shown in Figure 8.9. This loss function is intuitively appealing because it mimics reality. Note that the loss is zero and so is the slope when the characteristic value is on target. This is appropriate because the target is the best possible value of the characteristic. The loss increases slowly near the target and more rapidly away from the target, as it should. Loss occurs even when the characteristic value is within specification, thereby expressing a preference for on-target performance rather than just within-specification performance.

This assumed loss function is the simplest mathematical function with the desired behavior. This may be seen as follows. In general, the loss function $L(X)$ may be any continuous function of X with a minimum at the target value T. Such a function can be expanded about T using the Taylor series expansion:

$$L(X) = L(T) + \frac{L'(T)}{1!}(X - T) + \frac{L''(T)}{2!}(X - T)^2 + \ldots$$

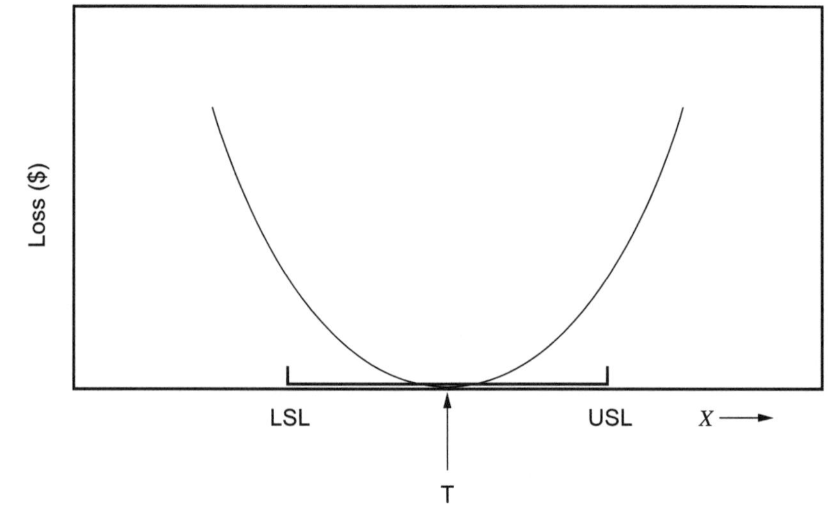

FIGURE 8.9
Quadratic loss function.

When $X = T$, $L(T) = 0$ and $L'(T) = 0$. When terms with power greater than 2 are neglected,

$$L(X) = \frac{L''(T)}{2!}(X - T)^2 = k(X - T)^2$$

Thus, the quadratic loss function reasonably approximates the real loss function within the region of interest.

It is important to determine the value for k that would best approximate the actual loss in the region of interest. This can be a difficult task. Since the loss function has only one unknown k, this coefficient can be estimated if the loss corresponding to any one value of X (other than target) can be specified. One convenient approach is to ask the question: What would happen if the customer received a product with characteristic value just outside specification? The product would not function well. Perhaps it could be repaired in the field, or might have to be replaced. There will be additional losses in terms of inconvenience, lack of availability of product, and so on. For food products, the eating quality and shelf life of the product may degrade, causing customer dissatisfaction and attendant losses. It is a common practice to replace the product if customers are not satisfied with it, so at a minimum, the loss will be equal to the scrap or repair cost and may be considerably higher.

Average Loss. In general, the quality characteristic X has a distribution $f(X)$ with mean μ and standard deviation σ as shown in Figure 8.10. The average loss per unit of product is computed by integrating $f(X)$ over the loss function as follows:

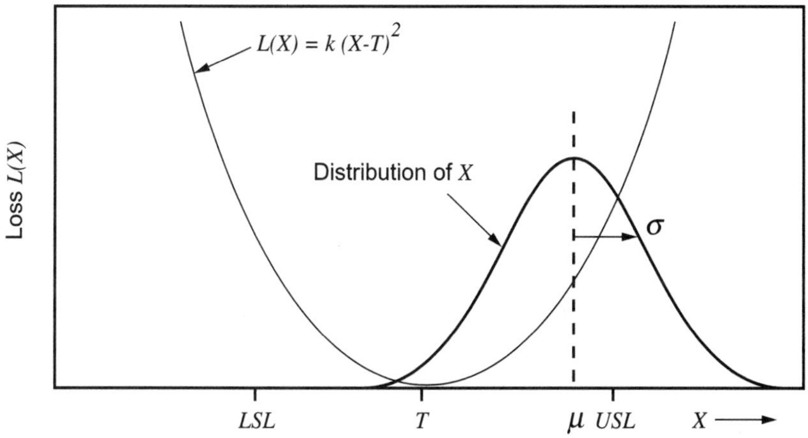

FIGURE 8.10 Average loss.

$$\text{Average Loss} = k\int (X - T)^2 f(X)dx = k\int (X - \mu + \mu - T)^2 f(X)dx$$

$$\text{Average Loss} = k[(\mu - T)^2 + \sigma^2] \tag{8.3}$$

The loss of $k(\mu - T)^2$ results from the process mean μ not being on target T. The loss of $k\sigma^2$ results from process variability around the process mean. Between the two components of loss, it is usually easier to eliminate the first component by bringing the process mean on target. Once this is done, the economic loss becomes directly proportional to the variance and can be reduced further only by reducing variance.

Quadratic Loss and C_p Index. With the process mean centered at the midpoint of the two-sided specification, $\mu = T$ and from Equation 8.3

$$\text{Average Loss} = k\sigma^2$$

Let L_{spec} be the loss at the specification limit. Then

$$L_{\text{spec}} = k\left(\frac{USL - LSL}{2}\right)^2$$

and

$$C_p = \frac{USL - LSL}{6\sigma}$$

The above two equations lead to

$$\text{Average Loss} = \frac{L_{\text{spec}}}{9C_p^2} \tag{8.4}$$

Thus, for a centered process, the quadratic loss is inversely proportional to the square of the C_p index. If the C_p index doubles, loss reduces by a factor of four for a centered process.

As an illustration, the example in Figure 8.8 is now reconsidered. The loss at the specification limit was \$100 per unit. Hence,

$$\text{Average Loss} = \frac{100}{9C_p^2}$$

Figure 8.11 shows the quadratic loss as a function of the C_p index, assuming the process to be centered. For comparison, the classical loss is also shown. Assuming equal loss at the specification limit, the quadratic loss model leads to higher loss estimates because loss is assumed to occur even inside the specification. This creates the need to achieve a high C_p index by continuously improving the process toward on-target performance.

Cp	Per Unit Economic Loss ($)
0.4	69.43
0.6	30.86
0.8	17.36
1.0	11.11
1.2	7.72
1.33	6.28
1.5	4.93
2.0	2.77

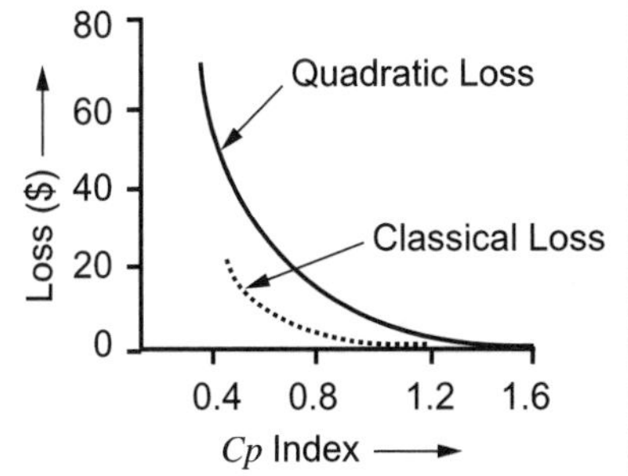

FIGURE 8.11 Economic loss as a function of *Cp* index (quadratic loss function).

Quadratic Loss and C_{pk} Index. There is not a one-to-one relationship between C_{pk} and average loss. Multiple values of C_{pk} may lead to the same average loss or the same value of C_{pk} may result in different values of average loss.

8.2.3 Variations of the Quadratic Loss Function

The quadratic loss function given by

$$L(X) = k(X - T)^2$$

is applicable when the characteristic X has a finite nonzero target, the specifications are two-sided, and the loss is symmetric about the target. Such a characteristic is called nominal-the-best type characteristic and occurs very commonly in practice.

Variations of this loss function are required to adequately cover some other common situations. These variations are described below and graphically illustrated in Figure 8.12.

Smaller-the-Better Characteristics. Some characteristics such as pollution, cycle time, and energy usage can never take negative values and have zero as their target value. Such characteristics are called smaller-the-better characteristics. The specification for such characteristics is one-sided with only an upper specification limit. Since the target value is zero ($T = 0$), the economic loss is approximated by

$$L(X) = kX^2 \tag{8.5}$$

and

$$\text{Average Loss} = k(\mu^2 + \sigma^2)$$

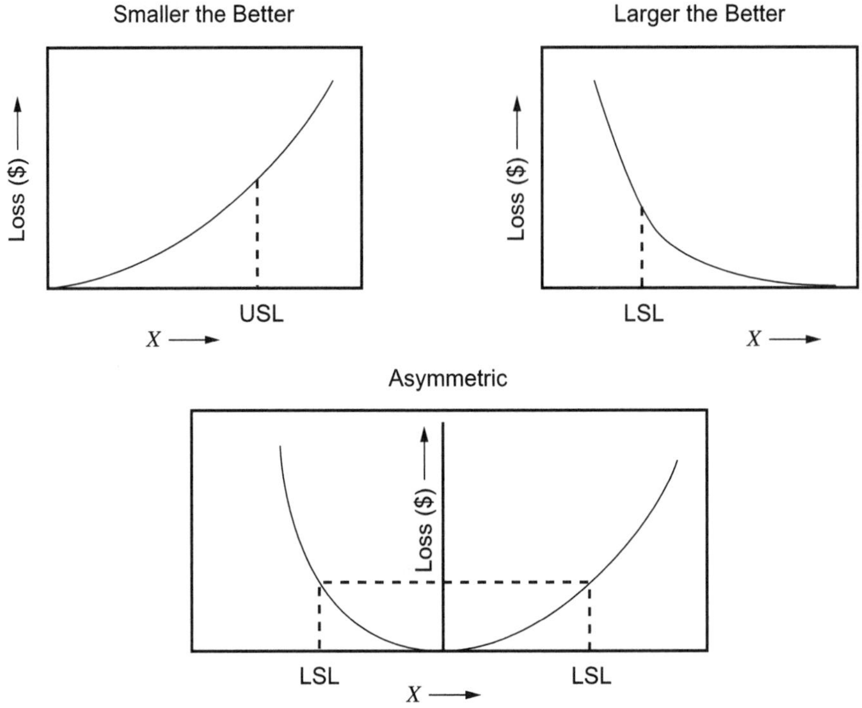

FIGURE 8.12
Variations of the quadratic loss function.

The value of k is usually determined by specifying the loss at the upper specification limit.

Larger-the-Better Characteristics. Characteristics such as bond strength also do not have negative values but larger values are preferred. The target value is infinity. Such characteristics are called larger-the-better characteristics. The specification is again one-sided but with only a lower specification limit. It is clear that reciprocal of such a characteristic is of the smaller-the-better type and the economic loss is approximated by

$$L(X) = \frac{k}{X^2} \tag{8.6}$$

The value of k is obtained by specifying the loss at the lower specification limit, and the average loss can be determined by numerically integrating the statistical distribution of X over the loss function.

Asymmetric Loss Function. In some instances, deviation of the characteristic in one direction is much more harmful than in the other direction. There is a finite target value, but the two-sided specification limits are set at unequal distances from the target. In such cases, a different coefficient may be used in the two directions. The economic loss is approximated by the following asymmetrical loss function:

$$L(X) = \begin{cases} k_1(X - T)^2 \text{ when } X > T \\ k_2(X - T)^2 \text{ when } X \leq T \end{cases} \tag{8.7}$$

As an example of the asymmetrical loss function, consider the case of the clearance between the cylinder and piston of an engine. The target value for the clearance is 4 μm and the specification limits are 1 μm to 8 μm. The loss at the specification limit is $100:

$$k_1(8 - 4)^2 = 100$$

Hence,

$$k_1 = 100/16$$

and

$$k_2(1 - 4)^2 = 100$$

Hence,

$$k_2 = 100/9$$

Given the observed statistical distribution of clearance, the average quality loss can be determined by numerical integration.

8.3 PLANNING FOR QUALITY IMPROVEMENT

The tools of variance components analysis, process capability, and economic loss can be integrated to plan quality improvement efforts and to make improvement decisions on an economic basis. Some examples of such decisions include:

1. Selecting a process to improve based upon high return on investment
2. Identifying where to focus the improvement effort and the extent of desired improvement
3. Making investment decisions intended to improve quality.
4. Deciding how to set specification limits and whether tightening tolerances is economical

5. Determining whether 100% inspection is justifiable or not
6. Evaluating suppliers on an economic basis

The following example, which is a continuation of the example in Section 8.1, illustrates how the concepts of variance components, process capability, and economic loss can be used to focus and plan quality improvement efforts. Table 8.3 shows the results of the current process and a sequence of "what if" analyses conducted to plan future quality improvement efforts. These results are discussed below.

1. Current Process. Column (a) in Table 8.3 shows the results for the current process discussed in Section 8.1. The target moisture is 10% with a two-sided specification of 9% to 11%. The current process mean is 10.27%. There are three sources of variance: supplier variance is 0.133, manufacturing variance is 0.254, and test variance is 0.023. The total variance is 0.410, which gives $\sigma_t = 0.64$. Assuming normality, out-of-specification product is calculated to be 15.45%. The P_{pk} index is calculated as (11.0 – 10.27)/(3 * 0.64) = 0.38.

To compute economic loss, the following additional assumptions are made. The economic loss at either specification limit is assumed to be $2, equal to the selling price of the product. The yearly sales are one million products. With these assumptions, the quadratic loss per year is obtained as follows:

$$\text{Quadratic loss at specification limit} = k(11 - 10)^2 = 2$$

TABLE 8.3 Planning for Quality Improvement

	(a) Current Process	(b) Mean On Target	(c) Reduction in Manufacturing Variance: (c1) 50%	(c) Reduction in Manufacturing Variance: (c2) 100%	(d) 90% Reduction in Supplier and Manufacturing Variance
Process mean	10.27	10.00	10.00	10.00	10.00
Supplier variance	0.133	0.133	0.133	0.133	0.013
Manufacturing variance	0.254	0.245	0.122	0.000	0.025
Test variance	0.023	0.023	0.023	0.023	0.023
Total variance	0.410	0.410	0.278	0.156	0.061
% out of specification	15.45	11.88	5.70	1.14	0.00
P_{pk} index	0.38	0.52	0.63	0.84	1.35
Quadratic loss per year ($k)	965	820	556	312	122
Classical loss per year ($k)	309	238	114	23	0

Hence, $k = 2$.

$$\text{Quadratic loss per product} = k[(\mu - T)^2 + \sigma^2] \approx 2[(0.27)^2 + (0.64)^2] = 0.965$$

$$\text{Quadratic loss per year} = \$965{,}000 = \$965\text{k}$$

The loss can also be computed using classical loss function:

$$\text{Classical loss per product} = 0.1545 * 2 = 0.309$$

$$\text{Classical loss per year is } \$309{,}000 = \$309\text{k}$$

The large difference between the quadratic and classical loss is due to the differing assumptions made by the two loss functions, principally the fact that the quadratic loss function assumes that significant losses can occur inside the specification, whereas classical loss function assumes no losses within specifications. For this reason, it is critical to ensure organizational acceptance of the selected loss function prior to using it for making decisions.

The P_{pk} index and the projected economic losses suggest the need for improvement. The yearly economic loss also provides a feel for the resources that could be committed to improve this process. A formal return on investment analysis can be done. Had the projected economic losses turned out to be small, attention would have focused on some other process in which there were larger gains to be made.

2. Mean on Target. Sometimes it is relatively easy to center the process on target. What if this could be achieved? The results would be as shown in Column (b) in Table 8.3. The percent out-of-specification product will decrease to 11.88%; the P_{pk} index will improve to 0.52 and will now be the same as the P_p index. The quadratic loss will decrease to \$820k and the classical loss will reduce to \$238k. If the quadratic loss is a reasonable approximation to the true loss, then this action of centering the process is approximately worth \$145k per year.

3. Reduced Manufacturing Variability. Further improvements can only be achieved by reducing variability, which is usually more difficult. The variance decomposition indicates that the initial focus should be on reducing manufacturing variability since it is the largest component. What if the manufacturing variance could be reduced by 50% if certain investments were made?

Column (c1) in Table 8.3 shows the predicted results. The quadratic loss will reduce to \$556k and the classical loss will reduce to \$114k. So this improvement action is worth between \$195k to \$410k per year, depending upon the selected loss function. If the projected investment is \$200k, then it may be justified on the basis of either loss function. If the investment is much higher, it may be justified under quadratic loss but not under classical loss, or it may not be justified under either loss function. Thus, the sensitivity of the decision to the various assumed loss functions can also be evaluated.

Note that even with this projected improvement, the P_{pk} index is still only 0.63. The target P_{pk} index may be 1.33. A reasonable question to ask is whether a P_{pk} index of 1.33 could be achieved by focusing on manufacturing process improvements alone. Column (c2) in Table 8.3 shows that even if the entire manufacturing process variability is eliminated, the P_{pk} index will only improve to 0.84.

If the manufacturing variability can be substantially reduced, the classical loss becomes very small and no further improvements may be justified on a financial basis. The quadratic loss continues to be over $300k, which is over 15% of sales volume, so further improvements are justifiable.

4. Reduced Supplier and Manufacturing Variability. To achieve a P_{pk} index of 1.33, a 90% reduction in both the supplier and manufacturing variance is necessary, as shown in Column (d) of Table 8.3. The out-of-specification product and the classical loss essentially become zero and the quadratic loss becomes small. Depending upon the selected loss function, expenditures between $300k and $850k appear justifiable to accomplish all of these improvements.

8.4 APPLICATION TO MULTILANE MANUFACTURING PROCESS

The Case of Pepperoni Pizza. A pizza manufacturer produces four million cases of pepperoni pizza annually. Each case contains 12 pizzas. The pepperoni component is expected to be no less than 10% of the declared weight of the pizza. A pizza with a declared weight of 12 ounces is expected to have a minimum 34 grams of pepperoni.

In this manufacturer's case, there was large variation in the weight of pepperoni from one pizza to another. This meant that the average weight of pepperoni per pizza had to be much greater than 34 grams to ensure that nearly all the pizzas would meet the minimum desired pepperoni weight. Since pepperoni is an expensive ingredient, large amounts of potential profits were lost due to this increased average weight. The solution was to implement a weight control strategy to reduce the variability of pepperoni weight and then retarget the average weight closer to 34 grams.

A study was designed to understand the pepperoni weight variability and to answer several specific questions:

1. What is the extent of variability in pepperoni weight?
2. What would the average weight have to be if the process is not changed?
3. What are the causes of variability? How much of the observed variability is due to each cause?
4. What actions are necessary to reduce variation?
5. If the variability is reduced, what will be the reduction in average pepperoni weight, while still meeting the desired minimum pepperoni weight per pizza? What is the resulting profit improvement?

Pizza Production Process. As shown in Figure 8.13, the production process simultaneously produces three lanes of pizzas and pepperoni is deposited on three pizzas at the same time. At periodic intervals, two consecutive pizzas were removed from each lane. The weight of pepperoni was measured in grams for each pizza. The data for 16 time periods are shown in Table 8.4.

This structured study is a two-variable crossed study with replicates. Lanes and times are crossed. Lane is a fixed factor; time is random. The consecutive weights of pepperoni are replicates. The variance components analysis for such a study is described in Table 7.7.

Histogram. Figure 8.14 shows the histogram of the collected data. The average weight of pepperoni is approximately 51 grams and the standard deviation from the raw data is approximately 8 grams. The individual pepperoni weight varies from 30 grams to 70 grams. For some piz-

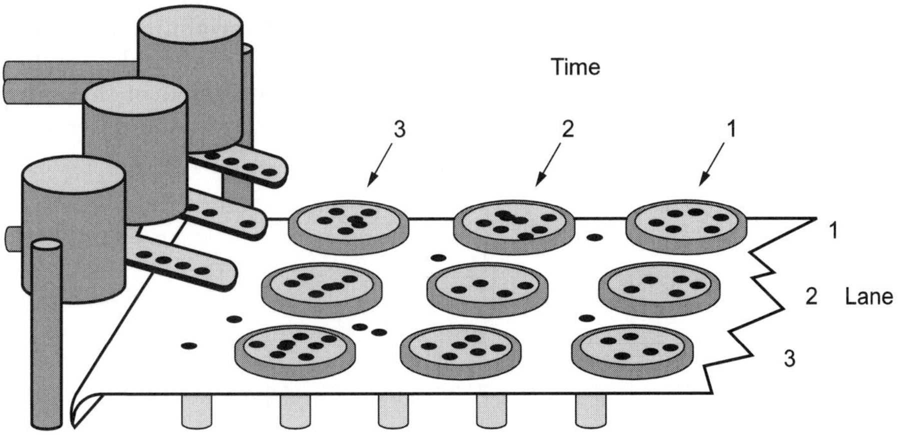

FIGURE 8.13 Pizza manufacturing process.

TABLE 8.4 Pepperoni Weight Data

Time	Lane 1		Lane 2		Lane 3	
1	45	48	54	52	51	54
2	62	61	52	56	45	42
3	46	52	43	47	44	39
4	48	44	52	48	50	52
5	44	42	47	50	36	34
6	56	57	52	54	34	32
7	58	62	57	60	46	50
8	63	60	58	62	44	43
9	54	56	52	56	45	48
10	51	48	64	55	52	59
11	50	54	54	58	34	31
12	40	46	56	57	48	42
13	63	70	66	60	60	58
14	57	58	52	58	51	46
15	58	62	49	52	51	45
16	51	46	51	52	40	40
Average	53.5		54.25		45.19	

zas, the pepperoni weight is less than the desired minimum 34 grams. For other pizzas, the pepperoni weight is twice as large as the minimum. The variation is very large, indicating that the process variability needs to be reduced so that the average could be retargeted closer to the minimum requirement.

Multi-Vari Chart. Figure 8.15 shows the multi-vari chart for pepperoni weight data. The variability between successive pizzas, represented by the vertical bar, is relatively small compared to the total range in weight from 30 grams to 70 grams. The principle source of variability appears to be lane-to-lane differences. Lane 3 often has considerably smaller weights than the other two lanes, which is also reflected in the much lower average weight for lane 3 in Table 8.4. However, the differences

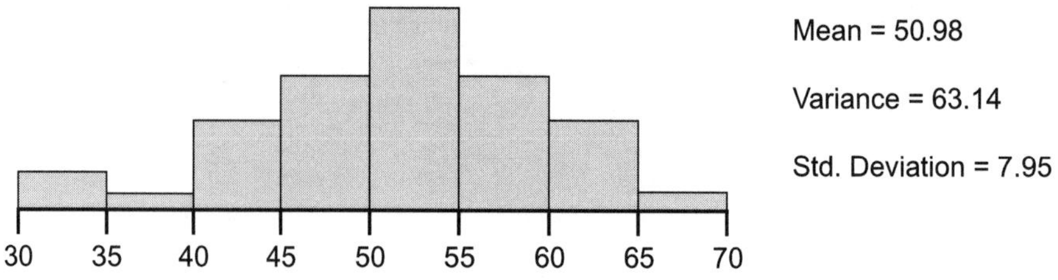

FIGURE 8.14 Histogram of pepperoni weight.

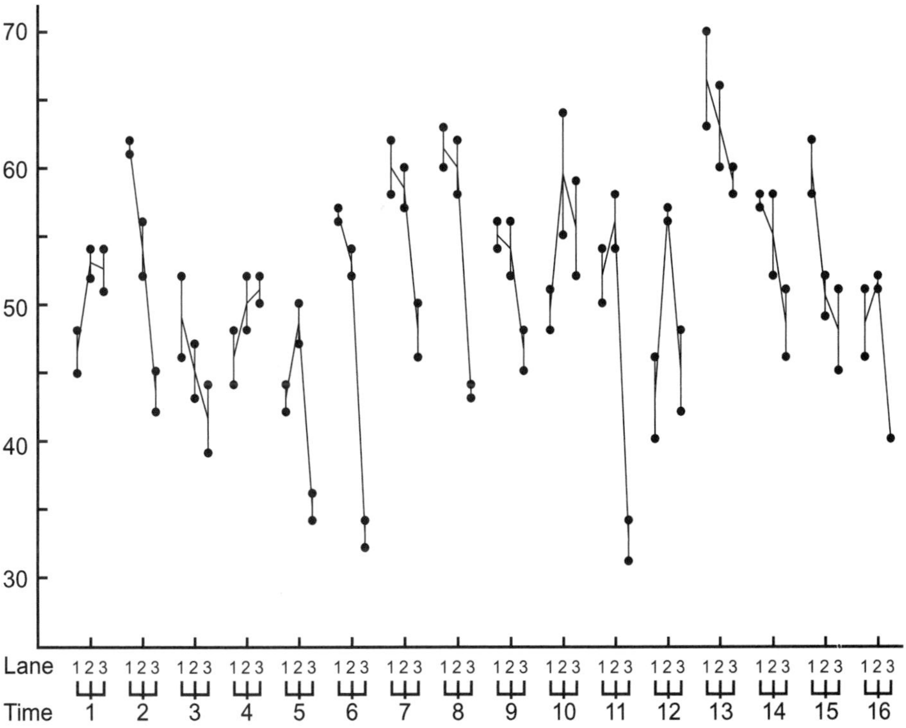

FIGURE 8.15 Multi-vari chart for pepperoni weight.

between the lanes are not always consistent: at time 2, lane 1 is high; at time 4, lane 3 is high; and at time 5, lane 2 is high. This suggests an interaction between lanes and times, namely, the differences between lanes change with time. The multi-vari chart could also be plotted by plotting all times for lane 1 first, then for lanes 2 and 3. Such a plot would provide a feel for time-to-time variability and lane-to-lane differences but will not provide a good feel for lane-by-time interaction. Variance components analysis makes the judgements regarding the key sources of variability easier by estimating the contribution of each source.

Variance Components Analysis. The easiest way to obtain variance components is to compute the mean square (MS) values in Table 7.7 by using one of the many available software packages and then equate these mean square values to the expected mean squares in Table 7.7. We use the last column of Table 7.7 because lanes are fixed and times are random, noting that $n = 2$, $b = 16$, and $a = 3$. Let L represent lanes, T represent times, and e represent the variability between replicated successive pizzas. Then,

$$\text{MS (Lanes)} = 809.5 = s_e^2 + 2s_{LT}^2 + 48s_L^2$$

$$\text{MS (Times)} = 151.5 = s_e^2 + 6s_T^2$$

$$\text{MS (Lane * Time)} = 57.3 = s_e^2 + 2s_{LT}^2$$

$$\text{MS (Replicates)} = 8.06 = s_e^2$$

Simultaneous solutions of these equations leads to the variance components in Table 8.5.

The following conclusions may be drawn:

1. There is a consistent difference between lanes, which accounts for 22% of the total weight variance. This is primarily caused by lane 3 average weight being considerably smaller than the average weight for the other two lanes.
2. The difference between lanes changes with time and this interaction accounts for 34% of the total weight variability. Thus, the consistent and inconsistent lane-to-lane differences account for 56% (= 22% + 34%) of the total variance.
3. Time-to-time variability accounts for 33% of the total variance. This is the special cause variability along the time axis and can be seen from the $\overline{X}$ control charts, separately designed for each lane, and shown in Figure 8.16. In all cases, the range charts (not shown) are in control.
4. The pepperoni weight differences between successive pizzas within a lane account for 11% of the total variance.
5. The total standard deviation of 8.5 is too large to be acceptable.

Planning Process Improvements. Let us suppose that we want the P_{pk} index to be at least 1 with respect to the 34 grams minimum specification. This will ensure that only rarely will an individual pizza have pepperoni less than 34 grams. We can then compute what the average

TABLE 8.5 Pepperoni Pizza Variance Components

Source	Degrees of Freedom	Variance	Standard Deviation	Percent Contribution
Lanes	2	15.67	3.96	22
Times	15	23.90	4.89	33
Lane * Time	30	24.62	4.96	34
Replicates	48	8.06	2.84	11
Total	95	72.25	8.5	100

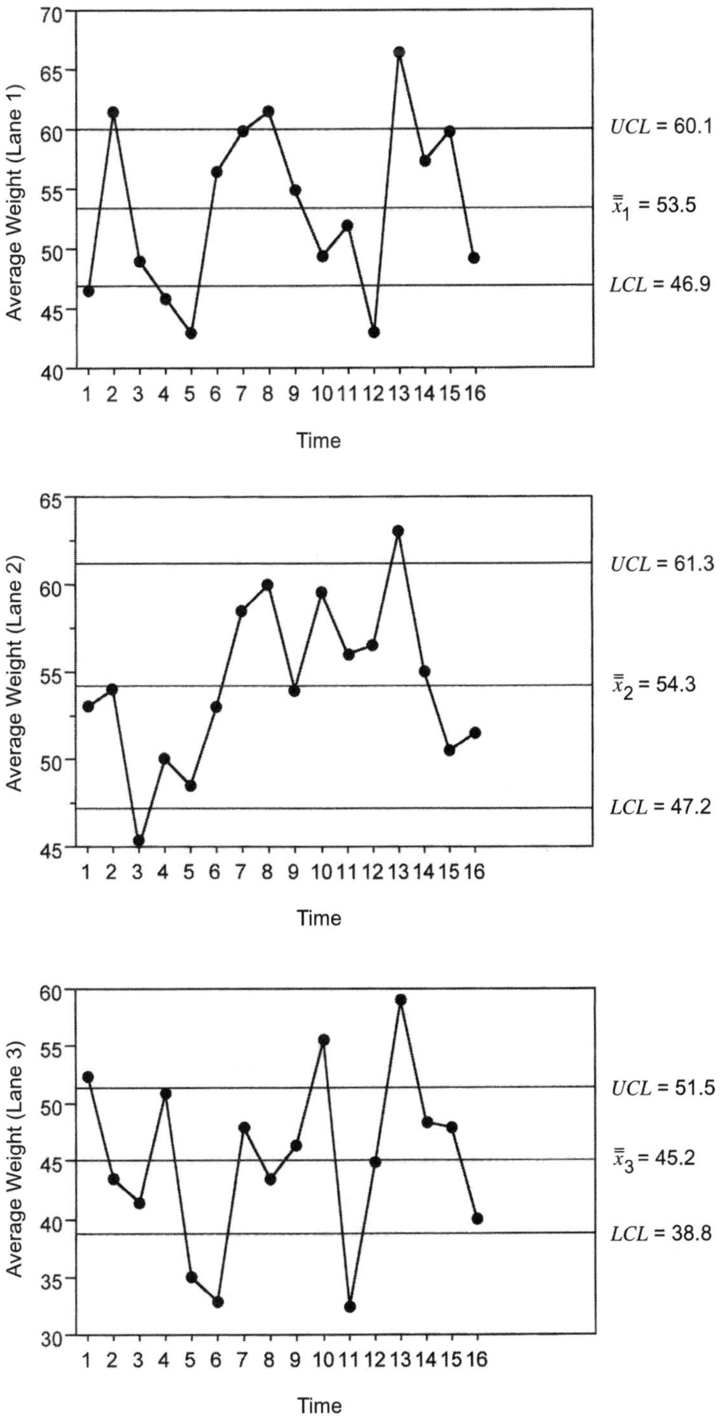

FIGURE 8.16 Lane-by-lane control charts.

pepperoni weight will have to be given the current variability and how much extra cost is being incurred because the average weight exceeds 34 grams. The impact of any proposed improvement could then be evaluated in economic terms as follows.

We want the P_{pk} to be 1 and P_{pk} is defined as:

$$P_{pk} = \frac{\text{Mean} - 34}{3\sigma_t} = 1$$

Hence,

$$\text{Mean} = 3\sigma_t + 34$$

where σ_t is the total standard deviation, currently equal to 8.5. If the process is not improved, process mean will have to be $(3 * 8.5 + 34) = 59.5$ grams.

If we assume the cost of pepperoni to be 0.3 cents per gram, then the extra cost of pepperoni due to variability is $(59.5 - 34) * 0.3$ cents = 7.65 cents per pizza. Since 48 million pizzas are produced per year, the cost of extra pepperoni is \$3.672 million/year. This represents the total potential cost reduction by complete elimination of pepperoni weight variability. The cost model considered here is purely from the viewpoint of meeting the declared weight.

If the consistent and inconsistent variability due to lanes could be eliminated, then from Table 8.5, the new $\sigma_t = \sqrt{23.90 + 8.06} = 5.65$ and the process mean could be retargeted at $(3 * 5.65 + 34) = 51$ grams. This would represent savings of $(59.5 - 51) * 0.3$ cents = 2.55 cents per pizza or \$1.224 million per year. If the special cause variability along the time axis could be eliminated as well, the savings would be \$2.4 million per year. In this manner the potential profit improvement due to variance reduction can be estimated. The costs of accomplishing these improvements can be evaluated against this potential profit improvement.

8.5 VARIANCE TRANSMISSION ANALYSIS

Before proceeding with further applications, we now discuss the subject of variance transmission analysis. Given a functional relationship between output and input factors, variance transmission analysis provides a method for determining how the variance in input factors is transmitted to the output.

For a simple linear relationship between output X and input A given by

$$X = \alpha_0 + \alpha_1 A + e$$

we can use the properties of variance in Chapter 2 to obtain variance of X:

$$\sigma_X^2 = \alpha_1^2\sigma_a^2 + \sigma_e^2$$

Similarly, if X is a linear function of several independent input factors,

$$X = \alpha_0 + \alpha_1 A + \alpha_2 B + \alpha_3 C + \ldots + e$$

Then

$$\sigma_X^2 = \alpha_1^2\sigma_a^2 + \alpha_2^2\sigma_b^2 + \alpha_3^2\sigma_c^2 + \ldots + \sigma_e^2$$

where σ_a, σ_b, . . . are standard deviations of A, B, . . . , etc.

The right-hand side of this equation represents the variance contribution due to each input factor. The contribution due to each factor is equal to the square of the coefficient times the variance of that factor, i.e.,

Variance component due to $A = \sigma_A^2 = \alpha_1^2\sigma_a^2$
Variance component due to $B = \sigma_B^2 = \alpha_2^2\sigma_b^2$
.
.
.
Variance component due to error $= \sigma_e^2$

What if the output is a nonlinear function of input factors? For example (Box, Hunter, and Hunter), if the volume V of a spherical bubble is computed by measuring its diameter D, how will the error in measuring diameter transmit to error in measuring the volume? The relationship between volume and diameter is given by

$$V = \frac{\pi}{6}D^3$$

Suppose the diameter of a particular bubble is 3 mm. Figure 8.17 shows how the standard deviation of measurement is transmitted to the standard deviation of volume at $D = 3$. If we locally approximate the relationship by a straight line, then the slope of the tangent is given by dV/dD and

$$\sigma_V = \left(\frac{dV}{dD}\right)\sigma_d$$

where σ_d is the standard deviation of diameter. Then

$$\sigma_V = \frac{\pi}{2}D^2\sigma_d$$

If σ_d is constant, then σ_V changes with D and increases as a quadratic function of D. Also

$$\frac{\sigma_V}{V} = 3\frac{\sigma_d}{D}$$

which means that the percentage error in measuring volume is approximately three times the percentage error in measuring diameter, or the CV for volume is three times the CV for diameter.

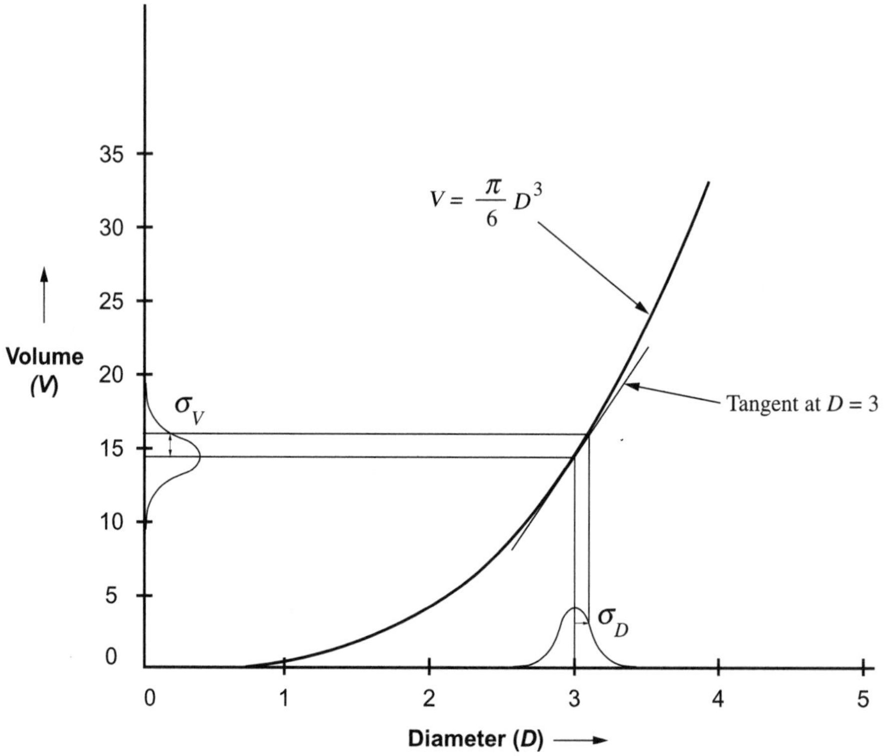

FIGURE 8.17 Variance transmission.

Similar results can be obtained for any general functional relationship:

$$X = f(A, B, C \ldots) + e$$

where A, B, C are independent factors. Using the locally linear approximation,

$$\sigma_X^2 \approx \left(\frac{df}{dA}\right)^2 \sigma_a^2 + \left(\frac{df}{dB}\right)^2 \sigma_b^2 \ldots + \sigma_e^2 \tag{8.8}$$

As an example, consider the case where the output is a product function of the input:

$$X = AB$$

Then

$$\sigma_X^2 \approx B^2\sigma_a^2 + A^2\sigma_b^2$$

Dividing both sides by X^2,

$$\frac{\sigma_X^2}{X^2} \approx \frac{\sigma_a^2}{A^2} + \frac{\sigma_b^2}{B^2}$$

If the linearization is done around the mean values of A and B, then

$$CV^2(X) \approx CV^2(A) + CV^2(B) \tag{8.9}$$

The above equation also holds for $X = A/B$. What if the linear approximation is inadequate? Then, higher-order approximations or a Monte Carlo simulation approach may be used to obtain the desired results.

8.6 APPLICATION TO A FACTORIAL DESIGN

This section illustrates the application of variance components analysis and variance transmission analysis to analyze designed experiments. To simplify the presentation, a simple 2^3 factorial experiment is considered, although the concepts extend to any situation in which an equation can be obtained to relate input and output factors, either on the basis of theory or on the basis of experiments.

Seal Strength Example. Consider the following experiment involving three factors, each at two levels. The three factors are days, lots, and machines. The output of interest is the seal strength of a package. Days is a surrogate for humidity. It was felt that ambient humidity may have a large effect on the seal strength and experiments were conducted on two days when the humidity was different. Lots represent the variation due to incoming lots of packaging material. Two different lots were used in the experiment. The production facility had two machines to seal the packages and both were used in the experiment. The experiment was designed to answer questions such as the following.

1. What are the effects of humidity (days), lots, and machines on the seal strength?
2. What proportion of the variability in seal strength is due to each source?
3. Can we meet the seal strength specification in future production? What will be the P_{pk} index?
4. If we cannot meet the specification, what actions should be taken?
   (a) Is it necessary to adjust the machines and make them more identical?
   (b) Is it necessary to improve supplier quality?
   (c) Is it necessary to control ambient humidity? If so, is the anticipated expenditure justified?

The designed 2^3 factorial experiment and data are shown in Table 8.6 and are graphically illustrated in Figure 8.18. A factorial experiment involves all possible combinations of the levels of the factors (Box, Hunter, and Hunter). Since there are 3 factors each at 2 levels, there are $2^3 = 8$ combinations that form a cube. The levels of each factor are shown in scaled units of –1 to +1 where –1 represents the low level and +1 represents the high level of each factor.

Model Building. The data in Table 8.6 can be analyzed to determine the effects of factors H, L, and M as well as all their interactions ,using regression analysis. Only the effects of factors H and L turn out to be large. The effects of M and all interactions are insignificant. The effect of factor H can be easily computed to be the difference between the average seal strength when H is at the +1 level and the average seal strength when H is at the –1 level. This difference is 12, meaning that the change in humidity from day 1 to day 2 causes the average seal strength to increase by 12. Similarly, the effect of lots is –9.5, indicating the difference in seal strength due to the two raw material lots.

These effects can be translated into an equation as follows:

$$x = 140.75 + 6H - 4.75L + e$$

with $s_e^2 = 15$. The equation is written in coded units, namely, H and L go from –1 to +1. The grand mean is 140.75, which corresponds to the predicted value of seal strength if $H = L = 0$ in coded units. The coefficients for H and L are half the effects because an effect corresponds to a change of two units in, say, H (from –1 to +1), whereas a coefficient corresponds to a change of one unit.

Fixed and Random Factors. In order to compute variance components, we need to decide whether the factors are fixed or random. Ma-

TABLE 8.6 Factorial Experiment

H: Days (Humidity)	L: Lots	M: Machines	x: Seal Strength
–1	–1	–1	141
–1	–1	+1	137
–1	+1	–1	134
–1	+1	+1	127
+1	–1	–1	155
+1	–1	+1	149
+1	+1	–1	138
+1	+1	+1	145
			Average = 140.75

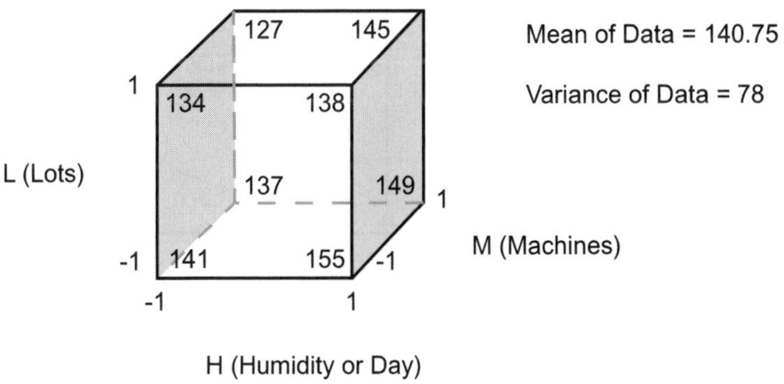

FIGURE 8.18 Factorial experiment.

chine (factor *M*) is a fixed factor. There are only two machines and both are in the experiment. The model indicates that the effect of machines is not significant, i.e., the machines may be considered to be equivalent. The variance component due to machines is essentially zero.

Humidity (days – factor *H*) is a random factor. It varies throughout the year. However, in this experiment only two values of humidity were used, designated –1 and +1, which on the two days in question actually were 30% and 40% relative humidity. If humidity is considered a random factor, then these two values would be taken to be random drawings from the population distribution of humidity. As shown in Figure 8.19, this population distribution of humidity can be estimated from the coded –1 and +1 values as follows:

$$\bar{\bar{x}} = 0 \text{ estimates the true mean humidity}$$

$$s^2 = \frac{\Sigma(-1-0)^2 + (1-0)^2}{2-1} = 2 \text{ estimates } \sigma_h^2\text{, the true variance of humidity}$$

In real humidity units, these translate to $\bar{\bar{x}} = 35$ and $s^2 = 50$. When humidity is considered a random factor; the variance component for humidity is calculated based upon this predicted distribution of humidity.

Unfortunately, as shown in Figure 8.19, this predicted distribution of humidity could be very different from the true humidity distribution obtained from historical records. This is so because we only have two observations on humidity, which is an extremely small sample size to estimate distribution. If we calculate the variance contribution of humidity, using the variance components analysis described earlier with humidity considered to be a random factor, the result could be much different from the truth.

Using Variance Transmission. A second approach uses variance transmission analysis as follows:

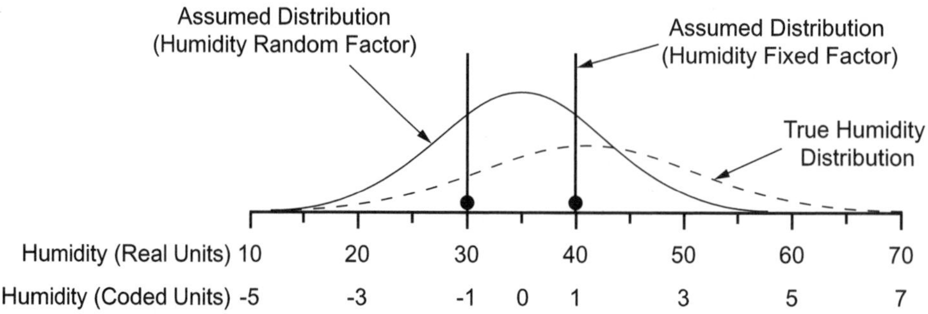

FIGURE 8.19 Humidity as a fixed and as a random factor.

1. Independently estimate the likely distribution of humidity. This could be done based upon meteorological records.
2. Estimate the variance component corresponding to humidity from the experimental results, assuming humidity to be a fixed factor.
3. Predict what the variance component would be for the likely distribution of humidity.

Let us assume that the records indicate that humidity varies between 10% to 70% throughout the year. The true humidity distribution may be as shown in Figure 8.19, with mean = 40% humidity (+1 in coded units) and standard deviation = 10% humidity (+2 in coded units).

If humidity is considered as a fixed factor in the experiment, it has a Bernoulli distribution as shown in Figure 8.19. The mean of this distribution is known to be 0 in coded units and variance of this distribution is

$$s_h^2 = \frac{(-1-0)^2 + (1-0)^2}{2} = 1 \text{ estimates } \sigma_h^2.$$

We divide by 2 and not (2 – 1) because as a fixed factor, the entire population of humidity is represented in the experiment. Note that $\sigma_h^2 = 1$ if humidity is a fixed factor and $\sigma_h^2 = 2$ if humidity is a random factor for a two-level experiment.

How does this variability in humidity transmit to variability in X? For the problem at hand the equation is

$$x = 140.75 + 6H - 4.75L + e$$

If both H and L are considered fixed, then $\sigma_h = 1$, $\sigma_l = 1$, and

$$\text{Variance component due to humidity} = \sigma_H^2 = (6)^2\sigma_h^2 = 36$$

$$\text{Variance component due to lot} = \sigma_L^2 = (-4.75)^2\sigma_l^2 = 22.6$$

$$\sigma_e^2 = 15$$

Notice that for simple additive model, in which each factor is considered fixed at two levels denoted by –1 and +1, the variance component due to each factor is equal to the square of its coefficient in the model. A very simple result indeed!

Projecting Realistic Variance Components. In order to project realistic variance components due to H and L, we must know the true distribution of H and L during actual manufacturing. Factor H is humidity and, as previously discussed, during manufacturing, it is expected to have a mean $\mu_h = 1$ and variance $\sigma_h^2 = 4$ in coded units. Factor L represents raw material lots. If the specific property of the lots that causes variability in output is known, then lots could be converted into a continuous factor (i.e., that specific property), just as days was converted into a continuous factor called humidity, and lots could be dealt with in the same manner as humidity. If such is not the case, then the best we can do is to assume lots to be a random factor with $\mu_l = 0$ and $\sigma_l^2 = 2$ in coded units. To summarize, during manufacturing

$$\text{Factor } H\text{: } \mu_h = 1,\ \sigma_h^2 = 4$$

$$\text{Factor } L\text{: } \mu_l = 0,\ \sigma_l^2 = 2$$

We are now ready to project the future mean and variance of X. Since

$$x = 140.75 + 6H - 4.75L + e$$

substituting the means of H and L in the above equation, we get

$$\mu_X = 146.75 \approx 147$$

The previously computed variance components $\sigma_H^2 = 36$ and $\sigma_L^2 = 22.6$ assumed H and L to be fixed factors with $\sigma_h^2 = \sigma_l^2 = 1$. We now project σ_h^2 to be 4 and σ_l^2 to be 2. The projected variance components can be calculated by multiplication as follows:

σ_H^2	36 * 4	=	144
σ_L^2	22.6 * 2	=	45
σ_e^2	15 * 1	=	15
σ_X^2		=	204

Thus, the projection is that seal strength will have a mean of approximately 147 and a variance of 204 in future production.

Making Decisions. Let us suppose that the specification for seal strength is 150 ± 30. We want to decide whether capital investments to control humidity should be made. Currently, $\mu_X = 147$ and $\sigma_X = \sqrt{204} = 14.3$. Therefore, $P_{pk} = 0.634$ and $P_p = 0.7$, leading to 3.5% to 4% of the product being outside specifications. If humidity is perfectly controlled, $\sigma_H^2 = 0$ and $\sigma_X^2 = 60$. The P_{pk} index will improve to 1.16 and very little product will be outside the specification. The cost of 4% out-of-specification product can be weighed against the cost of humidity control in order to make the capital investment decision on a financial basis.

8.7 SPECIFICATIONS AND VARIANCE COMPONENTS

This section addresses two topics related to specifications and variance components: how should specification be allocated to key sources of variability and how can R&D goals be set for within-lot and between-lot variability based upon lot acceptance specifications?

Products are usually accepted on the basis of specifications. For example, the two-sided specification for a certain dimension (X) of an extruded product may be $T \pm W$, where T is the target dimension and W is the half-width of the specification. Let raw material (M), analytical measurement method (A), and manufacturing process (P) be the important sources of variability in X. Then it is reasonable to ask the following questions:

1. What percent of the specification should be allocated to purchased material (or any one of the sources of variability)?
2. If analytical variability could be reduced, what will be the permissible increase in the percent of specification allocated to purchased material? (What are the trade-offs between the various sources of variability?)

In this case, the variance of X is the sum of the three variance components:

$$\sigma_X^2 = \sigma_M^2 + \sigma_A^2 + \sigma_P^2 \tag{8.9}$$

This addition may be graphically represented as shown in Figure 8.20, where the solid lines represent Equation 8.9 and the dotted lines suggest how this addition of variance components will continue if there were a larger number of components. The figure clearly illustrates how standard deviations add in a Pythagorian fashion. Variances add arithmatically.

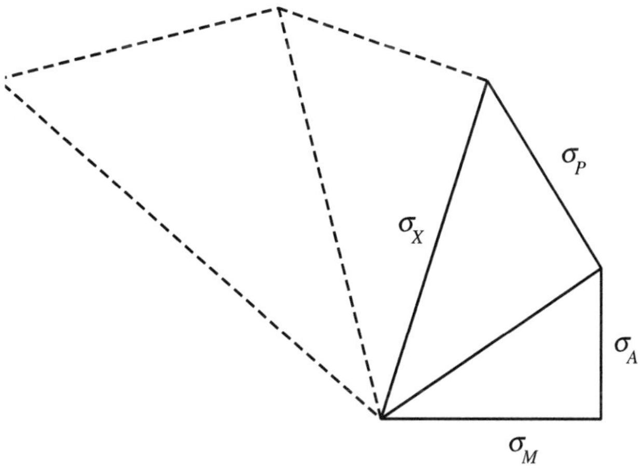

FIGURE 8.20 Adding variance components.

To answer the two questions posed earlier, for a given P_p

$$P_p = \frac{W}{3\sigma_X} \quad \text{or} \quad \sigma_X = \frac{W}{3P_p}$$

If we want the P_p index to be 1.33, then Equation 8.9 becomes

$$\sigma_X{}^2 = \frac{W^2}{16} = \sigma_M^2 + \sigma_A^2 + \sigma_P^2$$

Note that,

$$\sigma_X = \frac{W}{4} \neq \sigma_M + \sigma_A + \sigma_P$$

It is not the specification width that gets allocated to standard deviations of sources of variability; rather, it is the square of the specification width that gets allocated to the variance components in the usual summation fashion. If we arbitrarily assume $W = 40$, hence $\sigma_X^2 = 100$, and allocate it such that $\sigma_M^2 = 40$, $\sigma_A^2 = 20$, and $\sigma_P^2 = 40$, then $\sigma_M = 6.3$, $\sigma_A = 4.5$, and $\sigma_P = 6.3$. If analytical variability only needs $\sigma_A = 1$, how large can σ_M be? Here $\sigma_A{}^2 = 1$, hence $\sigma_M{}^2 = 59$ and $\sigma_M = 7.7$. This is how a reduced allocation to analytical variability trades off with an increased allocation to purchased material. In practice $(W/3P_p)^2$ should be allocated to various components of variability in proportion to their economically justifiable need so that the overall cost of the product is reduced.

For the rest of this section we turn to the common problem of lot-by-lot acceptance during manufacturing, with the idea of developing R&D goals for the following two variance components as a function of the specification:

σ_w^2 = Product-to-product variance within a lot

σ_b^2 = Variance between lots

These two variance components are influenced by different causes of variation. For example, analytical variability has a large influence on σ_w, whereas raw material batch-to-batch variability influences σ_b. Reducing both variance components is a key objective in R&D. Therefore, knowing the likely specification during future manufacturing, how can we set targets for σ_w and σ_b during the R&D phase?

Simple Specification. Suppose the lot acceptance criterion is that the lot will be accepted if all n randomly selected products from a lot are within the specification $T \pm W$. What constraints does this place on σ_w and σ_b?

In this case, we have

$$\sigma_t^2 = \sigma_w^2 + \sigma_b^2$$

where σ_t is the total standard deviation. If we minimally wish to have a process capability $P_p = 1.33$ (note that for all n products to be within specification with high probability, the P_p index needs to be high), then

$$P_p = \frac{W}{3\sigma_t} > 1.33 \qquad \text{or} \qquad \sigma_t < \frac{W}{4}$$

Hence

$$\sigma_w^2 + \sigma_b^2 < \frac{W^2}{16} \tag{8.10}$$

In the σ_w, σ_b domain, Equation 8.10 is a region enclosed by a circle of radius $W/4$ (generally $W/3P_p$) and is plotted in Figure 8.21 for various values of W. If the anticipated specification requires $W = 4$, then σ_w and σ_b need to be within the region enclosed by a circle of radius 1 centered at the origin. This means that if $\sigma_w = 1$, $\sigma_b = 0$; if $\sigma_w = 0.5$, $\sigma_b = 0.87$, and so on. Based upon the known or anticipated value of W, R&D goals for σ_w and σ_b can be set. Conversely, knowing the feasible values of σ_w and σ_b, a capability-based value of W can also be easily determined from Figure 8.21. For example, if $\sigma_w = 1$, $\sigma_b = 1$, then by inspection from Figure 8.21, $W \approx 6$ (the actual value of W is 5.66 from Equation 8.10). Figure 8.21 can be generalized by expressing W, σ_w, and σ_b as % of target T.

Simple Specification with s_w Constraint. Consider the case in which there are two requirements for the lot to be accepted:

1. Each of the $n = 10$ randomly selected products has to be within 100 ± 15.

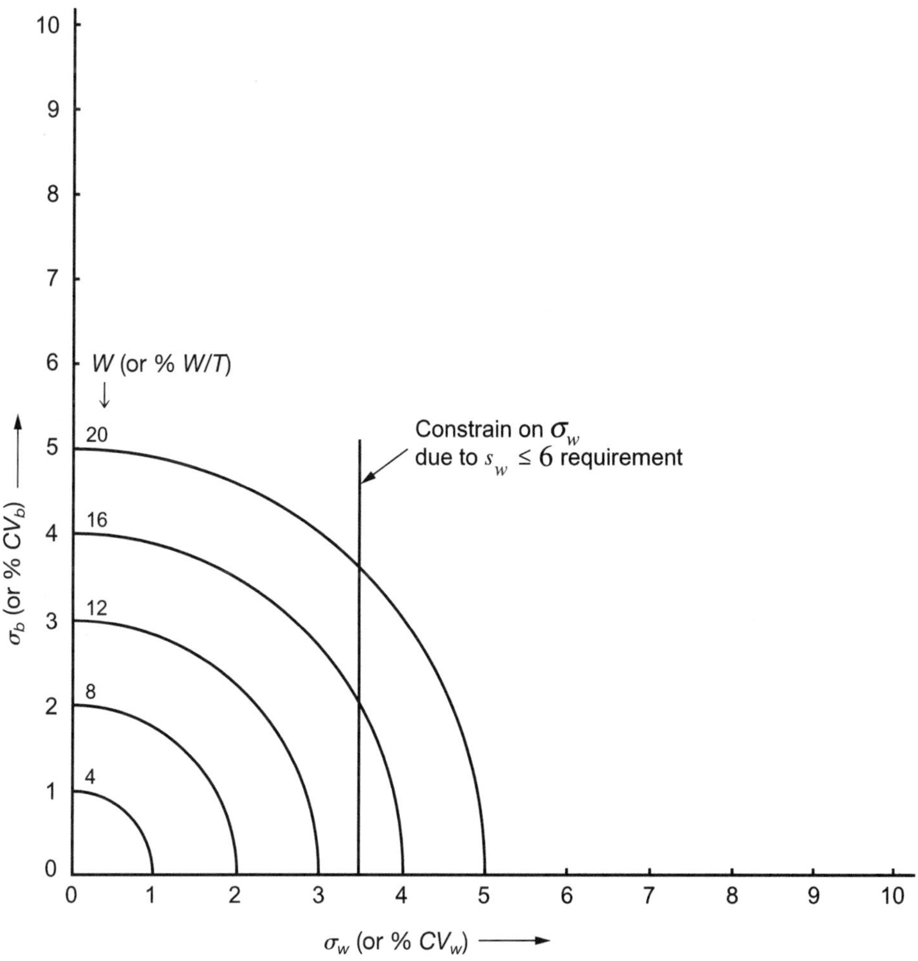

FIGURE 8.21 Specification and variance components (simple specification).

2. The standard deviation (s_w) calculated from these n = 10 observations must be less than six.

Drug content of a tablet generally has such specifications. Here W = 15 and the first requirement will result in a circle of radius $W/4$ = 3.75 in Figure 8.21.

The second requirement leads to a further constraint on σ_w as follows. We know that $(n-1)S_w^2/\sigma_w^2$ has a χ^2_{n-1} distribution. If we want $P[S_w > 6] \leq 0.001$ or $P[S_w^2 > 36] \leq 0.001$, then

$$P\left[\frac{(n-1)S_w^2}{\sigma_w^2} > \frac{36(n-1)}{\sigma_w^2}\right] \leq 0.001$$

From the chi-square table, for $n = 10$ $[36(9)/\sigma_w^2] \geq 27.9$ or $\sigma_w \leq 3.4$, which is shown by the vertical line in Figure 8.21. This narrows the permissible region for σ_w and σ_b by eliminating a portion of the circle of radius 3.75.

Multi-Level Specifications. Sometimes, lot acceptance involves multi-level specifications such as the specifications for drug release rate, which take the following form:

Level 1: Measure the individual release rates for six tablets. The average must be inside $T \pm W$. If all the individual values are also inside $T \pm W$, accept the lot.

Level 2: If some individual values are outside the Level 1 specification, measure another sample of six tablets. If all 12 individual release rates are inside $T \pm (W + 10)$, accept the lot.

Level 3: If some individual values are outside the Level 2 specification (usually no more than two values), measure an additional sample of 12 tablets. All 24 individual release rates must be inside $T \pm (W + 20)$, otherwise the lot is rejected.

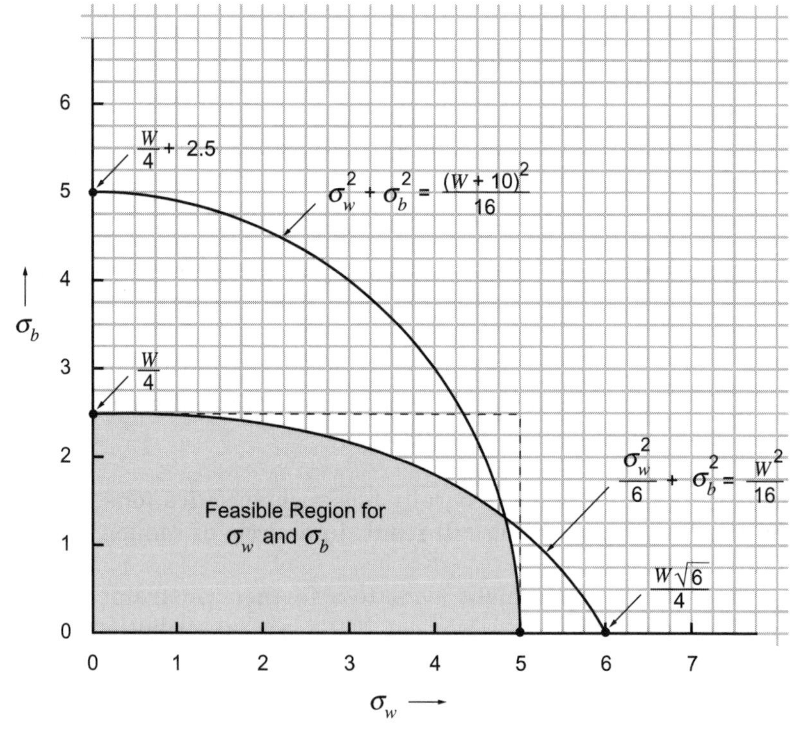

FIGURE 8.22 For $W = 10$, feasible region for σ_w and σ_b (multilevel specification).

The following calculates the constrains for σ_w and σ_b, assuming that we want to pass the lot with very high probability at Level 2.

Let $\overline{X}$ denote the average of six tablets. Assuming a desired P_p index of at least 1.33, for the Level 1 requirement, we have

$$P_p = \frac{W}{3\sigma_{\overline{X}}} > 1.33 \qquad \text{or} \qquad \sigma_{\overline{X}} < \frac{W}{4}$$

which leads to

$$\frac{\sigma_w^2}{6} + \sigma_b^2 < \frac{W^2}{16} \tag{8.11}$$

Let σ_t denote the total standard deviation. Then for the Level 2 requirement to be satisfied with high probability,

$$P_p = \frac{W + 10}{3\sigma_t} > 1.33 \qquad \text{or} \qquad \sigma_t < \frac{W + 10}{4}$$

Hence

$$\sigma_w^2 + \sigma_b^2 < \frac{(W + 10)^2}{16} \tag{8.12}$$

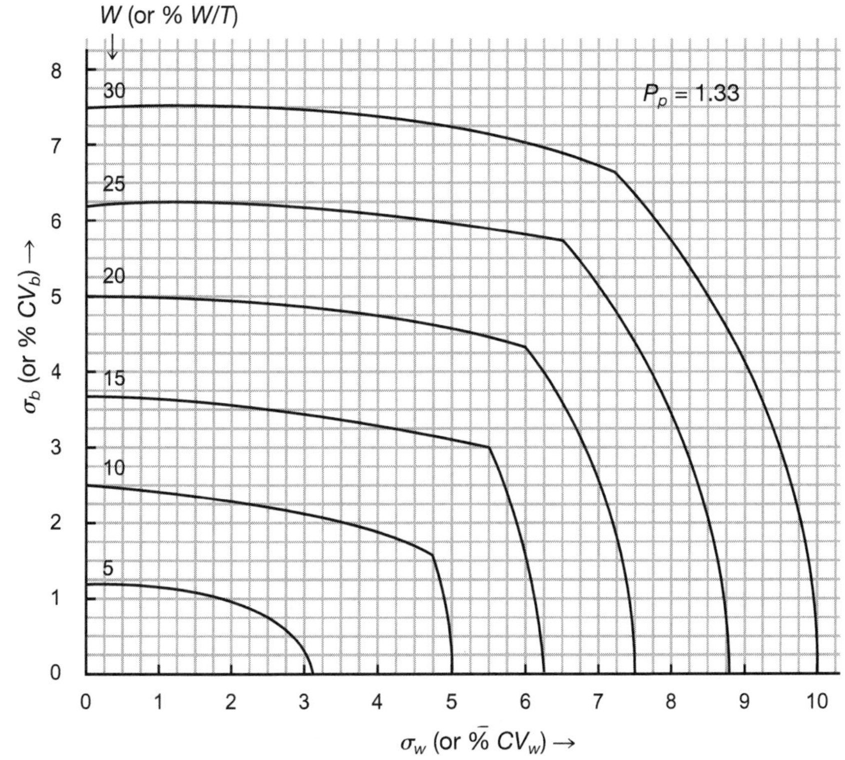

FIGURE 8.23
Feasible region for σ_w and σ_b (multilevel specification).

Equation 8.12 is a region enclosed by a circle of radius $(W + 10)/4$.

For $W = 10$, Figure 8.22 shows a plot of Equations 8.11 and 8.12, and the resultant feasible region for σ_w and σ_b.

Figure 8.23 shows the feasible regions for σ_w and σ_b for various values of W. As shown in figure 8.23, the graph can be generalized by expressing W, σ_w, and σ_b as % of tartet T. R&D can use this graph in a variety of ways as follows:

1. Given a value of W, to set goals for σ_w and σ_b.
2. Given the feasible values of σ_w and σ_b, find a satisfactory W, namely, to set a capability-based specification.
3. Given the values of W, σ_w, and σ_b, set improvement targets for σ_w and σ_b. For example, if $W = 10$, $\sigma_w = 6$, and $\sigma_b = 1$, then from Figure 8.23, an improvement target may be $\sigma_w = 4$. There is no specific need to reduce σ_b. In this connection, it should be noted that there is a trade-off between σ_w and σ_b. Figure 8.23 shows that smaller σ_w permits higher σ_b for any given value of W. However, the advantage of reducing σ_w is limited to a small region in which the Level 2 specification dominates.

For a P_p index of 2, corresponding to six sigma quality, Figure 8.24 shows the feasible regions for σ_w (or % CV_w) and σ_b (or % CV_b) for various values of W (or % W/T).

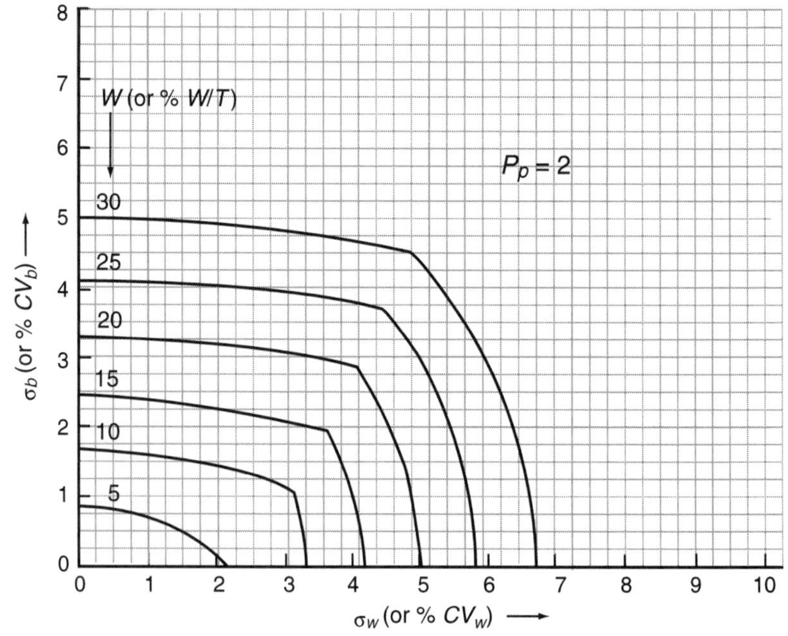

FIGURE 8.24. Feasible region for σ_w and σ_b (multilevel specification).

9

Measurement Systems Analysis

Measurements are often inexact. Decisions have to be made in the presence of measurement variability. Therefore, it is necessary to understand the extent and causes of measurement variability to assess if the measurement system is satisfactory for the task and, if not, how to improve it.

This chapter begins by defining the statistical properties of measurement systems. These properties are called stability, adequacy of measurement units, bias, repeatability, reproducibility, linearity, and constancy of variance. The detrimental effects of measurement variation are then considered, followed by acceptance criteria to judge the adequacy of a measurement system. The design and analysis of calibration studies, stability and bias studies, repeatability and reproducibility studies, intermediate precision and robustness studies, linearity studies, and method transfer studies are presented with examples to illustrate the assessment and improvement of measurement systems. Finally, an approach to determine the appropriate number of significant figures is presented.

9.1 STATISTICAL PROPERTIES OF MEASUREMENT SYSTEMS

A measurement system is a collection of gauges and equipment, procedures, people, and environment used to assign a number to the characteristic being measured. As shown in Figure 9.1, it is the complete process used to obtain measurements. When a measurement system is viewed as a process, it is easy to see the many causes of variation that lead to inexact measurements. An ideal measurement system is one that always produces a correct measurement (i.e., a measurement that agrees with a master standard) and may be said to have zero bias and zero variance. The statistical properties discussed below provide an assessment of how close a measurement system is to the ideal. They also provide guidance on how to improve the measurement system.

Stability. A measurement system is said to be stable if it is not affected by special causes over time. Stability of a measurement system is judged by a control chart of repeated measurements of a standard over a long duration of time. If the resultant average and range charts, or the individual and moving range charts, are in control, the measurement system is said to be stable. It is not necessary to have a master standard to judge stability. A product whose true measurement does not change over time is sufficient.

Adequacy of Measurement Units. The ability of the measurement system to measure a characteristic to an adequate number of decimal places is called the adequacy of measurement units. The determination of how many decimal places are adequate depends upon the task at hand and will be discussed later in this chapter.

Bias. Measurement system bias or inaccuracy is the systematic difference between the results of the measurement system and a master measurement method or a master standard. To determine bias, a master standard or a master measurement method must be available. Another possibility is to measure the same product several times using the best

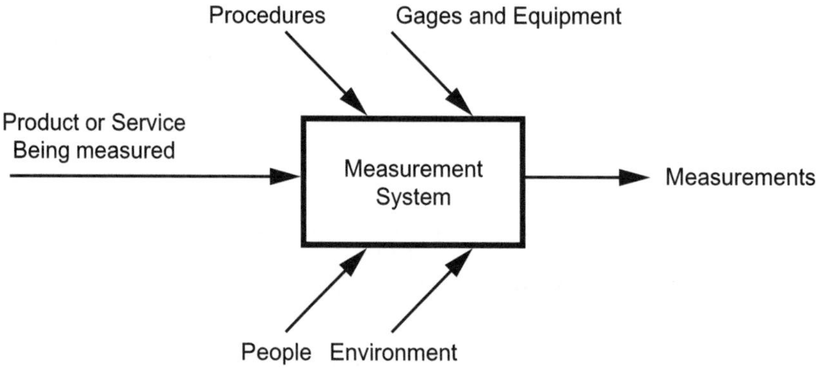

FIGURE 9.1
Measurement process.

available method and treat the average value as a master standard. Figure 9.2 shows the distribution of a large number of measurements of a standard for operator 1. For operator 1, bias is the difference between the average result and the master value.

Repeatability. As shown in Figure 9.2, repeatability is the variability in repeated measurements of the same product under identical operating conditions, i.e., using the same gauge, operator, and environment. The standard deviation of repeated measurements is a measure of repeatability. A reference standard is not necessary to estimate repeatability.

When a measurement system has poor repeatability, it means that some key factor in the measurement process is not being precisely repeated. For example, if the measurement process requires mixing, perhaps the mix time is not exactly the same in each repetition. Studies intended to assess the impact of such procedural factors are sometimes called robustness studies. Such studies can identify key factors that must be controlled to improve repeatability.

Reproducibility. As shown in Figure 9.2, the operator-to-operator variability, namely, the variability in the average of measurements made by different operators using the same gauge and product, is referred to as reproducibility. A reference standard is not necessary to measure reproducibility. Using variance components, the between-operator standard deviation can be estimated from the measurement of the same product by different operators and is a measure of reproducibility.

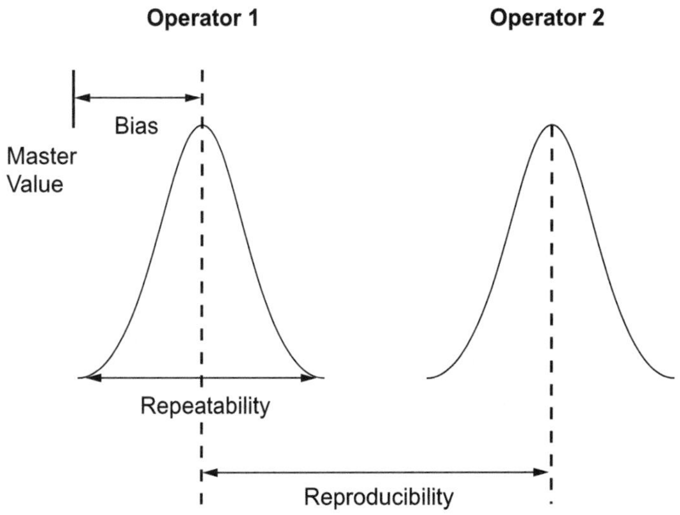

FIGURE 9.2 Bias, repeatability, and reproducibility.

We want a measurement system to be reproducible from operator to operator and also from day to day, laboratory to laboratory, and so on. The above definition of reproducibility can be generalized to include nonprocedural factors besides operators such as environment, instruments, and laboratories. Studies intended to assess the impact of such nonprocedural factors are sometimes called intermediate precision studies.

Linearity and Constancy of Variance. The bias and variability of the measurement system may or may not be constant throughout the operating range of the measurement system. If the bias is constant throughout the operating range, the measurement system is said to be linear. Similarly, if the variability remains constant throughout the operating range, the measurement system is said to have constant variance.

Linearity and constancy of variance may be determined by repeatedly measuring standards or products with known characteristic values that span the operating range of the measurement system. For each value of the standard, measurement bias can be determined, leading to a regression equation:

$$\text{Bias} = m(\text{standard value}) + b$$

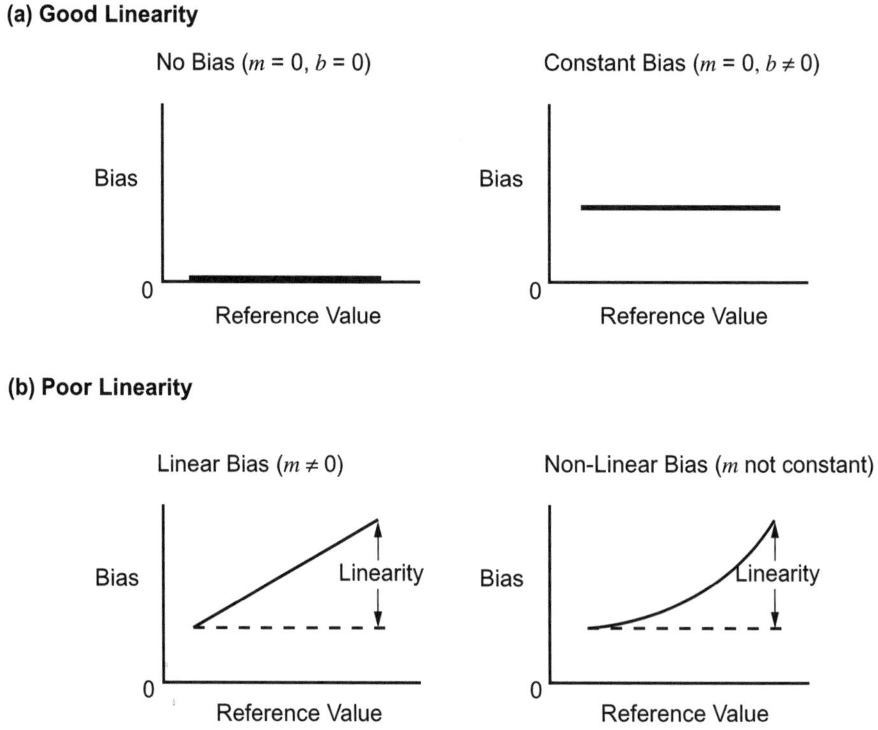

FIGURE 9.3
Linearity.

where m represents the slope and b is the intercept of the regression line. Linearity is the largest difference in bias values throughout the operating range, and for a fixed value of m, linearity may be quantified as

$$\text{Linearity} = |m| \text{ (operating range)}$$

Figure 9.3 graphically illustrates the differences between good and poor linearity.

Sometimes, the variability of the measurement system changes over the operating range. This is particularly so if the operating range is very large. The error often increases with the average value being measured, namely, the percent error is constant. Such data may require weighted least squares for calibration purposes or may need to be transformed, for example, by taking a log transform, prior to analysis.

9.2 ACCEPTANCE CRITERIA

The various statistical properties of measurement systems, described in the last section, ultimately influence measurement bias and variability. How much measurement bias and variability are acceptable? The answer depends upon the purpose of measurement and the detrimental effects of bias and variability.

Measurements serve two key purposes: improving products and processes, and demonstrating acceptability of products and processes. Control charts and various statistical studies such as comparative experiments, designed experiments and variance component studies are examples of using measurements for improvement purposes. Estimating process capability and lot-by-lot acceptance sampling are examples of using measurements for acceptance purposes. In making these assessments, it should be remembered that the observed variance is the sum of the true product-to-product variance and measurement variance as shown in Figure 9.4, and

$$\sigma_t^2 = \sigma_p^2 + \sigma_m^2 \tag{9.1}$$

where
σ_t^2 = total observed variance
σ_p^2 = true product-to-product variance
σ_m^2 = measurement variance

9.2.1 Acceptance Criteria for σ_m

The detrimental effects of measurement variability on control charts, statistical studies, process capability, and acceptance sampling plans

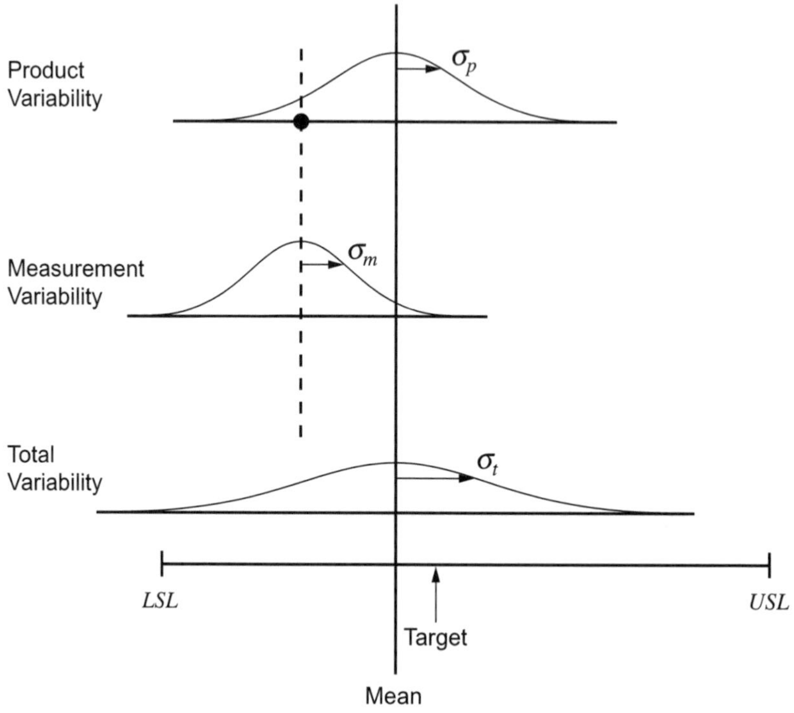

FIGURE 9.4 Product and measurement variability.

are now considered to develop acceptance criteria for measurement system standard deviation σ_m.

Control Charts. Large measurement errors increase the probability of wrong conclusions regarding process shifts and result in a need for larger subgroup size. To detect shifts or to make improvements, the measurement system must be able to discriminate between products. If the measurement variability is so large that the measurement system cannot statistically distinguish between products, it will be ineffective for process control.

The number of distinct categories into which products can be grouped using a single measurement per product are given by the following equation (Wheeler and Lyday, A.I.A.G.):

$$\text{Number of product categories} = \frac{1.41\sigma_p}{\sigma_m} \tag{9.2}$$

Clearly, the smaller the measurement error relative to product-to-product variation, the greater the number of categories. If the number of product (or data) categories is less than two, the measurement system cannot statistically distinguish between products on the basis of a single measurement per product. If the number of categories is two, then the measurement system is able to separate products into two groups, e.g., large and small. This is equivalent to an attribute measurement system. The number of categories should be at least three and preferably four or more for the measurement system to be acceptable. The acceptance criterion for σ_m may be derived as follows:

1. For the number of product categories to be three or more,

$$\frac{1.4\sigma_p}{\sigma_m} \geq 3$$

which, along with Equation 9.1, leads to

$$\sigma_m \leq \frac{\sigma_t}{2} \text{ (more correctly, } \sigma_m \leq 0.43\sigma_t)$$

2. For the number of product categories to be four or more,

$$\frac{1.4\sigma_p}{\sigma_m} \geq 4$$

which, along with Equation 9.1, leads to

$$\sigma_m \leq \frac{\sigma_t}{3}$$

Hence, one acceptance criterion for σ_m is

$$\sigma_m \leq \frac{\sigma_t}{2} \text{ and preferably less than } \frac{\sigma_t}{3} \tag{9.3}$$

Equation 9.3 implies that the measurement variance should be less than 25% of the total variance and should preferably be less than 10% of the total variance. It should be noted that the number of product categories in Equation 9.2 is based upon a single measurement per product. For nondestructive tests, if the product is measured n times and an average measurement is used, then the standard deviation of the average measurement will be $\sigma_m/\sqrt{n}$ and number of product categories will increase by a factor of $\sqrt{n}$. In this case,

$$\text{Number of product categories} = \frac{1.41\sqrt{n}\sigma_p}{\sigma_m}$$

Statistical Studies. These studies include experiments to estimate process mean and variance, compare two processing conditions, identify key factors for process improvement, implement control charts, and identify key sources of variation. Large measurement error can lead to wrong conclusions and results in larger sample size. The fact that the sample size n is proportional to the total variance σ_t^2 leads to the following conclusions that strongly support the acceptance criterion in Equation 9.3:

1. If $\sigma_m = 0$, then $\sigma_t = \sigma_p$ and $n \propto \sigma_p^2$. If $\sigma_m = \sigma_t/2$, then from Equation 9.1,

$$\sigma_t^2 = \left(\frac{\sigma_t}{2}\right)^2 + \sigma_p^2$$

or

$$\sigma_t^2 = 1.33\sigma_p^2$$

and

$$n \propto 1.33\sigma_p^2$$

This means that compared to the case of no measurement error, $\sigma_m = \sigma_t/2$ leads to a 33% increase in sample size and, correspondingly, an approximately equal increase in the cost of experimentation.

2. If $\sigma_m = \sigma_t/3$, then

$$\sigma_t^2 = 1.13\sigma_p^2$$

and compared to the case of no measurement error, $\sigma_m = \sigma_t/3$ leads to a 13% increase in sample size and cost.

Process Capability and Six Sigma. Large measurement variability leads to an underestimate of true process capability. If $\sigma_m = 0$, then $\sigma_t = \sigma_p$ and C_p = specification width/$6\sigma_p$. If $\sigma_m = \sigma_t/2$, then $\sigma_t = \sqrt{1.33}\sigma_p$ and C_p = specification width/$6.9\sigma_p$ so that C_p is underestimated by 15%. Similarly, if $\sigma_m = \sigma_t/3$, C_p is underestimated by 5%.

From another point of view, we do not want the measurement variability to get in the way of accomplishing demonstrated high quality levels, namely, a C_p index of at least 1.33, which corresponds to four sigma quality, or, for the six sigma objective, a C_p index of 2.

$$C_p = \frac{USL - LSL}{6\sigma_t} = 2$$

If $\sigma_m = \sigma_t/2$, then

$$C_{p\text{measurement}} = \frac{USL - LSL}{6\sigma_m} = 4 \tag{9.4}$$

The six-sigma goal requires the measurement system C_p index to be at least 4. This requirement may be restated as follows. Let W represent half the width of the specification for a two-sided specification or the distance between process mean and specification limit for a one-sided specification, i.e.,

$$W = \frac{USL - LSL}{2}$$

for a two-sided specification and $W = |\text{specification limit} - \text{mean}|$ for a one-sided specification.

Then substituting W in Equation 9.4 results in

$$\begin{aligned} \sigma_m &\le \frac{W}{12} \quad \text{for } C_p = 2 \\ \sigma_m &\le \frac{W}{8} \quad \text{for } C_p = 1.33 \end{aligned} \tag{9.5}$$

Equation 9.5 shows the values of the measurement standard deviation necessary to achieve high quality levels relative to specifications.

Acceptance Sampling Plans. Measurement errors can lead to misclassification of products. Out-of-specification products may be recorded as in-specification and products within specification may be recorded as out-of-specification. Figure 9.5 shows how the probability of product acceptance changes as a function of the true value of the measured characteristic. If the measurement error is known, then a guard banding approach can be taken to develop an acceptance region such that the probability of accepting an out-of-specification product will be very

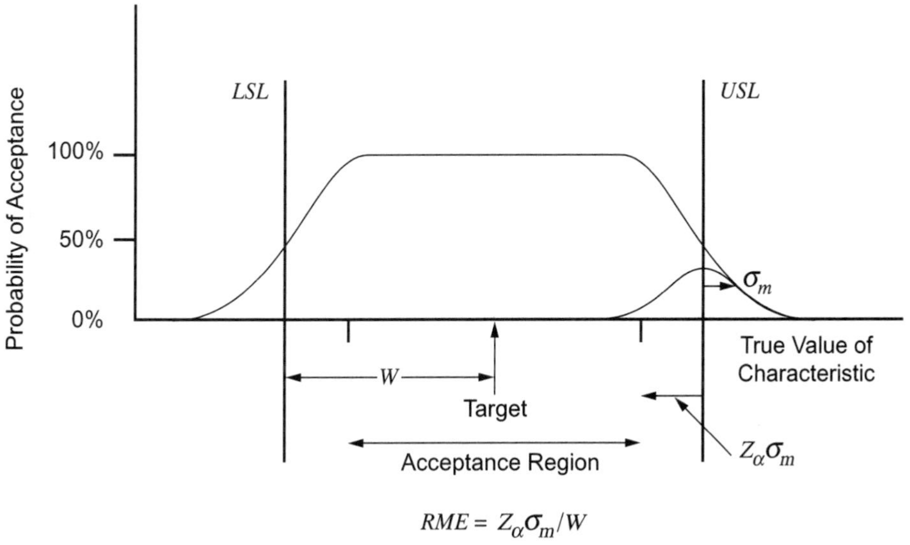

FIGURE 9.5 Relative measurement error (RME).

small. This approach also ensures a very low probability of an accepted product being classified as rejected upon retest. As shown in Figure 9.5, these acceptance limits may be set $Z_\alpha \sigma_m$ inside the specification limits. The percent relative measurement error (*%RME*) may be defined to be

$$\%RME = \frac{100\, Z_\alpha \sigma_m}{W} \tag{9.6}$$

%RME denotes the percent of the specification allocated to the measurement error. For $\alpha = 5\%$, $Z_\alpha = 1.64$, and for $\sigma_m = W/12$, $\%RME = 14\%$. For $\sigma_m = W/8$, $\%RME = 20\%$.

What *%RME* is acceptable? The answer partly depends upon the ease of getting precise measurements. In the metal parts industry, a recommended value is 10%, with 20% as a barely acceptable number and 30% to be definitely considered very poor. The σ_m suggested by Equation 9.5 leads to a *%RME* of 15% to 20%, which seems reasonable for moderately complex measurements.

For nondestructive tests, Equation 9.3 provides the acceptance criterion when the measurement system is to be used for improvement purposes and Equation 9.5 shows the acceptance criterion when the measurement system is to be used for specification-related purposes such as product acceptance. These criteria are summarized in Table 9.1.

σ_m includes all sources of variability that could occur in replicated measurements. For example, if replicated observations are taken over a long period of time, the observations may have been taken with different instruments, operators, environmental conditions, and so on, and their effects will appear as replication error. Therefore,

$$\sigma_m = \sqrt{\sigma^2_{\text{repeatability}} + \sigma^2_{\text{reproducibility}} + \sigma^2_{\text{other}}} \tag{9.7}$$

where σ^2_{other} may include instrument-to-instrument variability, environmental variability, etc.

9.2.2 Acceptance Criteria for Bias

Acceptance criteria for bias apply to the maximum bias within the operating range of the measurement system. Following the requirements

TABLE 9.1 Acceptance Criteria for σ_m (Nondestructive tests)

Purpose	Acceptable	Preferred
Improvement-related	$\sigma_m < \frac{\sigma_t}{2}$	$\sigma_m < \frac{\sigma_t}{3}$
Specification-related	$\sigma_m < \frac{W}{8}$	$\sigma_m < \frac{W}{12}$

TABLE 9.2 Acceptance Criteria for Bias (Nondestructive tests)

Purpose	Acceptable	Preferred
Improvement-related	$\lvert \text{Bias} \rvert < \dfrac{\sigma_t}{2}$	$\lvert \text{Bias} \rvert < \dfrac{\sigma_t}{3}$
Specification-related	$\lvert \text{Bias} \rvert < \dfrac{W}{8}$	$\lvert \text{Bias} \rvert < \dfrac{W}{12}$

of six-sigma quality, which permits the process mean to be centered within $\pm 1.5\sigma_{\text{short}}$ of the target, the acceptance criterion for bias may be stated as

$$\text{Bias} \leq 1.5\,(\text{acceptable } \sigma_{\text{repeatability}}) \tag{9.8}$$

If we assume acceptable $\sigma^2_{\text{repeatability}}$ to be half of σ^2_m, then $\sigma_{\text{repeatability}} = 0.7\sigma_m$ and Bias $\leq \sigma_m$, which leads to the acceptance criteria for bias summarized in Table 9.2.

9.2.3 Acceptance Criteria for Measurement Units

A measurement system reports the measured values to a certain number of decimal places. If measurements are recorded to the nearest hundredth of a gram, the reported values may be 1.02, 1.03, 1.04, etc. In this case, the measurement unit, denoted by d, is the difference between consecutive reportable values and equals 0.01. Is this measurement unit adequate or do we need to measure to a greater number of decimal places?

The effect of the measurement unit is to produce rounding errors. Rounding is the difference between the true value and the nearest reportable value. If the true value is 1.02357, it will be reported as 1.02 and will result in a rounding error of 0.00357. Any true value that is within $\pm d/2$ away from a reportable value will be rounded to that reportable value. Rounding errors have a uniform (rectangular) distribution with zero mean and width d. For such a uniform distribution, the standard deviation of rounding error is (Johnson and Kotz)

$$\sigma_{\text{rounding}} = \frac{d}{\sqrt{12}} = 0.2887d$$

Rounding error is one of the many sources of measurement error and it should be a small portion. If we somewhat arbitrarily assume σ_{rounding} to preferably be less than $\sigma_m/3$ or be at least less than $\sigma_m/2$, then, with acceptable σ_m ranging from $W/12$ to $W/8$ (see Table 9.1), we obtain the following acceptance criteria for d:

1. $\sigma_{\text{rounding}} < \frac{\sigma_m}{3}$ with $\sigma_m = \frac{W}{12}$ leads to $d < \frac{W}{10}$

2. $\sigma_{\text{rounding}} < \frac{\sigma_m}{2}$ with $\sigma_m = \frac{W}{8}$ leads to $d < \frac{W}{5}$

Hence, one acceptance criterion for the measurement unit d is

$$d < \frac{W}{5} \text{ or preferably less than } \frac{W}{10} \tag{9.8}$$

This means that the measurement unit should be small enough to divide the distance between process mean and specification limit into at least five regions and, preferably, 10 or more regions.

As one example, if the specification is 10 ± 1, then $W = 1$ and d should be at least 0.2 and preferably less than 0.1 for specification-related applications. As another example, if the characteristic values are being reported as whole numbers, then $d = 1$, and $\sigma_{\text{rounding}} = d/\sqrt{12} = 0.29$. If the measurement standard deviation was estimated to be 0.35, then rounding is the largest source of measurement error and reporting the values to one decimal place should significantly reduce the measurement variation.

Measurement units can also have a significant impact on control charts (Wheeler and Lyday) in terms of creating artificial out-of-control points. For $\overline{X}$ and R charts, this may be seen as follows. Recall that $\overline{R} = d_2\sigma_{\text{short}}$, where σ_{short} is the short-term standard deviation of the process. For small subgroup sizes $d_2 \approx 1$ and $\overline{R} \approx \sigma_{\text{short}}$. If $d = \sigma_{\text{short}}$, then many ranges will be zero and $\overline{R}$ will decrease, thus tightening all control limits. Simultaneously, the greater discreteness of measurements will spread out the running record of the data, causing artificial out-of-control points. Therefore, another acceptance criterion for d is

$$d < \sigma_{\text{short}} \tag{9.9}$$

Based upon the design of range charts in Chapter 3, it follows that for the range chart, $(UCL - LCL)/\sigma_{\text{short}}$ is estimated to be equal to $d_2 + 3d_3$ for $n < 7$ and equal to $6d_3$ for $n > 7$. Substituting the values of d_2 and d_3 from Appendix G, for the usual subgroup sizes, Equation 9.9 generally translates to having five possible values of range inside the range chart control limits (Wheeler and Lyday). The acceptance criteria for measurement units are summarized in Table 9.3.

9.2.4 Acceptance Criteria for Destructive Testing

Repeated measurements of the same product are not possible with destructive tests and $\sigma_{\text{repeatability}}$ is confounded with product-to-product

TABLE 9.3 Acceptance Criteria for Measurement Units

Purpose	Acceptable	Preferred
Improvement-related	$d < \sigma_{\text{short}}$	
Specification-related	$d < \frac{W}{5}$	$d < \frac{W}{10}$

variability. A direct estimate of repeatability cannot be obtained. The usual approach is to try to create a situation in which the products (or samples) being tested are as homogeneous as possible so that the tests could be interpreted as repeated measurements of the same product, producing an estimate of $\sigma_{\text{repeatability}}$. Note that under the assumption of random sampling, $\sigma_{\text{reproducibility}}$ can be estimated for destructive tests. If this approach works, then the acceptance criteria in Tables 9.1–9.3 will apply.

Examples of the above approach include homogenizing the product prior to taking samples, creating homogeneous products with known characteristic values, measuring completely different products that are very homogeneous, and selecting consecutively produced products because they may vary little from each other.

What if such solutions could not be found? Then σ_m^2 would include product-to-product variability and would equal σ_{short}^2. If the target values of C_p are in the range of 1.33 to 2, then an acceptance criterion is

$$\sigma_m = \sigma_{\text{short}} < \frac{W}{4} \quad \text{to} \quad \frac{W}{6}$$
$$|\text{Bias}| < \frac{W}{4} \tag{9.10}$$

We now consider studies necessary to validate and use a measurement system. These studies are aimed at demonstrating the adequacy of the measurement system and identifying areas of potential improvement. They are known as calibration studies, stability and bias studies, repeatability and reproducibility studies, intermediate precision and robustness studies, linearity studies, and method transfer studies. We begin by considering the calibration study.

9.3 CALIBRATION STUDY

The purpose of a calibration study is to establish a relationship between the true value of the characteristic of interest and the measured response. For example, the characteristic of interest may be drug concentration and the measured response may be peak height ratio of drug and internal standard. For an unknown sample, the peak height ratio is determined and the calibration relationship is used to back-calculate the drug concentration.

Table 9.4 shows an example of a typical calibration study. Calibration standards range from 5 ng/mL to 3000 ng/mL, a rather broad range of concentration (x). Ten standards were analyzed, each in triplicate, and the resultant peak height ratio (y) responses are shown in Table 9.4. It is not necessary to have 10 standards for a calibration study. For ex-

TABLE 9.4 Calibration Study

Concentration (x)	Peak Height Ratio (y)	Unweighted Regression		Weighted Regression	
		$\hat{x}$	% RE	$\hat{x}$	% RE
5	0.048	12.6	151.8	4.8	–4.3
5	0.053	13.1	162.5	5.3	6.5
5	0.063	14.4	187.6	6.6	31.8
10	0.090	17.6	76.0	9.8	–1.5
10	0.089	17.4	74.2	9.7	–3.3
10	0.098	18.5	85.0	10.8	7.5
25	0.215	32.5	29.8	24.9	–0.5
25	0.215	32.5	29.8	24.9	–0.5
25	0.217	32.8	31.3	25.2	0.9
50	0.416	56.4	12.9	49.1	–1.8
50	0.411	55.9	11.8	48.6	–2.9
50	0.414	56.3	12.5	48.9	–2.1
100	0.795	101.7	1.7	94.9	–5.1
100	0.825	105.3	5.3	98.5	–1.5
100	0.828	105.7	5.7	98.9	–1.1
250	1.953	239.9	–4.0	234.6	–6.2
250	2.015	247.3	–1.1	242.0	–3.2
250	1.996	245.1	–2.0	239.8	–4.1
500	3.963	479.8	–4.0	477.0	–4.6
500	4.025	487.1	–2.6	484.5	–3.1
500	4.110	497.3	–0.5	494.8	–1.0
1000	7.942	954.7	–4.5	957.1	–4.3
1000	8.040	966.3	–3.4	968.8	–3.1
1000	8.502	1021.4	2.1	1024.6	2.5
2000	16.946	2029.0	1.4	2043.1	2.2
2000	17.028	2038.8	1.9	2053.0	2.7
2000	15.753	1886.7	–5.7	1899.2	–5.0
3000	23.971	2867.4	–4.4	2890.6	–3.6
3000	25.929	3101.0	3.4	3126.7	4.2
3000	25.852	3091.9	3.1	3117.5	3.9

ample, if the relationship between the true value and the measured response is known to be linear, two standards will do. Ideally, the two standards should be positioned at the edges of the range of interest. Furthermore, if it is also known that the linear relationship goes through the origin, a single standard, positioned at the end of the range of interest, will be sufficient. To assess whether the relationship is linear or not, three to five equispaced standards should be sufficient.

Unweighted Regression. Figure 9.6 shows a plot of the data with concentration on the x-axis and peak height ratio on the y-axis (note that the replicated observations are often too close to each other to see them as separate dots). A straight line can be fitted to the data using the regression analysis approach described in Chapter 3. Since each observation is given equal weight, this approach is also known as unweighted regression. The fitted equation is

$$y = -0.0575 + 0.00838x + e \qquad (9.11)$$

where
y = Peak height ratio
x = Drug concentration
e = Error

The intercept in Equation 9.11 is statistically insignificant, meaning that the true intercept may be zero. The slope is significant with 99.99% confidence. The R^2 value is 0.998.

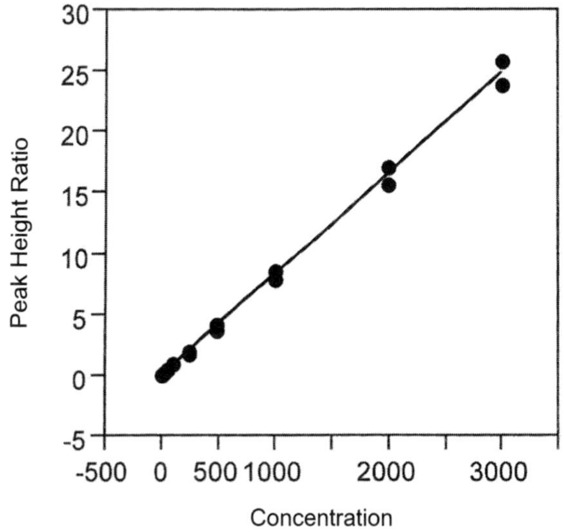

FIGURE 9.6 Peak height ratio versus concentration.

Equation 9.11 can be used to calculate the predicted drug concentration ($\hat{x}$) as follows:

$$\hat{x} = \frac{y + 0.0575}{0.00838} \tag{9.12}$$

If we know the peak height ratio for an unknown sample, the corresponding drug concentration can be determined from Equation 9.12. Table 9.4 shows these predicted drug concentrations based upon the observed peak height ratios. The percent relative error (*%RE*) is computed as

$$\%RE = 100\left(\frac{\hat{x} - x}{x}\right)$$

%RE is shown in Table 9.4 and provides a measure of error associated with the use of Equation 9.12. The *%RE* values in Table 9.4 for unweighted regression are very large for drug concentrations below 100 ng/mL. For a 5 ng/mL sample, the error exceeds 150%. Even though the regression equation has a very high R^2 value, the prediction errors have turned out to be unacceptably large.

Weighted Regression. Figure 9.7 shows a plot of errors (e) or residuals against concentration. The error variance increases as concentration increases, implying that the variance of peak height ratio increases as concentration increases. Since each observation does not have the same variance, each observation cannot be given equal weight and weighted regression is necessary. Weighted regression attaches a weight to each observation that is inversely proportional to the variance of the observation. Intuitively, this is as it should be because an observation with large variance tells us less about where the true value is compared to an observation with small variance. In unweighted regression, the coefficients are estimated using the criterion

$$\text{Minimize } \Sigma e_i^2 = \text{Minimize } \Sigma(y_i - \hat{y}_i)^2$$

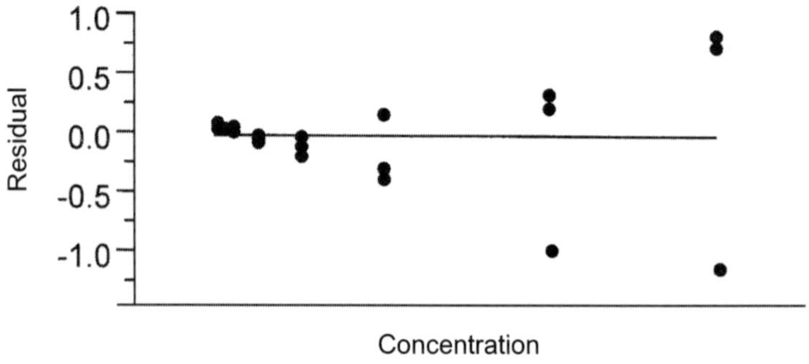

FIGURE 9.7 Residuals versus concentration.

In weighted regression the coefficients are estimated using the criterion

$$\text{Minimize } \Sigma w_i e_i^2 = \text{Minimize } \Sigma w_i (y_i - \hat{y}_i)^2$$

where $w_i = 1/\sigma_i^2$ and $\hat{y}_i$ is the estimated value of y_i. Presently, we assume the weights to be inversely proportional to x_i (similar results are obtained by assuming the weights to be inversely proportional to the square of x_i, which would be the case if *CV* were constant) and estimate the coefficients to

$$\text{Minimize } \Sigma[(y_i - \hat{y}_i)^2/x_i]$$

The fitted equation is

$$y = 0.00835 + 0.00829x + e \qquad (9.13)$$

The R^2 value is 0.9986. Given a value of y, the drug concentration x can be back-calculated as:

$$\hat{x} = \frac{y - 0.00835}{0.00829} \qquad (9.14)$$

These predicted values of drug concentration and the calculated *%RE* are shown in Table 9.4. The *%RE* is small throughout the entire range of drug concentration. For concentration less than 100 ng/mL, the *%RE* has decreased dramatically, whereas for higher concentrations there is a slight increase in *%RE*. Compared to Equation 9.12, Equation 9.14 provides a better calibration throughout the operating range.

9.4 STABILITY AND BIAS STUDY

The primary purpose of a stability study is to demonstrate that the measurement system is stable over time. The study consists of repeated testing of a single product over time by a single operator. $\overline{X} - R$ or $X - mR$ control charts are constructed to demonstrate stability. The data can also be analyzed to estimate short-term variation, which is a measure of repeatability. If the true characteristic value for the product is known, or if measurements of a known standard are taken, then bias can also be estimated. The calculated short-term variation and bias can be compared to the acceptance criteria to assess the initial acceptability of the measurement system. These ideas are now illustrated using an example.

Stability. At the beginning of each day, three independent measurements were taken on a master standard. Table 9.5 shows the deviations of the measured values from the standard. By considering the three dai-

TABLE 9.5 Measured Deviations from Standard

Day	Deviation from Standard			Day	Deviation from Standard		
1	–0.02	0.03	–0.03	11	–0.03	0.02	–0.05
2	–0.02	0.03	0.04	12	0.06	0.09	–0.03
3	0.01	0.07	0.09	13	0.02	–0.01	0.06
4	0.03	–0.03	0.01	14	0.05	–0.08	–0.05
5	0.02	0.04	–0.04	15	0.04	0.08	0.02
6	–0.05	0.05	0.03	16	–0.02	0.04	0.04
7	0.07	–0.03	–0.02	17	–0.04	0.07	0.03
8	0.02	–0.06	0.01	18	–0.03	0.02	0.04
9	0.02	0.07	–0.02	19	–0.09	0.01	–0.05
10	–0.03	–0.02	0.04	20	–0.02	0.02	0.03

ly measurements to be a subgroup, the data may be plotted as $\overline{X}$ and R control charts, as shown in Figure 9.8. Since no out-of-control rules are violated, the measurement system is said to be stable.

This concept of the stability of a measurement system has important consequences for measurement system calibration. When should this measurement system be recalibrated? If the system is recalibrated every day based upon the average daily variation from standard, then this recalibration is based upon common cause deviation and will only serve to increase the measurement error. Therefore, recalibration should be considered only when the control chart indicates a special cause. A second reason for recalibration is when the overall mean deviation from standard, estimated by $\overline{\overline{x}}$, significantly differs from zero, indicating a bias.

Repeatability. The standard deviation due to repeatability may be calculated as the short-term standard deviation $\overline{R}/d_2$:

$$\sigma_{\text{repeatability}} = \frac{\overline{R}}{d_2} = \frac{0.081}{1.693} = 0.0478$$

The specification for the product characteristic being measured was 10 ± 1. Also, from the control charts maintained by the production department, the total standard deviation was estimated to be $\sigma_t = 0.2$. Based upon this information, the acceptance criteria in Table 9.1 can be applied with the following results:

1. $\sigma_{\text{repeatability}}$ is much less than $\sigma_t/3$, hence the measurement system is initially judged to be satisfactory for improvement purposes.
2. $W = 1$ and $\sigma_{\text{repeatability}}$ is much less than $W/12$, hence the measurement system is initially judged to be satisfactory for specification-related purposes, including achieving six-sigma quality.

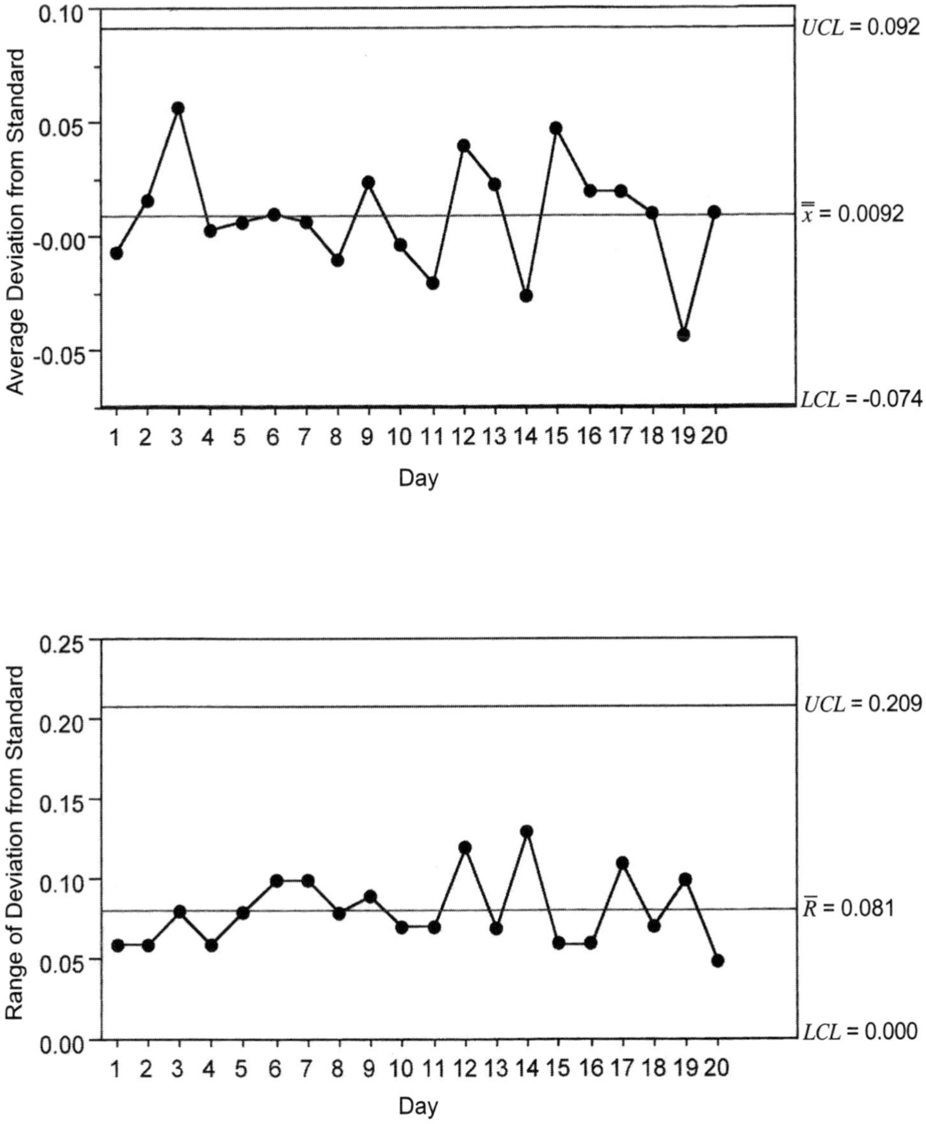

FIGURE 9.8 $\overline{X}$ and R charts for measurement stability.

These are initial judgements because $\sigma_{\text{repeatability}}$ does not include all sources of measurement variability included in σ_m.

Bias. Data in Table 9.5, which are differences from a master standard, are suitable for determining bias. The mean of 60 measurements is 0.0092 and represents the average difference between the standard and the results of the measurement system. Hence, bias = 0.0092. A 95% confidence interval for true bias may be constructed as

$$0.0092 \pm 2(0.0478/\sqrt{60}) = 0.0092 \pm 0.0123$$

This means that we are 95% sure that the true bias is somewhere between –0.0031 and +0.0215. Since zero is included in the confidence interval, the observed bias is statistically insignificant. The confidence interval also suggests that the true bias could be as large as 0.0215. The observed bias of 0.0092 and also its 95% limit of 0.0215 can be compared to the bias acceptance criteria in Table 9.2. Since the observed bias and its 95% confidence limit are smaller than $\sigma_t/3 = 0.067$ and $W/12 = 0.083$, the bias is acceptable.

Measurement Units. σ_{short}, the short-term standard deviation of the process, will not be less than $\sigma_{\text{repeatability}} = 0.0478$. Since the measurement unit $d = 0.01$, $d < \sigma_{\text{short}}$. Also, the specification was 10 ± 1, hence $W = 1$ and $d < W/10$. Therefore, the acceptance criteria for measurement units, shown in Table 9.3, are satisfied.

To illustrate the effect of inadequate measurement units if, instead of measuring to the second decimal place, measurements were made to one decimal place, the data would appear as in Table 9.6. The resultant $\overline{X}$ and R charts are shown in Figure 9.9. Note that compared to Figure 9.8, there are several zero ranges and $\overline{R}$ is reduced. There are only two possible values of range within R chart control limits. The control limits

TABLE 9.6 Results of Inadequate Measurement Units

Subgroup	Measured Deviations			Subgroup	Measured Deviations		
1	0.0	0.0	0.0	11	0.0	0.0	–0.1
2	0.0	0.0	0.0	12	0.1	0.1	0.0
3	0.0	0.1	0.1	13	0.0	0.0	0.1
4	0.0	0.0	0.0	14	0.1	–0.1	–0.1
5	0.0	0.0	0.0	15	0.0	0.1	0.0
6	–0.1	0.1	0.0	16	0.0	0.0	0.0
7	0.1	0.0	0.0	17	0.0	0.1	0.0
8	0.0	–0.1	0.0	18	0.0	0.0	0.0
9	0.0	0.1	0.0	19	–0.1	0.0	–0.1
10	0.0	0.0	0.0	20	0.0	0.0	0.0

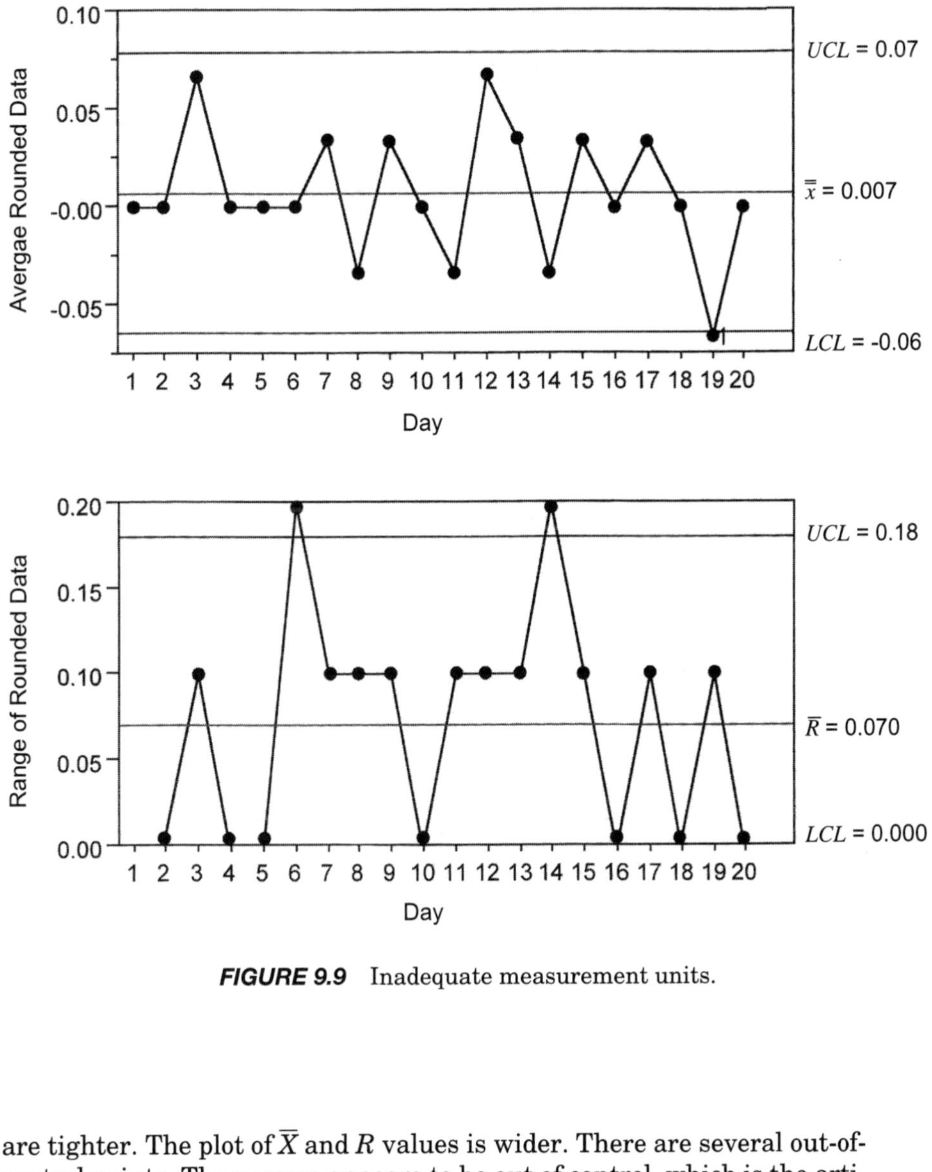

FIGURE 9.9 Inadequate measurement units.

are tighter. The plot of $\overline{X}$ and R values is wider. There are several out-of-control points. The process appears to be out of control, which is the artificial result of inadequate measurement units. In this case, $d = 0.1$ and exceeds measurement system $\sigma_{\text{short}} = 0.07/1.693 = 0.041$. Since $W = 1$, the criterion $d < W/10$ is just met. The implication is that the measurement unit of 0.1 is satisfactory for specification-related purposes but, depending upon the σ_{short} for the manufacturing processs, may not be satisfactory for improvement purpose, and is certainly not satisfactory for judging the stability of the measurement system.

9.5 REPEATABILITY AND REPRODUCIBILITY (R&R) STUDY

The purpose of conducting a measurement system R&R study is to assess the extent of measurement variability and the percent contributions of the various sources of variability to the total measurement variance. The structured studies discussed in Chapters 7 and 8 help in this regard. For example, if there is a single operator and multiple products to be measured, a nested study with duplicate measurements nested inside products may be appropriate. If there are multiple products and multiple operators, a two-variable crossed study may be used, and so on. What is necessary is to identify potential sources of measurement variation and then select the appropriate structured study to determine the variance contribution due to each source.

The following example illustrates the typical gauge repeatability and reproducibility study, using a "two variables crossed with duplicates" structure to estimate product-to-product variability, operator-to-operator variability, product-by-operator interaction, and test error (A.I.A.G.).

For the study, 10 products were randomly selected from the production line and measured twice by three different operators. The measurements were conducted in a random sequence. The collected data are shown in Table 9.7 after subtracting 99 from the observed values. In this study, products and operators are crossed and the duplicate measurements are nested.

Table 9.8 shows the analysis of variance (ANOVA) for this study. Product-to-product variability, operator-to-operator variability, and product-by-operator interaction are all highly statistically significant.

The variance components can be determined by equating the expected mean squares in Table 7.7 to the computed mean squares in Table 9.8. Products, operators, and duplicates are all random factors under the assumption that the three operators were randomly selected from a larger pool of operators. If the three operators were the only ones to take

TABLE 9.7 Data for Gauge R&R Study (Measurement – 99)

		Products									
Operator	Trial	1	2	3	4	5	6	7	8	9	10
A	1	0.64	1.02	0.87	0.85	0.55	0.97	1.06	0.84	1.11	0.65
	2	0.60	0.99	0.80	0.95	0.49	0.91	0.98	0.89	1.09	0.60
B	1	0.52	1.05	0.78	0.82	0.37	1.03	0.87	0.75	1.02	0.77
	2	0.55	0.94	0.75	0.76	0.39	1.05	0.95	0.72	0.95	0.74
C	1	0.48	1.02	0.89	0.88	0.48	1.08	0.98	0.69	1.02	0.85
	2	0.55	1.00	0.81	0.77	0.59	1.02	0.90	0.76	1.05	0.87

TABLE 9.8 ANOVA for Gauge R&R Study

Source	Degrees of Freedom	Sum of Squares	Mean Square	F Value	Percent Confidence
Product	9	1.9830	0.22000	111.1	99.9%
Operator	2	0.0341	0.01710	8.6	99.9%
Interaction	18	0.1510	0.00841	4.2	99.9%
Test	30	0.0595	0.00198		
Total	59	2.2276			

such measurements, then operators will be a fixed factor. Presently, all factors are assumed to be random. Then by equating the mean squares in Table 9.8 to their expectations in Table 7.7, the following equations are obtained, where p represents products, o represents operators, po represents the product-by-operator interaction, and e represents errors in duplicate measurements. In Table 7.7, $a = 10$ since there are 10 products, $b = 3$ since there are three operators, and $n = 2$ since there are two duplicates.

$$\sigma_e^2 + 2\sigma_{\text{po}}^2 + 6\sigma_p^2 = 0.2200$$

$$\sigma_e^2 + 2\sigma_{\text{po}}^2 + 20\sigma_o^2 = 0.01710$$

$$\sigma_e^2 + 2\sigma_{\text{po}}^2 = 0.00841$$

$$\sigma_e^2 = 0.00198$$

The simultaneous solution of these equations leads to the estimated variance components in Table 9.9.

The following conclusions may be drawn from this gauge R&R study:

1. The total variance (σ_t^2) is 0.04090. This partitions into product-to-product variance (σ_p^2) and measurement system variance (σ_m^2) as follows:

$$\sigma_t^2 = \sigma_p^2 + \sigma_m^2$$

TABLE 9.9 Variance Components from Gauge R&R Study

Source	Degrees of Freedom	Variance	Standard Deviation	Percent Contribution
Product	9	0.03530	0.1880	86.3%
Operator	2	0.00043	0.0208	1.1%
Interaction	18	0.00320	0.0567	7.8%
Test	30	0.00198	0.0445	4.8%
Total	59	0.04090	0.2020	100.0%

where $\sigma_p^2 = 0.03530$ and $\sigma_m^2 = 0.00562$ being the sum of the variances due to operators, product-by-operator interaction, and test method. Thus, 86% of the total variance is due to product-to-product differences and the remaining 14% is due to the measurement system. Note that it is the individual variances that add to the total variance. Standard deviations do not add to the total standard deviation.

2. Let us suppose that the specification for the measured characteristic is 100 ± 0.5. Since the data in Table 9.7 were reported after subtracting 99, the specification becomes 1.0 ± 0.5. Is the measurement system adequate? From Table 9.1, for improvement purposes, $\sigma_m < \sigma_t/2$ is acceptable, with $\sigma_m < \sigma_t/3$ being preferred. Presently, $\sigma_m = 0.075$ and $\sigma_t = 0.202$ so that the measurement system is acceptable for improvement purposes but not at the preferred level. For specification-related purposes, the criteria are $\sigma_m < W/8$ to be acceptable and $\sigma_m < W/12$ to be preferred. Presently, $W = 0.5$ and to be acceptable σ_m needs to be 0.0625. Hence, the conclusion is that the measurement system is acceptable for improvement purposes but not for specification-related purposes such as meeting the six-sigma goal.

3. Is the measurement unit acceptable? The measurement unit $d = 0.01$ For control chart applications, the within-subgroup standard deviation σ_{short} will be at least as large as the test standard deviation of 0.0445. Since $d < \sigma_{\text{short}}$ and also $d < W/10 = 0.05$, the measurement unit is adequate.

4. The total measurement variance can be further partitioned into variances due to repeatability and reproducibility, and from Table 9.9,

$$\sigma_m^2 = \sigma_{\text{repeatability}}^2 + \sigma_{\text{reproducibility}}^2$$

$$\sigma_{\text{repeatability}}^2 = 0.00198$$

$$\sigma_{\text{reproducibility}}^2 = 0.00043 + 0.00320 = 0.00363$$

The variance due to repeatability is the same as the test variance. The variance due to reproducibility is the sum of the operator and interaction variances. Both these components are due to the differences between operators and will disappear if a single operator is used to take measurements or if all operators measure alike. Repeatability accounts for 5% of the total variance and reproducibility accounts for 9% of the total variance, for the total measurement system variance of 14%.

5. Variance due to reproducibility is further divided into consistent

and inconsistent differences between operators. Consistent difference between operators measures the variability due to the true differences between the average measurements of each operator. In Table 9.9, the source of consistent differences is referred to as operator. If the differences between operators are not always consistent and depend upon the product being measured, then a product-by-operator interaction exists, referred to in Table 9.9 as interaction. For example, if for some products operator *A* measurements are higher than operator *B* measurements, but for other products the reverse is true, then this inconsistency signifies interaction.

Figure 9.10 shows a multi-vari chart for the measurement data. It is clear that there is large product-to-product variability.

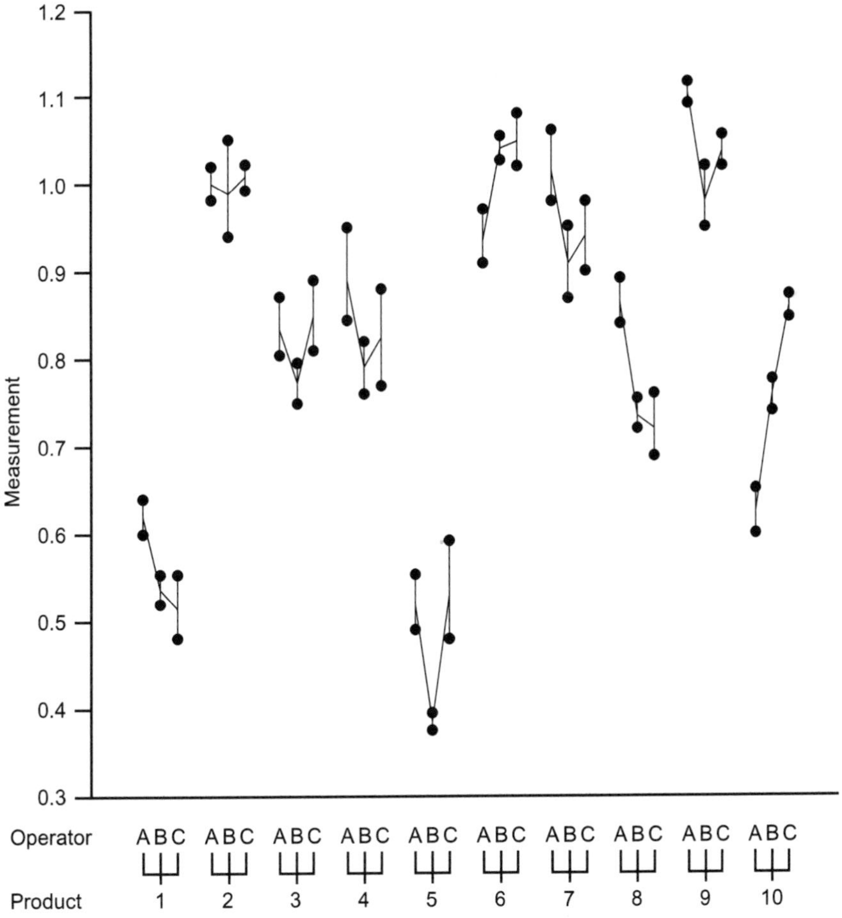

FIGURE 9.10
Multi-vari chart for measurement data.

The operator-to-operator variability within a product is clearly seen for product 5 and product 10. However, this operator-to-operator variability is not consistent from product to product. For example, for product 5, the measurements by operator A are higher than those by B. For product 10, measurements by operator B are higher than measurements by operator A. This inconsistency is called the product-by-operator interaction. From Table 9.9, the consistent variability between operators is 1% and the inconsistent variability is 8%, for a total reproducibility of 9% of σ_t^2.

6. If by providing further training regarding measurement procedures, by providing appropriate jigs and fixtures, etc., the consistent and inconsistent operator differences could be eliminated, then σ_m would become equal to 0.0445, the test–retest standard deviation. The reason for partitioning reproducibility into consistent and inconsistent differences is to help identify corrective actions. Generally, consistent operator differences are easier to correct than inconsistent differences.

   If σ_m reduces to 0.0445, the new σ_t will be 0.1931. W will continue to be 0.5. Now $\sigma_m < \sigma_t/3$ and the measurement system will work well for improvement purposes. Also, the new $\sigma_m < W/8$ and the improvement will be sufficient for specification-related purposes as well. Further optimization of the measurement system will not be necessary to reduce $\sigma_{\text{repeatability}}$.

Sensory Tests. The typical sensory tests in the food and other industries have a structure similar to the gauge R&R structure discussed above. Sensory tests usually involve multiple products, and multiple testers evaluate each product one or more times. If the sensory panel consists of a small number of testers to be routinely used, the testers should be considered fixed. The data could be analyzed to determine the variances due to products, testers, product-by-tester interaction, and duplicates. The sum of tester and interaction variances measures reproducibility and the duplicate variance measures repeatability. $\sigma^2_{\text{reproducibility}}$ provides a measure of how close the testers are to each other, namely, the effectiveness of prior training sessions.

If a single product is evaluated by k testers with n repeat measurements per tester, then the variance of the average of kn measurements is

$$\sigma^2_{\text{average}} = \frac{\sigma^2_{\text{reproducibility}}}{k} + \frac{\sigma^2_{\text{repeatability}}}{kn}$$

If a particular single tester is used, then

$$\sigma^2_{\text{average}} = \frac{\sigma^2_{\text{repeatability}}}{n}$$

These variances can be used to construct confidence intervals, determine sample size, and assess the detectable degree of difference between products.

9.6 ROBUSTNESS AND INTERMEDIATE PRECISION STUDIES

The purpose of these studies is to demonstrate that the measurement system is insensitive to routine changes in analytical factors. If measurements are susceptible to routine variations in analytical factors (or conditions), the analytical factors should be suitably controlled, or precautionary statements should be included. Ideally, the sensitivity of measurements to routine changes in analytical factors should be assessed during the design of the measurement system when it may be possible to change the design to make the measurement system less sensitive to such changes.

Uncontrolled factors that influence measurements are known as noise factors. A distinction may be made between internal and external noise factors, as is done in the pharmaceutical industry. Internal noise factors are associated with the method of measurement and lead to poor repeatability. Studies intended to evaluate the sensitivity of measurements to changes in internal noise factors are called robustness studies. In the case of liquid chromatography, examples of typical internal noise factors are:

1. Variation of pH in the mobile phase
2. Variation of mobile phase composition
3. Different columns
4. Temperature variation
5. Flow rate variation

External noise factors are external to the measurement method and lead to poor reproducibility. Studies intended to evaluate the sensitivity of measurements to changes in external noise factors are called intermediate precision studies. Examples include:

1. Differences between analysts
2. Day-to-day variation
3. Changes in ambient temperature and humidity

4. Differences between instruments
5. Differences between laboratories

Robustness and intermediate precision studies are best structured as designed experiments, particularly as factorial and fractional factorial designs. Such experiments can be analyzed to identify key noise factors and also to develop appropriate control schemes. This approach, using variance transmission analysis, was exemplified in Section 8.3 and is not repeated here.

9.7 LINEARITY STUDY

The purpose of a linearity study is to determine if measurement bias is constant over the operating range and, if not, the manner in which the bias varies over the operating range. To determine linearity, either products with known reference values or standards that cover the operating range of the measurement system are necessary. For each product, bias is determined as the difference between the measured value and the reference value. An equation relating bias to the reference value is derived using regression analysis and linearity can be estimated from the fitted equation.

Table 9.10 shows data from a typical linearity study. Five products with known reference values were measured four times each by a single operator. The products are equispaced over the operating range. Bias was calculated by subtracting the reference value from the measured value.

Figure 9.11 shows a plot of bias as a function of reference value. Bias is changing throughout the operating range. For small reference values, bias is positive, meaning that the measured value is larger than the true value. For large reference values, the reverse is true. The plot suggests a linear relationship between bias and reference value. The fitted equation using regression analysis is

$$\text{Bias} = 1.1005 - 0.2305\,(\text{Reference Value}) + e$$

Over the operating range from 3 to 7, estimated bias goes from +0.409 to −0.513. Linearity is estimated as

$$\text{Linearity} = (\text{slope})\,(\text{operating range}) = 0.2305\,(7 - 3) = 0.922$$

and can also be obtained as 0.409 − (−0.513) = 0.922. The maximum absolute value of bias over the operating range is

$$\text{Max}\,|\,\text{bias}\,| = 0.513$$

which can be compared to the acceptance criteria for bias.

TABLE 9.10 Data for Linearity Study

Product	Reference Value	Measurement	Bias
1	3	3.51	0.51
1	3	3.42	0.42
1	3	3.30	0.30
1	3	3.44	0.44
2	4	4.21	0.21
2	4	4.11	0.11
2	4	3.98	–0.02
2	4	4.25	0.25
3	5	4.91	–0.09
3	5	5.15	0.15
3	5	5.03	0.03
3	5	4.85	–0.15
4	6	5.82	–0.18
4	6	5.74	–0.26
4	6	5.66	–0.34
4	6	5.71	–0.29
5	7	6.34	–0.66
5	7	6.55	–0.45
5	7	6.41	–0.59
5	7	6.57	–0.43

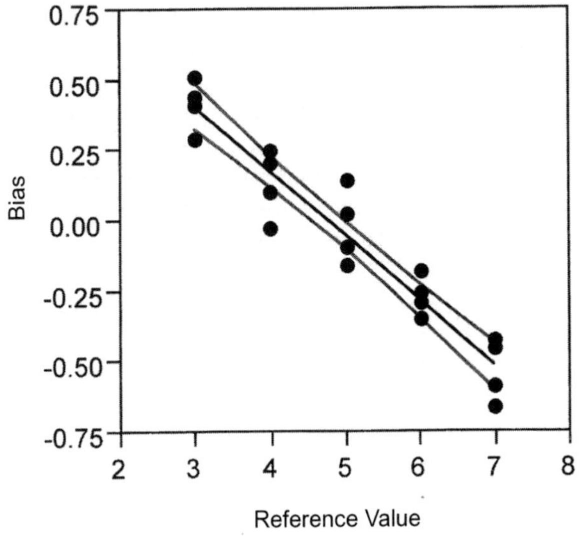

FIGURE 9.11 Linearity study.

The curved lines in Figure 9.11 show the 95% confidence limits for the regression line, from which the confidence limits for linearity and maximum bias can be determined. Should the relationship turn out not to be linear, curvilinear equations (e.g., quadratic) can be fitted and used to estimate linearity and bias over the operating range. Potential causes of curvilinearity such as improper calibration, worn instrumentation, and inaccurate reference values should be investigated and corrected wherever possible.

9.8 METHOD TRANSFER STUDY

Often, a measurement method is developed in R&D and is eventually transferred to production. A similar situation occurs when the same method is used in two different laboratories, perhaps the supplier and producer laboratories. The key question to be answered in method transfer is whether the method was correctly transferred, i.e., if the same products are measured in R&D and manufacturing, are the results identical in terms of the mean and variance? This question of comparing two populations was dealt with in Chapter 3 and is now answered for the following method transfer example.

Impurity Example. A method to measure the amount of a certain impurity, in mg per tablet, was developed in R&D. The one-sided upper specification limit for the impurity was 2 mg per tablet. Prior R&D results suggested a mean impurity level of approximately 0.4 mg per tablet with a standard deviation of approximately 0.02 mg per tablet. The method was transferred to production and a side-by-side comparison was planned to determine whether the method was correctly transferred. The test consisted of R&D evaluating a randomly selected set of n tablets and production evaluating another randomly selected set of n tablets, the test being a destructive test. The differences between R&D and production were primarily expected to be in the mean.

What should be the sample size n? From Chapter 3, n is given by

$$n = \frac{(\sigma_1^2 + \sigma_2^2)(Z_{\alpha/2} + Z_{\beta})^2}{\Delta^2}$$

where

σ_1 = standard deviation in R&D

σ_2 = standard deviation in production

α = probability of concluding that the method transfer is incorrect when it is correct

β = probability of concluding that the method transfer is correct when it is incorrect

Δ = smallest meaningful difference in mean that will make the method transfer incorrect

Presently, we assume that $\sigma_1 = \sigma_2 = 0.02$. The β risk has worse consequences than the risk. We assume $\beta = 0.01$ and $\alpha = 0.05$. The smallest meaningful difference Δ is taken to be 0.05 because a mistake as big as Δ is not likely to result in adverse consequences with respect to the specification:

$$n = \frac{(0.02^2 + 0.02^2)(1.96 + 2.33)^2}{(0.05)^2} = 6$$

It was decided to take 10 observations per laboratory. The results are in Table 9.11.

The data may be analyzed as follows:

1. We first conduct an F-test to determine if the two variances are different. This F-test is done prior to conducting the t-test for difference in means because if the variances turn out to be not different, then a more sensitive t-test can be done by pooling the two variances. Presently,

$$F = \frac{s_2^2}{s_1^2} = 1.53$$

   The critical value of F with 9 and 9 degrees of freedom and 95% confidence is 3.18. The conclusion is that the hypothesis of equal variance cannot be rejected.

TABLE 9.11 Method Transfer Study

R&D	Production
0.4430	0.4567
0.4295	0.4658
0.4332	0.4781
0.4139	0.4694
0.4606	0.4428
0.4350	0.4574
0.4240	0.4798
0.4335	0.4304
0.4608	0.4299
0.4427	0.4420
$\bar{x}_1 = 0.43762$	$\bar{x}_2 = 0.45523$
$s_1 = 0.01482$	$s_2 = 0.01836$

2. We can now pool the two variances and conduct a t-test to determine if the two means are statistically different. A pooled estimate of variance is

$$s^2_{\text{pooled}} = \frac{s_1^2 + s_2^2}{2} = 0.000278$$

and $s_{\text{pooled}} = 0.01668$. The calculated value of t is

$$t = \frac{\bar{x}_2 - \bar{x}_1}{s_{\text{pooled}}\sqrt{2/n}} = 2.36$$

The critical value of t is $t_{18,0.025} = 2.1$. The conclusion is that we are more than 95% sure that the two means are statistically different.

3. Even though the means have been shown to be statistically significantly different, the real question is how different are they? This is answered by constructing a confidence interval for difference in means. The 95% confidence interval for $(\mu_2 - \mu_1)$ is

$$(\bar{x}_2 - \bar{x}_1) \pm 2.1(s_{\text{pooled}}\sqrt{2/n}) = 0.00193 \text{ to } 0.03329$$

A practically meaningful difference was previously defined to be ±0.05. Since the entire confidence interval is contained within the meaningful difference, the conclusion is that although the difference in means is statistically significant, it is practically unimportant and the method transfer may be considered to be satisfactory as long as the definition of practically meaningful difference is correct.

9.9 CALCULATING SIGNIFICANT FIGURES

The number of significant figures in a measured quantity is defined to be the number of certain digits plus one estimated digit. If a number is measured as 1.234, it has four significant figures. The first three digits are certain and the last digit is estimated. 1.02 has three, 0.021 has two (initial zeros do not count), and 4.1 also has two significant figures. In the following, we assume that when a number is measured as 4.1, the uncertainty associated with this number is ±0.05 so that numbers between 4.05 and 4.15 have been rounded off to 4.1. Questions arise regarding the number of significant figures a calculated answer should be reported to. If we add all the above numbers, or subtract, multiply, or divide them, to how many significant figures should the answer be reported? Or alternately, to how many decimal places should the answer be reported? In general, if x_i denotes the measured numbers, each measured to some number of significant figures, and if y is any arbitrary function of x_i (such as $y = \log x_i$), then to how many significant figures (or decimal places) should y be reported? Some rules are often used to deal with cases such as additions and multiplication. For addition and subtraction, the rule is that the number of decimal places in the final result should equal the number of decimal places in the number with the least number of decimal places. A product or quotient may have only as many significant figures as the measurement with the fewest significant figures. A general approach may be formulated as follows by recognizing that the essential question has to do with how the uncertainty in x_i is transmitted to y. The approach is based upon an understanding of two statistical tools: distribution of rounding errors and variance transmission analysis.

1. If x_i is reported to be 4.1, then the reportable measurement unit $d_i = 0.1$, with the measurement uncertainty being $\pm d_i/2$. When the true values are rounded off to 4.1, the round-off errors have a uniform distribution with zero mean and width d_i. The variance of such a distribution is given by (Johnson and Kotz)

$$\sigma_{xi}^2 = \frac{d_i^2}{12}$$

2. If y is an arbitrary function of $x_1, x_2 \ldots x_n$ independent measurements,

$$y = f(x_1, x_2 \ldots x_n)$$

then using variance transmission analysis and assuming local linearity,

$$\sigma_y^2 \approx \left(\frac{dy}{dx_1}\right)^2 \sigma_{x1}^2 + \left(\frac{dy}{dx_2}\right)^2 \sigma_{x2}^2 \ldots$$

3. The uncertainty in y may be approximated by $\pm 2\sigma_y$ and determines the permissible number of decimal places and hence significant figures for y.

Consider the following examples to illustrate these ideas.

Example 1. $y = \ln x$. If $x = 24$, then $y = 3.178$. To how many significant figures (or decimal places) should y be reported? x has two significant figures. $d = 1$, hence, $\sigma_x = 1/\sqrt{12}$.

Since $(dy/dx) = 1/x$,

$$\sigma_y \approx \left(\frac{dy}{dx}\right)\sigma_x = \left(\frac{1}{x}\right)\frac{1}{\sqrt{12}} = \frac{1}{24\sqrt{12}}$$

and $\pm 2\sigma_y = \pm 0.024$.

The relationship between the uncertainty in y approximated by $\pm 2\sigma_y$ and the correct number of decimal places for y is summarized in Table 9.12.

Presently, since the uncertainty in y is $\pm$ 0.024, y should be reported to one (at most two) decimal places, i.e., it should be reported as 3.2 (two significant figures).

Example 2. % Difference is defined to be

$$\% \text{ Difference} = 100\left[\frac{\text{Observed value} - \text{Actual value}}{\text{Actual value}}\right]$$

or

$$y = 100\left[\frac{x_1 - x_2}{x_2}\right] = 100\left(\frac{x_1}{x_2} - 1\right)$$

If $x_1 = 5.182$ and $x_2 = 5.154$, to how many significant figures should % Difference be reported?

$$\sigma_y^2 \approx \left(\frac{100}{x_2}\right)^2 \sigma_{x_1}^2 + \left(-\frac{100x_1}{x_2^2}\right)^2 \sigma_{x_2}^2$$

Since x_1 and x_2 are approximately five,

$$\sigma_y^2 \approx (20)^2\,\sigma_{x_1}^2 + (20)^2\,\sigma_{x_2}^2$$

Also, since $d = 0.001$,

$$\sigma_{x_1}^2 = \sigma_{x_2}^2 = \frac{(0.001)^2}{12}$$

Hence,

$$\sigma_y^2 \approx \frac{800(0.001)^2}{12}$$

and

$$\pm 2\sigma_y \approx \pm 0.016$$

TABLE 9.12 Uncertainty and Reportable Decimal Places

Uncertainty in $y = \pm 2\sigma_y$	Number of Decimal Places for y
± 0.5	0
± 0.05	1
± 0.005	2
± 0.0005	3
± 0.00005	4

Based upon Table 9.12, y should be reported to no more than two decimal places. The calculated value of y is 0.54327%, which should be rounded off to 0.54%. Note that the original data have four significant figures but the final answer has at most two significant figures.

Example 3. $\bar{x}$ is the average of 100 observations, each observation being on the order of 4.1 with two significant figures and one decimal place. To how many significant figures should $\bar{x}$ be reported?

In general, if there are n observations, then

$$\bar{x} = \frac{x_1}{n} + \frac{x_2}{n} + \ldots + \frac{x_n}{n}$$

Hence

$$\sigma_{\bar{x}}^2 = \frac{\sigma_x^2}{n} = \frac{d^2}{12n}$$

Presently, $d = 0.1$ and $n = 100$. Hence,

$$\sigma_{\bar{x}}^2 = \frac{(0.1)^2}{1200}$$

and

$$\pm 2\sigma_{\bar{x}} = \pm 0.006$$

From Table 9.12, $\bar{x}$ should be reported to two decimal places and will have three significant figures, one more than in the data.

The above approach can be used by engineers and technical personnel to calculate the appropriate number of decimal places and significant figures. The answers need to be translated in an easy-to-follow fashion for implementation by operators; for example, by indicating the appropriate significant figures using spaces (-. - -).

10

What Color Is Your Belt?

This chapter consists of a test and answers to the test. You could use this test as a pretest, namely, take the test before reading this book. If you missed a few (or many) answers, do not be discouraged, because this book is for you. Please either read this book from beginning to end or read those portions of the book for which you missed the answers. The sequence of questions generally follows the sequence in this book. You could then take the test again until you are able to answer all questions correctly.

One aspect of the six sigma movement is the training of individuals to green belt, black belt, and master black belt levels. Some companies have even considered additional colors! These belts represent different levels of expertise, and different roles and responsibilities. A green belt, working with data in an R&D and manufacturing environment, should be able to answer most of the test questions correctly and should be able to make simple applications. A black belt should not only be able to answer the test questions correctly but also be able to apply the statistical methods to solve more difficult industrial problems. A master black belt would be expected to be able to teach these methods, get people enthused about the use of these methods, and answer new questions that may arise.

This multiple-choice test contains 64 questions. Please do not guess. Before you answer a question, make sure that you truly do know the answer and can explain it to others. There is only one correct answer for each question.

Good luck!

10.1 TEST

1. The heights of three randomly selected trees in a certain area turned out to be 4, 5, and 6 feet. What is the estimated sample variance?
   a. 1
   b. 2
   c. $\sqrt{2}$
   d. 4
   e. Data are too few to calculate variance
2. A doctor prescribed two tablets per day for a patient. The drug content of each tablet has a normal distribution with $\mu = 50$ mg and $\sigma =$ 4 mg. What is the standard deviation of drug taken by the patient per day?
   a. 4
   b. $2\sqrt{2}$
   c. $4\sqrt{2}$
   d. 8
   e. $8\sqrt{2}$
3. What is the approximate sample size to estimate σ within 10% of the true value with 95% confidence?
   a. 10
   b. 50
   c. 100
   d. 200
   e. 500
4. The histogram of certain X values was positively skewed. Which of the following is **not** a likely reason for this positive skewness?
   a. Data are close to zero
   b. X could be particle size
   c. Sorting is taking place
   d. Successfully trying to maximize X
   e. Both a and b above
5. The standard deviation of individual values σ was known to be 4. The average of 4 observations turned out to be 10. What is the approximate 95% confidence interval for population mean μ?
   a. 10 ± 4
   b. 10 ± 8
   c. 10 ± 2
   d. $10 \pm 2\sqrt{2}$
   e. $10 \pm 4\sqrt{2}$
6. Seven candy bars were randomly selected from a production line and their thicknesses were measured. The sample average $\bar{x}$ was found to be 3 mm and the sample standard deviation s was calculated to be

0.1 mm. With 95% confidence, what is the approximate interval within which 95% of the candy bar thicknesses for the entire population of candy bars are expected to be?
a. 3 ± 0.1
b. 3 ± 0.2
c. 3 ± 0.3
d. 3 ± 0.4
e. 3 ± 0.5

7. Moisture content of a certain product was routinely measured as the difference between wet weight of the product (*X*) and dry weight of the product (*Y*). *X* had a normal distribution with $\mu_X = 100$ and $\sigma_X = 2$. *Y* had a normal distribution with $\mu_Y = 90$ and $\sigma_Y = 2$. What is the approximate coefficient of variation (*CV*) or relative standard deviation (*RSD*) for moisture?
a. 8%
b. 16%
c. 30%
d. 80%
e. 4%

8. If a fair coin is tossed four times, what is the approximate chance of getting two heads?
a. 30%
b. 40%
c. 50%
d. 60%
e. 70%

9. If the data (*X*) show that standard deviation of *X* is proportional to the mean of *X*, which is the right transformation to use prior to data analysis?
a. $Y = 1/X$
b. $Y = \sin X$
c. $Y = \sqrt{X}$
d. $Y = \log X$
e. $Y = X^2$

10. Raw material hardness is an important property of interest. There are two suppliers of this raw material and we have collected 10 hardness observations on material from each supplier. We wish to determine if the two suppliers meaningfully differ in terms of average hardness. What is the **best** way to do so?
a. Conduct a *t*-test
b. Conduct an *F*-test
c. Conduct a paired *t*-test
d. Check if the distribution is normal
e. Construct a confidence interval for difference in mean

11. To compare two treatments in terms of their mean, what is the ap-

proximate sample size per treatment if the difference in mean to be detected is equal to the standard deviation σ, the α risk is 5%, and the β risk is 10%?
a. 10
b. 20
c. 30
d. 35
e. 40

12. Old and new product designs are to be compared with respect to variability. We wish to know if the standard deviation for the new design is half that for the old design with an α risk of 5% and a β risk of 10%. How many observations are necessary per design?
a. 25
b. 40
c. 50
d. 60
e. 100

13. In interpreting a scatter diagram and the associated correlation coefficient between an input factor X and an output response Y, we do **not** have to worry about which of the following?
a. Correlation without causation
b. Stratification
c. Inadequate range of Y
d. Inadequate range of X
e. Parabolic curvature

14. Regression analysis was used to obtain the following equation relating response Y to input factors X_1 and X_2:

$$y = 100 + 3x_1 - 0.5x_2 + \text{error}$$

The percent confidence associated with both coefficients in the equation exceeded 95% and the R^2 value turned out to be 20%. Which of the following statements is **not** true?
a. X_2 has a small but statistically significant effect on Y
b. Measurement error must be small to find this small effect of X_2
c. X_1 has a statistically significant effect on Y
d. Some other factor is influencing Y
e. Measurement error may be large

15. A local newspaper stated that less than 10% of apartments did not allow renters with children. The city council conducted a random survey of 100 units and found 13 units that excluded children. Which of the following statements is true?
a. The newspaper statement must be wrong
b. The newspaper survey must be wrong
c. There is not enough data to decide if the newspaper statement is right or wrong

d. The newspaper statement is right
e. Both a and b above

16. For variable data, the assumption of normality is very important for which of the following tasks?
a. Designing an $\overline{X}$ chart
b. Estimating variance components
c. Conducting a t-test
d. Establishing a tolerance interval
e. Plotting a histogram

17. Which of the following concepts is **not** used in determining $\overline{X}$-chart control limits with a subgroup size of two?
a. Variance of $\overline{X}$ reduces inversely proportional to sample size
b. X is normally distributed
c. False alarm probability is approximately 3 in 1000
d. Between subgroup standard deviation is used to calculate the limits
e. The centerline is often taken to be the grand average

18. A moving average chart is an appropriate one to use when which of the following statements is true?
a. A small sustained shift in the mean is to be detected
b. The mean is expected to move considerably
c. Variability changes from subgroup to subgroup
d. Both b and c are true
e. Large sporadic shifts in mean are expected and data need to be smoothed out

19. Which of the following is not an out-of-control rule?
a. One point outside control limits
b. Six points in a row above the centerline
c. Fifteen points in a row within $\pm 1\sigma$ from the centerline
d. 2 out of 3 points more than 2σ away on the same side of the centerline
e. 14 points alternating up and down

20. Why is the np chart considered inappropriate for variable sample size?
a. Expected fraction defective will vary with sample size
b. The centerline will keep varying
c. Each subgroup will have a different amount of information
d. The number of defectives will change with sample size
e. None of the above

21. When may a modified limit chart be used?
a. The control limits are too narrow
b. The cost of correcting a process is high
c. The current process capability is high
d. Both b and c are true
e. The control limits are too wide

22. A job shop produces different products using the same manufactur-

ing process. Each product is produced over small time durations. The products have different target values and different variabilities. Which is the right control chart to use?

a. $\overline{X}$ and R
b. $\Delta\overline{X}$ and mR
c. X and mR
d. $\bar{Z}$ and W
e. None of the above

23. On a control chart, one plotted point had a value of +2. Which of the following charts is **not** the one being used?
    a. $\overline{X}$ chart
    b. c chart
    c. u chart
    d. np chart
    e. p chart
24. A manufacturer wishes to set up a control chart for workmanship defects on a refrigerator. For the past 20 days, different numbers of refrigerators were inspected each day and the number of defects found were recorded each day. Which is the appropriate chart to use?
    a. u chart
    b. p chart
    c. $\overline{X}$ chart
    d. c chart
    e. None of the above
25. A baseball player claimed that he was retiring because his batting average had eroded that year. His baseball statistics were available for the past 15 years in terms of the number of times at bat per year and the number of hits per year. In order to assess if his batting average has in fact eroded, which control chart should be used?
    a. $\overline{X}$ chart
    b. c chart
    c. p chart
    d. u chart
    e. np chart
26. Which of the following is **not** a consideration in defining a rational subgroup?
    a. Understanding the target value
    b. Minimizing within-subgroup variance
    c. Maximizing the opportunity for variation between subgroups
    d. Understanding the causes of variation
    e. Understanding the process
27. Why do variable charts occur in pairs (e.g., $\overline{X}$ and R) and attribute charts occur singly (e.g., p)?
    a. Variable data are more precise

b. Both mean and variance can be computed for variable data but not for attribute data
c. Variance is a deterministic function of mean for the attribute data
d. Attribute data have only two categories, so that by knowing one you know the other
e. None of the above

28. When should a properly designed control chart be redesigned?
a. When two points fall outside control limits
b. When a deliberate process change is made and the data indicate that the process has changed
c. When 20 subgroups worth of additional data is available
d. Both a and c above
e. The control limits should be updated as each new data point becomes available to keep the control limits current

29. Which is the **least** important factor in determining the sampling frequency?
a. Process average
b. Expected frequency of process shifts
c. Cost of sampling and testing
d. Process capability index
e. Lack of sufficient data on process characteristic being measured

30. How will you interpret the $\overline{X}$ chart when there are 15 or more points in a row within $\pm 1\sigma$ from the centerline?
a. The process variability has decreased
b. The subgroup consists of samples from distributions with different means
c. There is another factor causing the points to go consistently up and down
d. Could be a and/or b above
e. Could be a and/or c above

31. Why do you never put specifications on an $\overline{X}$ chart?
a. Specifications can be determined from the control limits
b. Specifications are unnecessary when you already have control limits
c. Such reference points are never used
d. Misleading conclusions may be drawn
e. None of the above

32. For an $\overline{X}$ chart with a subgroup size of 2, the lower control limit is 6 and the upper control limit is 14. If the subgroup size is increased to 8, what will be the control limits?
a. $LCL = 8$ and $UCL = 12$
b. $LCL = 6$ and $UCL = 14$
c. $LCL = 9$ and $UCL = 11$
d. $LCL = 4$ and $UCL = 16$
e. None of the above

33. In the context of an $\overline{X}$ chart, what does β risk mean?
    a. Risk of false alarm
    b. Risk of producing bad product
    c. Risk of over correcting the process
    d. Risk of concluding that the process is out of control when it is not
    e. Risk of not detecting a shift in process average
34. For a conventional $\overline{X}$ chart with 3-sigma control limits, what is the result of increasing the subgroup size?
    a. α risk increases and β risk decreases
    b. β risk increases and α risk decreases
    c. α risk decreases
    d. β risk decreases
    e. Anything is possible
35. Given the same data, five students computed the values of C_p, C_{pk}, P_p, and P_{pk} indices. Which computations are likely to be correct?

	C_p	C_{pk}	P_p	P_{pk}
a.	2	1.5	1.33	1.0
b.	1.5	1.7	1.42	1.0
c.	2.0	1.0	0.62	0.5
d.	1.82	1.0	1.47	1.2
e.	Both a and c above			

36. Given $C_p = 1.5$, $C_{pk} = 1.4$, and $P_{pk} = 0.7$, which corrective action should be taken?
    a. Stabilizing the process and centering the process are equally important.
    b. Stabilize, process is essentially centered
    c. Center, process is essentially stable
    d. Process needs to be fundamentally changed
    e. No action is necessary because C_p and C_{pk} are essentially equal and both exceed the target of 1.33.
37. Which of the following assumptions is necessary for the C_p index but not for the C_{pk} index?
    a. The sample size should be large
    b. The specification must be known
    c. The distribution should be normal
    d. The process must be stable
    e. The specification must be two-sided
38. What combination of values will yield $C_p = 1.0$?
    a. $\overline{X} = 10$ $\sigma = 3$ Tolerance = 12
    b. $\overline{X} = 20$ $\sigma = 1$ Tolerance = 6
    c. $\overline{X} = 20$ $\sigma = 2$ Tolerance = 6

d. $\overline{X} = 10$ $\sigma = 1$ Tolerance = 12
e. $\overline{X} = 20$ $\sigma = 2$ Tolerance = 16

39. If $LSL = 2$, $USL = 20$, Process Mean = 14, and $\sigma = 2$, what is the value of C_{pk}?
 a. $C_{pk} = 1.33$
 b. $C_{pk} = 2.0$
 c. $C_{pk} = 0.5$
 d. $C_{pk} = 1.5$
 e. C_{pk} = None of the above
40. C_{pk} was computed to be 1.0 on the basis of 10 observations. What is the 95% confidence interval for true C_{pk}?
 a. 0.5 to 1.5
 b. 0.8 to 1.2
 c. 0 to 2
 d. 0.9 to 1.1
 e. 0.2 to 1.8
41. What will you conclude if C_{pk} is negative?
 a. Standard deviation is negative
 b. Mean is negative
 c. Process mean is outside the specifications
 d. One of the specification limits must be negative
 e. Calculations are wrong because C_{pk} cannot be negative
42. Which of the following statements is **incorrect?**
 a. To be predictable, a process must be stable
 b. A stable process is a capable process
 c. A process may have $C_p = 1.0$ and $C_{pk} = 1.7$
 d. None of the above
 e. Both b and c above
43. If the process is improved by centering the mean and reducing variability, what will happen?
 a. C_p index will increase
 b. C_{pk} index will increase
 c. C_p and C_{pk} will be equal
 d. b and c above
 e. All of the above
44. From an $\overline{X}$ chart, variance within subgroup was found to be 2 and variance between subgroups was found to be 0.1. What action should be taken to significantly improve the process?
 a. Eliminate time trend
 b. Eliminate shift-to-shift differences
 c. Supplier needs to improve batch-to-batch variability
 d. Measurement system variability should be evaluated
 e. Remove cycling along the time axis

45. Analysis of a centered process showed the within-lot standard deviation to be $\sigma_w = 4$ and the between lot standard deviation to be $\sigma_b = 3$. The specification was 100 ± 20. What is the P_p index?
    a. Less than 1
    b. 1.33
    c. 1.67
    d. 2.00
    e. 2.22
46. Which of the following statements is true for a nested structure with factor B nested inside factor A?
    a. A is a random factor
    b. Both A and B must be fixed factors
    c. B is a random factor
    d. Levels of B must be different for each level of A
    e. Both a and c above
47. Ambient temperature is a random factor that influences a key output of interest. To find the variability of output due to varying ambient temperature, experiments were conducted on a very hot day and a very cold day, and the variance contribution of temperature was determined assuming it to be a random factor. Which of the following statements is true?
    a. Variance contribution due to temperature will be underestimated
    b. Variance contribution due to temperature will be overestimated
    c. Variance contribution due to temperature will be correctly estimated due to the central limit theorem
    d. Variance contribution due to temperature will be correctly estimated due to the randomness of temperature
    e. Variance contribution due to temperature will be correctly estimated because average temperature was close to the true average
48. A and B are fixed factors in a two-level factorial experiment, the two levels of each factor being -1 and $+1$. Analysis of the factorial experiment resulted in the following equation:

$$Y = 10 + 4A + 3B + \text{error}$$

    What is the variance component due to factor A?
    a. 40
    b. $8\sqrt{2}$
    c. 12
    d. 4
    e. 16
49. The specification for X is $T \pm \Delta$. For a centered process, the quadratic loss function for X is proportional to which of the following?
    a. $(X - T)^2$
    b. X^2

c. $1/X^2$
d. Δ^2
e. $(T + \Delta)^2$

50. Average quadratic loss due to a perfectly centered process is \$4 million/year. If the variance is reduced by a factor of 4, what will be the new average loss?
a. \$2 Million
b. \$4 Million
c. \$0.50 Million
d. \$0.25 Million
e. None of the above

51. There were three sources of variation in a structured study: batches of raw material, manufacturing process, and test method. Based upon a large amount of data, the grand average turned out to be 48 and the three variance components were estimated to be $\sigma^2_{\text{batch}} = 5$, $\sigma^2_{\text{manufacturing}} = 7$, and $\sigma^2_{\text{test}} = 4$. The specification was 50 ± 10. What is the P_{pk} index?
a. 1/3
b. 2/3
c. 1
d. 4/3
e. 5/3

52. Variance components analysis of a multilane process showed a large lane-by-time interaction. Lane was a fixed factor and time was a random factor. What is the interpretation of this large interaction?
a. The analysis is incorrect because a random factor cannot interact
b. Lane-to-lane differences are a function of time
c. There are consistent, large differences between lanes
d. There is a large time trend
e. None of the above

53. In order to determine differences between the three shifts, 100 products were evaluated from each shift. What is the structure for this study?
a. Shifts are nested inside products
b. Products are nested inside shifts
c. Shifts and products are crossed
d. Both a and b above
e. It does not matter because there are only two factors

54. A hamburger chain had five fast food restaurants. A different group of 10 randomly selected people were sent to each restaurant (a total of 50 people) and each person evaluated the quality of two hamburgers. What is the structure for this data collections scheme?
a. Restaurants and people are crossed and hamburgers are nested
b. All three variables are crossed

c. People and hamburgers are crossed but are nested inside restaurants
d. Hamburgers are nested inside people who are nested inside restaurants
e. None of the above

55. How will you interpret the following variance components table for an $\overline{X}$ chart?

Source	Degrees of Freedom	Variance	Percent Contribution
Between subgroups	20	0.1	25%
Within subgroup	80	0.3	75%
Total	100	0.4	100%

a. The total variance is small
b. The common cause variance is 0.3
c. The common cause variance is 0.1
d. Bulk of the variation can be eliminated by controlling the process
e. The degrees of freedom are insufficient

56. If W represents half the width of a two-sided specification, to achieve 6-σ quality, the standard deviation of measurement σ_m should be less than which of the following?
a. W
b. $W/2$
c. $W/4$
d. $W/8$
e. $W/12$

57. If σ_m represents measurement standard deviation and σ_t represents the total standard deviation of the process, for the measurement system to be acceptable for improvement purposes, σ_m should be less than which of the following?
a. σ_t
b. $\sigma_t/2$
c. $\sigma_t/4$
d. $\sigma_t/8$
e. $\sigma_t/10$

58. Suppliers X, Y, and Z have the same $P_{pk} = 1.33$. Supplier X has large measurement error, Y has medium measurement error, and Z has small measurement error. Who will you buy the product from?
a. Supplier X
b. Supplier Y
c. Supplier Z
d. All are equally good because they have the same P_{pk}
e. None. The P_{pk} does not meet the six-sigma target.

59. In a measurement study, the same six products were measured by three operators. Each operator made duplicate measurements on each product. What is the study structure?
 a. Duplicates are nested inside operators and operators are nested inside products
 b. Duplicates and products are crossed and operators are nested
 c. Products and operators are crossed and duplicates are nested
 d. Duplicates and operators are crossed and products are nested
 e. All three factors are crossed
60. A master standard is necessary to evaluate which of the following properties of measurement systems?
 a. Repeatability
 b. Adequacy of measurement unit
 c. Stability
 d. Reproducibility
 e. Bias
61. When should a measurement system be recalibrated (adjusted)?
 a. As often as economically feasible
 b. Whenever a different operator conducts measurements
 c. When the control chart of repeated measurements of a standard signals out of control
 d. At a predetermined fixed interval from a prevention viewpoint
 e. Very often if the measurement capability is low
62. In a measurement study, the interaction between the operators and products turned out to be the largest variance component. What is the right corrective action?
 a. Reduce measurement bias
 b. Repeat the measurement study with products that are more alike
 c. Provide training for the operators so they measure alike
 d. Optimize the measuring equipment to reduce inherent variation
 e. Improve the linearity of the measurement system
63. What indicates that the measurement system measures to an inadequate number of decimal places?
 a. The $\overline{X}$ chart shows a steady trend going up or down
 b. The $\overline{X}$ chart shows a small sustained shift
 c. There are only two possible values of range inside the range chart control limits
 d. Either a or b above
 e. The range chart shows cyclic behavior
64. If $y = x^{1/3}$, to how many significant figures should y be reported if $x =$ 4.8?
 a. 0
 b. 1
 c. 2

d. 3
e. 4

10.2 ANSWERS

The correct answers to the test questions are in parentheses following the question numbers; related discussions follow.

1. (a) Sample variance can be calculated so long as the sample size n is greater than one. Here $n = 3$ and since $\bar{x}$ is 5 feet,

$$s^2 = \frac{(4-5)^2 + (5-5)^2 + (6-5)^2}{(3-1)} = 1$$

The units of variance are (feet)^2.

2. (c) If X_1 and X_2 denote the drug content of the first and second tablets taken each day, respectively, then

$$\text{Variance}(X_1 + X_2) = \text{Variance}(X_1) + \text{Variance}(X_2) = 16 + 16 = 32$$

Hence, standard deviation of drug per day = $\sqrt{32} = 4\sqrt{2}$ mg. The key point is that it is the variances that add; standard deviations do not add arithmetically. Also, the assumption of normality is not necessary to compute variance, but it is useful in interpreting standard deviation.

3. (d) The sample size to estimate σ within δ% of the true value with 95% confidence is approximately given by

$$n = 2\left(\frac{100}{\delta}\right)^2$$

So for $\delta = 10\%$, $n = 200$. It takes a large number of observations to estimate σ precisely.

4. (d) Data are usually positive and if they are close to zero, the histogram could become positively skewed. Particle size often has a lognormal distribution that is positively skewed. If the lower portion of a normal distribution is sorted out, the remaining distribution will look positively skewed. On the other hand, if we are successful in maximizing X, the histogram may look negatively skewed, not positively skewed, because large values of X will occur often and small values occasionally.

5. (a) The 95% confidence interval for μ is

$$\bar{x} \pm 2\sigma/\sqrt{n}$$

which, upon substitution becomes

$$10 \pm 2(4/\sqrt{4}) = 10 \pm 4$$

6. (d) This question pertains to tolerance intervals, not confidence intervals, since we are looking for an interval which, with 95% confidence, encloses 95% of the population. This is a 95:95 tolerance interval given by $\bar{x} \pm ks$ Since $n = 7$ is very small, the value of k is not 2 but much larger; around 4, and the answer is

$$3 \pm 4\,(0.1) = 3 \pm 0.4$$

The key point is that for small n, with μ and σ unknown, the tolerance interval becomes much wider than $\bar{x} \pm 2s$.

7. (c) If Z denotes moisture content then

$$Z = X - Y$$

and

$$\mu_Z = \mu_X - \mu_Y = 100 - 90 = 10$$

$$\sigma_Z^2 = \sigma_X^2 + \sigma_Y^2 = 4 + 4 = 8$$

Hence, $\sigma_Z = \sqrt{8}$ and

$$\%RSD_Z = \frac{100\sigma_Z}{\mu_Z} = \frac{100\sqrt{8}}{10} = 28.3\%$$

The key point is that if a quantity of interest is determined by subtracting two large numbers, it is likely to have a large RSD.

8. (b) The probability of getting a head in a single toss of a fair coin is $p = 0.5$. The probability of getting x heads in n tosses is given by the binomial distribution as

$$\binom{n}{x} p^x (1 - p)^{n-x}$$

For $n = 4$ and $x = 2$,

$$\binom{4}{2} p^2 (1 - p)^2 = \frac{4!}{2!2!}(0.5)^2(0.5)^2 = 37.5\%$$

9. (d) If σ_X is proportional to μ_X, then $CV_X = \sigma_X/\mu_X$ = constant. In such a situation, the correct transformation is

$$Y = \log X$$

This is so because, using variance transmission,

$$\sigma_Y{}^2 \approx \left(\frac{1}{x}\right)^2 \sigma_X{}^2 = \left(\frac{\sigma_X}{x}\right)^2$$

or

$$\sigma_Y^2 \approx CV_X^2 = \text{constant}$$

With the log transformation, Y has essentially constant variance, independent of μ_Y.

10. (e) The question has to do with whether the two suppliers meaningfully differ in terms of average hardness. A paired t-test cannot be conducted here because the data cannot be appropriately paired. Checking normality is useful in terms of checking the underlying assumptions (not crucial here because of the central limit theorem) but sheds no light on the difference between suppliers. We could do an F-test first to assess if the variances are equal and then conduct the appropriate t-test to see if the means are different. However, the t-test will not tell us the amount by which the means are different to judge if the difference is practically meaningful or not. The best course of action is to construct a confidence interval for the difference in mean.

11. (b) For α and β risks in the neighborhood of 5 to 10%, the sample size per treatment is approximately given by

$$n = \frac{20}{d^2}$$

where

$$d = \frac{\text{difference in mean to be detected}}{\sigma}$$

Since $d = 1$ in this case, $n = 20$. It is useful to remember such formulae so that one can make quick judgements without having to access computers and textbooks.

12. (a) The operating characteristic curves necessary to answer the question are in Chapter 3. However, it is useful to remember the following numbers:

$n_1 = n_2$ = Number of observations	σ_1/σ_2 that can be detected
10	3/1
20 – 25	2/1
50 – 100	1.5/1

So the answer is that we need approximately 25 observations per design to be able to detect a two-to-one ratio of standard deviations with high probability.

13. (c) X and Y could be correlated without having any causal relationship, particularly when a third factor causes both X and Y to increase and decrease together. Stratification is important because looking at the data in groups may reveal or mask the relationship. If the range of X is not practically large enough, the relationship between X and Y may be masked by error. Parabolic curvature makes the correlation coefficient zero, which

could lead to the wrong conclusion that no relationship exists. The answer is inadequate range of Y. Assuming that the range of X is correctly chosen, range of Y will be a function of the relationship.

14. (b) Both X_1 and X_2 have statistically significant effects. The low R^2 could be caused by large measurement error, or an unknown key factor that varied during data collection. What is not true is that measurement error has to be small to detect small effects. Even when measurement error is large, small effects can be detected by taking a large number of observations.

15. (c) If p denotes the fraction of apartments that do not allow renters with children, then the estimated value of p is $\bar{p} = 13/100 = 0.13$. The approximate 95% confidence interval for p is

$$\bar{p} \pm 2\sqrt{\bar{p}(1-\bar{p})/n} = 0.13 \pm 2\sqrt{0.13(0.87)/100} = 0.13 \pm 0.07$$

Based upon the data, the true value of p could be anywhere between 6% to 20%, which is a very wide range to decide whether the newspaper statement is right or wrong.

16. (d) The assumption of normality is not very important when decisions are being made regarding the center of the distribution, as is the case with the $\bar{X}$ chart and the t-test. In calculating variance components, no distribution assumptions are made. A histogram can be plotted for any distribution. A tolerance interval requires estimating the tails of the distribution, which are very sensitive to changes from normality.

17. (d) The $\bar{X}$ chart control limits are calculated as

$$\bar{\bar{x}} \pm 3\frac{\sigma_{\text{short}}}{\sqrt{n}}$$

The centerline is usually taken to be the grand average. The division by $\sqrt{n}$ recognizes that variance of $\bar{X}$ reduces inversely proportional to the sample size. The multiplier three is based upon a false alarm probability of 3/1000 and the assumption of normal distribution, particularly necessary for very small subgroup size. σ_{short}, the standard deviation used to calculate limits, is within-subgroup standard deviation, not between-subgroup standard deviation.

18. (a) A moving average chart, particularly one with a large span, averages a large number of subgroups together and is, therefore, able to detect small sustained shifts more rapidly using the "single point outside control limits" rule.

19. (b) Six points in a row above the centerline is not an out-of-control rule. All the out-of-control rules are chosen such that the probability that a rule violation will occur just by chance is less than

about five in 1000. The probability of getting six points in a row above the centerline is

$$\left(\frac{1}{2}\right)^6 \approx 16 \text{ in } 1000$$

The rule is eight points in a row above the centerline, for which the probability is

$$\left(\frac{1}{2}\right)^8 \approx 4 \text{ in } 1000$$

20. (b) The control limits for the *np* chart are

$$n\bar{p} \pm 3\sqrt{n\bar{p}(1-\bar{p})}$$

If the subgroup size n changes, both the centerline and the control limits will change from subgroup to subgroup. The changing control limits are not a problem (as happens with the p-chart) but the changing centerline will make it very difficult to appreciate the running record of the process.

21. (d) When the cost of correcting a shift in the process is very high, the modified limit chart widens the usual control limits to let the process drift over a wider range to reduce the need for process corrections. This strategy is possible without producing an out-of-specification product only if the process capability is very high, usually a C_p index of two or more.

22. (d) Since different products are being produced using the same manufacturing process, a short-run chart should be used. The products have different targets and variability; therefore, data need to be standardized with respect to both the target and the standard deviation by using the Z transform:

$$Z = \frac{X - T}{\sigma}$$

The Z values for all products can be plotted on the same chart. Hence, the right chart to use is Z and W.

23. (e) A p-chart deals with fraction defective, which can only take values between zero and one. The plotted points on all other charts can have values between 0 and ∞ either on a continuous or integer scale. Points on variable charts may have negative values as well.

24. (a) Workmanship defects are count data so an attribute chart is necessary. The number of refrigerators per day, which constitutes the subgroup size, is changing. Hence, the number of defects per day needs to be standardized by dividing by the subgroup size. A u chart is necessary.

25. (c) The primary data are: number of hits per year (x_i) and number of times at bat per year (n_i). Number of hits per year are integer or attribute data. Number of times at bat per year is subgroup size. The batting average (x_i/n_i) is conceptually similar to fraction defective (except that bigger numbers are better). The subgroup size will change from year to year. A p chart should be used.
26. (a) The rational subgroup deals with determining which data should constitute a subgroup on a control chart. This requires an understanding of the process and, in particular, an understanding of the causes of variation. Also, the purpose of a control chart is to identify special causes, which is often accomplished by minimizing within-subgroup variation (resulting in tighter control limits) and maximizing the opportunity for variation between subgroups so that out-of-control subgroups could be more easily identified and their reasons found and corrected. On the other hand, the target value has to do with the centerline on a control chart but has nothing to do with defining subgroups.
27. (c) The commonly used variable charts are based upon the normal distribution, which is characterized by two independent parameters μ and σ. Because knowledge of μ tells us nothing about σ, we need two separate charts to monitor mean and standard deviation. Attribute charts are based upon binomial and Poisson distributions, in which the variance is a deterministic function of the mean. For example, for a Poisson distribution variance = mean. Hence, only one chart is necessary for attribute data.
28. (b) A properly designed control chart should not be redesigned just because some additional data have been collected. If the process is drifting and the chart is redesigned every time new data are available, then, instead of identifying the drift, the chart will tend to follow the drift, defeating the purpose of the chart. Nor should the chart be redesigned just because it signals an out-of-control situation. A properly designed control chart should be redesigned when a deliberate process change is made or a permanent known process change has occurred and the control chart indicates that the process has changed.
29. (a) The process average has nothing to do with deciding sampling frequency. The frequency of sampling is predominantly determined by the frequency of anticipated process shifts. Additionally, high cost of sampling and testing, high capability index, and large amount of available data will tend to reduce the sampling frequency.

30. (d) This would suggest that the process variability has decreased, perhaps because of a deliberate improvement that was made, or because a certain batch of raw material happens to be extra homogeneous, or because the measurement system malfunctioned. Another possibility is that the within-subgroup data came from, say, two distributions with different means. In this case, the within-subgroup range will be large, causing the control limits to inflate. However, the $\overline{X}$ values will stay close to the centerline. In this case, hugging of the centerline will occur immediately after the implementation of the chart.

31. (d) The reason is that specifications are usually on individual values and $\overline{X}$ charts deal with averages. The $\overline{X}$ chart control limits are a function of subgroup size n, and by changing n, the limits can be made tighter or wider. Therefore, the control limits cannot be compared to specifications. If the specifications are wider than the control limits, it would not necessarily mean that the process was producing good products. It could just be that the subgroup size was large. If the specification was on the average of n observations, then the control limits of an $\overline{X}$ chart with subgroup size n would be comparable to the specification. But even in this case, they could result in misleading actions. For example, if a point fell outside the control limits but inside the specification limits, perhaps no action would be taken, which is wrong. So putting specifications on the $\overline{X}$ chart will be misleading in any case.

32. (a) The current control limits are 10 ± 4. If the subgroup size is increased by a factor of four, the width of the $\overline{X}$ chart control limits will narrow by a factor of two and they will become 10 ± 2.

33. (e) The $\overline{X}$ chart deals with the process mean. The null hypothesis is that the process mean is at the centerline, the alternate hypothesis being that the process mean has shifted due to a special cause. β risk is the probability of accepting the null hypothesis when it is false; namely, the probability of saying that the process mean is at the centerline when it has shifted. It is the probability of not detecting a shift in mean. α risk is the probability of rejecting the null hypothesis when it is true; namely, the probability of saying that the mean has shifted when it has not. It is the probability of false alarm.

34. (d) So long as we use 3-sigma limit charts, the α risk is fixed at 3/1000. If we increase the subgroup size, the $\overline{X}$ chart control limits will tighten and it will become easier to detect a mean shift. β risk will decrease.

35. (a) The capability and performance indices have the following relationships:

$$C_p \geq C_{pk} \geq P_{pk}$$

$$P_p = \frac{C_p P_{pk}}{C_{pk}}$$

Only the answers in row (a) agree with these relationships.

36. (b) The fact that C_p and C_{pk} are close means that the process is centered. The large difference between P_{pk} and C_{pk} means that the process is unstable. We need to bring the process in control.

37. (e) The C_p index assumes that the process is perfectly centered in the middle of a two-sided specification. The C_{pk} index does not make this assumption. Both C_p and C_{pk} assume a stable process that is normally distributed.

38. (b) The C_p index is defined as Tolerance/6σ. Information regarding $\overline{X}$ is irrelevant to this question. When the tolerance is 6 and σ is 1, $C_p = 1$.

39. (e) Since the process mean is closer to the upper specification limit, C_{pk} is estimated as

$$C_{pk} = \frac{USL - \text{Mean}}{3\sigma} = \frac{20 - 14}{6} = 1$$

So the answer is none of the above.

40. (a) With ten observations, the confidence interval is ±50% of the calculated C_{pk} value (see Chapter 5). So the answer is 1 ± 0.5.

41. (c) C_{pk} is defined as the smaller of

$$\frac{USL - \text{Mean}}{3\sigma} \quad \text{or} \quad \frac{\text{Mean} - LSL}{3\sigma}$$

σ is always positive. For the C_{pk} to be negative, the numerator must be negative, which will happen if Mean > USL or Mean < LSL. So, C_{pk} will be negative if the process mean is outside the specification limits.

42. (e) Statement (a) is correct. Unless a process is stable, we cannot make predictions regarding the future. Statement (b) is incorrect. A stable process may or may not be capable, depending upon process variability, centering, and specifications. Statement (c) is also incorrect. C_{pk} cannot be greater than C_p.

43. (e) The C_p index will increase due to reduced variability. C_{pk} will increase both due to centering and reduction in variability, and the two indices will become equal: $C_p = C_{pk}$.

44. (d) Variance within subgroup is common cause variance and variance between subgroups is special cause variance. Here, special cause variance is small and common cause variance is large. Time trend, shift-to-shift differences, batch-to-batch variability, and cycling along the time axis are examples of special causes.

Measurement system variability is one source of common cause variance and should be evaluated.

45. (b) The total standard deviation σ_t may be calculated as

$$\sigma_t^2 = \sigma_w^2 + \sigma_b^2 = 16 + 9 = 25$$

$$\sigma_t = 5$$

and

$$P_p = \frac{20}{3\sigma_t} = \frac{20}{15} = 1.33$$

46. (d) For factor B to be nested inside factor A, levels of B must be different for each level of A. Both A and B may be random or fixed factors.

47. (b) When ambient temperature is considered a random factor and experiments are conducted on a very hot day and a very cold day, these two hot and cold temperatures are considered to be random drawings from the distribution of ambient temperature. The result is that the variance of temperature is overestimated and the variance contribution due to temperature is overestimated.

48. (e) Using variance transmission, the variance component due to factor A is

$$\left(\frac{dy}{dA}\right)^2 \sigma_a^2 = 16\sigma_a^2$$

where σ_a^2 is the variance of factor A, which is a fixed factor with levels –1 and 1. These two values are the entire population of A. Hence, $\mu_a = 0$ and

$$\sigma_a^2 = \frac{(-1-0)^2 + (1-0)^2}{2} = 1$$

The variance contribution due to factor A is 16.

49. (a) The quadratic loss function says that the loss is zero when X is on target T and as X deviates from target, the loss increases in proportion to the square of the deviation. Mathematically, loss is proportional to $(X - T)^2$.

50. (e) For a centered process, quadratic loss is directly proportional to variance. If the variance is reduced by a factor of four, the quadratic loss will become $1 Million.

51. (b) $\sigma_{\text{total}}^2 = 16$ because it is the sum of the three variance components. Hence, $\sigma_{\text{total}} = 4$. The mean was 48, with the nearest specification limit being the lower specification of 40. Therefore,

$$P_{pk} = \frac{48 - 40}{3(4)} = \frac{2}{3}$$

52. (b) When the effect of one factor depends upon the level of another factor, the two factors are said to interact. If lane and time interact, it means that the differences between lanes are not constant; they vary with time. If the interaction is small, it means that the differences between lanes are essentially constant and their cause may be easier to find than when the interaction is large.
53. (b) 100 different products are evaluated for each shift. Therefore, products are nested inside shifts. If the same products were evaluated by each shift, then products and shifts would form a crossed structure.
54. (d) Since a different group of people went to each restaurant, people are nested inside restaurant. Since each person evaluated different hamburgers, hamburgers are nested inside people.
55. (b) The between-subgroup variance is due to special causes and the within-subgroup variance is due to common causes. The common cause variance is 0.3. Since bulk of the variance is common cause, it cannot be eliminated by controlling the process. Whether the total variance is small or big cannot be judged unless we know the specification. Finally, 20 degrees of freedom are minimally necessary to estimate variance. The larger the degrees of freedom, the better.
56. (e) To meet six sigma quality, the C_p index for the measurement system should be four or more. Hence,

$$\frac{2W}{6\sigma_m} \geq 4 \qquad \text{or} \qquad \sigma_m \leq \frac{W}{12}$$

57. (b) For a measurement system to be useful for improvement purposes, it should be able to distinguish between products. If a measurement system is able to classify products into only two statistically different categories, then it is no better than an attribute system. So, minimally, three or more categories are required. The number of categories into which products can be grouped using a single measurement per product are $1.4\sigma_p/\sigma_m$, where σ_p and σ_m are the product and measurement standard deviations, respectively. So we need

$$\frac{1.4\sigma_p}{\sigma_m} \geq 3$$

Squaring both sides gives

$$2\sigma_p^2 \geq 9\sigma_m^2$$

Also,

$$\sigma_p^2 = \sigma_t^2 - \sigma_m^2$$

and, upon substitution,

$$\sigma_m \le 0.43\sigma_t$$

which is rounded to

$$\sigma_m \le \sigma_t/2$$

58. (a) The three suppliers have the same P_{pk}, so they must have the same σ_t (total standard deviation). Since

$$\sigma_t^2 = \sigma_p^2 + \sigma_m^2$$

the supplier with the largest measurement variance must have the smallest true product-to-product variance and should be the one to buy the product from.

59. (c) The same three operators measured each product. Therefore, products and operators are crossed. The levels of duplicate measurements are all different for each combination of products and operators, hence, duplicates are nested inside products and operators.

60. (e) A master standard is only absolutely necessary to determine the bias or accuracy of a measurement system. Properties such as repeatability, reproducibility, and stability can be determined if the same product can be repeatably measured without changing it. Such a product need not be a master standard.

61. (c) A measurement system should be recalibrated only when it goes out of calibration. Otherwise, we will be recalibrating the system on the basis of noise, which will only serve to inflate measurement variability. A good way to judge whether the measurement system is out of calibration or not is by using a control chart of repeated measurements of a standard or a product that does not change over time. An out-of-control message signals the need to recalibrate.

62. (c) Interaction between operators and products means that there are operator-to-operator differences and these differences change as a function of products. The solution is to reduce operator-to-operator differences by finding out why these differences exist and by providing training for the operators to reduce the differences.

63. (c) If the measurement system measures to inadequate number of decimal places, then it will not be able to distinguish between products. Many within-subgroup ranges may be zero, causing $\overline{R}$ to decrease and control limits to tighten. The number of possible values for range within the range chart control limits will become very small. When the number of possible values of range within the range chart control limits is four or less, the

measurement unit is inadequate. A lack of adequate number of decimal places will not cause deterministic shifts in the $\overline{\overline{X}}$ or R charts.

64. (d) If $y = x^{1/3}$, then

$$\sigma_y^2 = \left(\frac{1}{3x^{2/3}}\right)^2 \sigma_x^2$$

When $x = 4.8$ (two significant figures), the measurement unit is 0.1 Therefore,

$$\sigma_x^2 = \frac{(0.1)^2}{12}$$

and

$$\sigma_y^2 = \left[\frac{1}{3(4.8)^{2/3}}\right]^2 \frac{(0.1)^2}{12} = 0.000011$$

Hence, $\pm 2\sigma_y = \pm 0.0068$, which means that y should be reported to 2 decimal places. The computed value of y is 1.6868653, which should be reported as 1.69 with three significant figures.

Appendix A

Tail Area of Unit Normal Distribution

Z	0.00	0.01	0.02	0.03	0.04	0.05	0.06	0.07	0.08	0.09
0.0	0.5000	0.4960	0.4920	0.4880	0.4840	0.4801	0.4761	0.4721	0.4681	0.4641
0.1	0.4602	0.4562	0.4522	0.4483	0.4443	0.4404	0.4364	0.4325	0.4286	0.4247
0.2	0.4207	0.4168	0.4129	0.4090	0.4052	0.4013	0.3974	0.3936	0.3897	0.3859
0.3	0.3821	0.3783	0.3745	0.3707	0.3669	0.3632	0.3594	0.3557	0.3520	0.3483
0.4	0.3446	0.3409	0.3372	0.3336	0.3300	0.3264	0.3228	0.3192	0.3156	0.3121
0.5	0.3085	0.3050	0.3015	0.2981	0.2946	0.2912	0.2877	0.2843	0.2810	0.2776
0.6	0.2743	0.2709	0.2676	0.2643	0.2611	0.2578	0.2546	0.2514	0.2483	0.2451
0.7	0.2420	0.2389	0.2358	0.2327	0.2296	0.2266	0.2236	0.2206	0.2177	0.2148
0.8	0.2119	0.2090	0.2061	0.2033	0.2005	0.1977	0.1949	0.1922	0.1894	0.1867
0.9	0.1841	0.1814	0.1788	0.1762	0.1736	0.1711	0.1685	0.1660	0.1635	0.1611
1.0	0.1587	0.1562	0.1539	0.1515	0.1492	0.1469	0.1446	0.1423	0.1401	0.1379
1.1	0.1357	0.1335	0.1314	0.1292	0.1271	0.1251	0.1230	0.1210	0.1190	0.1170
1.2	0.1151	0.1131	0.1112	0.1093	0.1075	0.1056	0.1038	0.1020	0.1003	0.0985
1.3	0.0968	0.0951	0.0934	0.0918	0.0901	0.0885	0.0869	0.0853	0.0838	0.0823
1.4	0.0808	0.0793	0.0778	0.0764	0.0749	0.0735	0.0721	0.0708	0.0694	0.0681
1.5	0.0668	0.0655	0.0643	0.0630	0.0618	0.0606	0.0594	0.0582	0.0571	0.0559
1.6	0.0548	0.0537	0.0526	0.0516	0.0505	0.0495	0.0485	0.0475	0.0465	0.0455
1.7	0.0446	0.0436	0.0427	0.0418	0.0409	0.0401	0.0392	0.0384	0.0375	0.0367
1.8	0.0359	0.0351	0.0344	0.0336	0.0329	0.0322	0.0314	0.0307	0.0301	0.0294
1.9	0.0287	0.0281	0.0274	0.0268	0.0262	0.0256	0.0250	0.0244	0.0239	0.0233
2.0	0.0228	0.0222	0.0217	0.0212	0.0207	0.0202	0.0197	0.0192	0.0188	0.0183
2.1	0.0179	0.0174	0.0170	0.0166	0.0162	0.0158	0.0154	0.0150	0.0146	0.0143
2.2	0.0139	0.0136	0.0132	0.0129	0.0125	0.0122	0.0119	0.0116	0.0113	0.0110
2.3	0.0107	0.0104	0.0102	0.0099	0.0096	0.0094	0.0091	0.0089	0.0087	0.0084
2.4	0.0082	0.0080	0.0078	0.0075	0.0073	0.0071	0.0069	0.0068	0.0066	0.0064
2.5	0.0062	0.0060	0.0059	0.0057	0.0055	0.0054	0.0052	0.0051	0.0049	0.0048
2.6	0.0047	0.0045	0.0044	0.0043	0.0041	0.0040	0.0039	0.0038	0.0037	0.0036
2.7	0.0035	0.0034	0.0033	0.0032	0.0031	0.0030	0.0029	0.0028	0.0027	0.0026
2.8	0.0026	0.0025	0.0024	0.0023	0.0023	0.0022	0.0021	0.0021	0.0020	0.0019
2.9	0.0019	0.0018	0.0018	0.0017	0.0016	0.0016	0.0015	0.0015	0.0014	0.0014
3.0	0.0013	0.0013	0.0013	0.0012	0.0012	0.0011	0.0011	0.0011	0.0010	0.0010
3.1	0.0010	0.0009	0.0009	0.0009	0.0008	0.0008	0.0008	0.0008	0.0007	0.0007
3.2	0.0007	0.0007	0.0006	0.0006	0.0006	0.0006	0.0006	0.0005	0.0005	0.0005
3.3	0.0005	0.0005	0.0005	0.0004	0.0004	0.0004	0.0004	0.0004	0.0004	0.0003
3.4	0.0003	0.0003	0.0003	0.0003	0.0003	0.0003	0.0003	0.0003	0.0003	0.0002
3.5	0.0002	0.0002	0.0002	0.0002	0.0002	0.0002	0.0002	0.0002	0.0002	0.0002
3.6	0.0002	0.0002	0.0001	0.0001	0.0001	0.0001	0.0001	0.0001	0.0001	0.0001
3.7	0.0001	0.0001	0.0001	0.0001	0.0001	0.0001	0.0001	0.0001	0.0001	0.0001
3.8	0.0001	0.0001	0.0001	0.0001	0.0001	0.0001	0.0001	0.0001	0.0001	0.0001
3.9	0.0000	0.0000	0.0000	0.0000	0.0000	0.0000	0.0000	0.0000	0.0000	0.0000

Appendix B

Probability Points of the t Distribution with ν Degrees of Freedom

	Tail Area Probability									
ν	0.4	0.25	0.1	0.05	0.025	0.01	0.005	0.0025	0.001	0.0005
1	0.325	1.000	3.078	6.314	12.706	31.821	63.657	127.32	318.31	636.62
2	0.289	0.816	1.886	2.920	4.303	6.965	9.925	14.089	22.326	31.598
3	0.277	0.765	1.638	2.353	3.182	4.541	5.841	7.453	10.213	12.924
4	0.271	0.741	1.533	2.132	2.776	3.747	4.604	5.598	7.173	8.610
5	0.267	0.727	1.476	2.015	2.571	3.365	4.032	4.773	5.893	6.869
6	0.265	0.718	1.440	1.943	2.447	3.143	3.707	4.317	5.208	5.959
7	0.263	0.711	1.415	1.895	2.365	2.998	3.499	4.029	4.785	5.408
8	0.262	0.706	1.397	1.860	2.306	2.896	3.355	3.833	4.501	5.041
9	0.261	0.703	1.383	1.833	2.262	2.821	3.250	3.690	4.297	4.781
10	0.260	0.700	1.372	1.812	2.228	2.764	3.169	3.581	4.144	4.587
11	0.260	0.697	1.363	1.796	2.201	2.718	3.106	3.497	4.025	4.437
12	0.259	0.695	1.356	1.782	2.179	2.681	3.055	3.428	3.930	4.318
13	0.259	0.694	1.350	1.771	2.160	2.650	3.012	3.372	3.852	4.221
14	0.258	0.692	1.345	1.761	2.145	2.624	2.977	3.326	3.787	4.140
15	0.258	0.691	1.341	1.753	2.131	2.602	2.947	3.286	3.733	4.073
16	0.258	0.690	1.337	1.746	2.120	2.583	2.921	3.252	3.686	4.015
17	0.257	0.689	1.333	1.740	2.110	2.567	2.898	3.222	3.646	3.965
18	0.257	0.688	1.330	1.734	2.101	2.552	2.878	3.197	3.610	3.922
19	0.257	0.688	1.328	1.729	2.093	2.539	2.861	3.174	3.579	3.883
20	0.257	0.687	1.325	1.725	2.086	2.528	2.845	3.153	3.552	3.850
21	0.257	0.686	1.323	1.721	2.080	2.518	2.831	3.135	3.527	3.819
22	0.256	0.686	1.321	1.717	2.074	2.508	2.819	3.119	3.505	3.792
23	0.256	0.685	1.319	1.714	2.069	2.500	2.807	3.104	3.485	3.767
24	0.256	0.685	1.318	1.711	2.064	2.492	2.797	3.091	3.467	3.745
25	0.256	0.684	1.316	1.708	2.060	2.485	2.787	3.078	3.450	3.725
26	0.256	0.684	1.315	1.706	2.056	2.479	2.779	3.067	3.435	3.707
27	0.256	0.684	1.314	1.703	2.052	2.473	2.771	3.057	3.421	3.690
28	0.256	0.683	1.313	1.701	2.048	2.467	2.763	3.047	3.408	3.674
29	0.256	0.683	1.311	1.699	2.045	2.462	2.756	3.038	3.396	3.659
30	0.256	0.683	1.310	1.697	2.042	2.457	2.750	3.030	3.385	3.646
40	0.255	0.681	1.303	1.684	2.021	2.423	2.704	2.971	3.307	3.551
60	0.254	0.679	1.296	1.671	2.000	2.390	2.660	2.915	2.232	3.460
120	2.254	0.677	1.289	1.658	1.980	2.358	2.617	2.860	3.160	3.373
∞	0.253	0.674	1.282	1.645	1.960	2.326	2.576	2.807	3.090	3.291

From E. S. Pearson and H. O. Hartley (Eds.) (1958), *Biometrika Tables for Statisticians,* Vol. 1, used by permission of Oxford University Press.

Appendix C

Probability Points of the χ^2 Distribution with ν Degrees of Freedom

	Tail Area Probability								
ν	0.995	0.990	0.975	0.950	0.500	0.050	0.025	0.010	0.005
1	0.00+	0.00+	0.00+	0.00+	0.45	3.84	5.02	6.63	7.88
2	0.01	0.02	0.05	0.10	1.39	5.99	7.38	9.21	10.60
3	0.07	0.11	0.22	0.35	2.37	7.81	9.35	11.34	12.84
4	0.21	0.30	0.48	0.71	3.36	9.49	11.14	13.28	14.86
5	0.41	0.55	0.83	1.15	4.35	11.07	12.83	15.09	16.75
6	0.68	0.87	1.24	1.64	5.35	12.59	14.45	16.81	18.55
7	0.99	1.24	1.69	2.17	6.35	14.07	16.01	18.48	20.28
8	1.34	1.65	2.18	2.73	7.34	15.51	17.53	20.09	21.96
9	1.73	2.09	2.70	3.33	8.34	16.92	19.02	21.67	23.59
10	2.16	2.56	3.25	3.94	9.34	18.31	20.48	23.21	25.19
11	2.60	3.05	3.82	4.57	10.34	19.68	21.92	24.72	26.76
12	3.07	3.57	4.40	5.23	11.34	21.03	23.34	26.22	28.30
13	3.57	4.11	5.01	5.89	12.34	22.36	24.74	27.69	29.82
14	4.07	4.66	5.63	6.57	13.34	23.68	26.12	29.14	31.32
15	4.60	5.23	6.27	7.26	14.34	25.00	27.49	30.58	32.80
16	5.14	5.81	6.91	7.96	15.34	26.30	28.85	32.00	34.27
17	5.70	6.41	7.56	8.67	16.34	27.59	30.19	33.41	35.72
18	6.26	7.01	8.23	9.39	17.34	28.87	31.53	34.81	37.16
19	6.84	7.63	8.91	10.12	18.34	30.14	32.85	36.19	38.58
20	7.43	8.26	9.59	10.85	19.34	31.41	34.17	37.57	40.00
25	10.52	11.52	13.12	14.61	24.34	37.65	40.65	44.31	46.93
30	13.79	14.95	16.79	18.49	29.34	43.77	46.98	50.89	53.67
40	20.71	22.16	24.43	26.51	39.34	55.76	59.34	63.69	66.77
50	27.99	29.71	32.36	34.76	49.33	67.50	71.42	76.15	79.49
60	35.53	37.48	40.48	43.19	59.33	79.08	83.30	88.38	91.95
70	43.28	45.44	48.76	51.74	69.33	90.53	95.02	100.42	104.22
80	51.17	53.54	57.15	60.39	79.33	101.88	106.63	112.33	116.32
90	59.20	61.75	65.65	69.13	89.33	113.14	118.14	124.12	128.30
100	67.33	70.06	74.22	77.93	99.33	124.34	129.56	135.81	140.17

From E. S. Pearson and H. O. Hartley (Eds.) (1966), *Biometrika Tables for Statisticians,* Vol. 1, used by permission of Oxford University Press

Appendix D1

k Values for Two-Sided Normal Tolerance Limits

	90% Confidence that Percentage of Population between Limits Is			95% Confidence that Percentage of Population between Limits Is			99% Confidence that Percentage of Population between Limits Is		
n	90%	95%	99%	90%	95%	99%	90%	95%	99%
2	15.98	18.80	24.17	32.02	37.67	48.43	160.2	188.5	242.3
3	5.847	6.919	8.974	8.380	9.916	12.86	18.93	22.40	29.06
4	4.166	4.943	6.440	5.369	6.370	8.299	9.398	11.15	14.53
5	3.494	4.152	5.423	4.275	5.079	6.634	6.612	7.855	10.26
6	3.131	3.723	4.870	3.712	4.414	5.775	5.337	6.345	8.301
7	2.902	3.452	4.521	3.369	4.007	5.248	4.613	5.448	7.187
8	2.743	3.264	4.278	3.136	3.732	4.891	4.147	4.936	6.468
9	2.626	3.125	4.098	2.967	3.532	4.631	3.822	4.550	5.966
10	2.535	3.018	3.959	2.829	3.379	4.433	3.582	4.265	5.594
11	2.463	2.933	3.849	2.737	3.259	4.277	3.397	4.045	5.308
12	2.404	2.863	3.758	2.655	3.162	4.150	3.250	3.870	5.079
13	2.355	2.805	3.682	2.587	3.081	4.044	3.130	3.727	4.893
14	2.314	2.756	3.618	2.529	3.012	3.955	3.029	3.608	4.737
15	2.278	2.713	3.562	2.480	2.954	3.878	2.945	3.507	4.605
16	2.246	2.676	3.514	2.437	2.903	3.812	2.872	3.421	4.492
17	2.219	2.643	3.471	2.400	2.858	3.754	2.808	3.345	4.393
18	2.194	2.614	3.433	2.366	2.819	3.702	2.753	3.279	4.307
19	2.172	2.588	3.399	2.337	2.784	3.656	2.703	3.221	4.230
20	2.152	2.564	3.368	2.310	2.752	3.615	2.659	3.168	4.161
21	2.135	2.543	3.340	2.286	2.723	3.577	2.620	3.121	4.100
22	2.118	2.524	3.315	2.264	2.697	3.543	2.584	3.078	4.044
23	2.103	2.506	3.292	2.244	2.673	3.512	2.551	3.040	3.993
24	2.089	2.489	3.270	2.225	2.651	3.483	2.522	3.004	3.947
25	2.077	2.474	3.251	2.208	2.631	3.457	2.494	2.972	3.904
26	2.065	2.460	3.232	2.193	2.612	3.432	2.469	2.941	3.865
27	2.054	2.447	3.215	2.178	2.595	3.409	2.446	2.914	3.828
28	2.044	2.435	3.199	2.164	2.579	3.388	2.424	2.888	3.794
29	2.034	2.424	3.184	2.152	2.554	3.368	2.404	2.864	3.763
30	2.025	2.413	3.170	2.140	2.549	3.350	2.385	2.841	3.733
35	1.988	2.368	3.112	2.090	2.490	3.272	2.306	2.748	3.611
40	1.959	2.334	3.066	2.052	2.445	3.213	2.247	2.677	3.518
50	1.916	2.284	3.001	1.996	2.379	3.126	2.162	2.576	3.385
60	1.887	2.248	2.955	1.958	2.333	3.066	2.103	2.506	3.293
80	1.848	2.202	2.894	1.907	2.272	2.986	2.026	2.414	3.173
100	1.822	2.172	2.854	1.874	2.233	2.934	1.977	2.355	3.096
200	1.764	2.102	2.762	1.798	2.143	2.816	1.865	2.222	2.921
500	1.717	2.046	2.689	1.737	2.070	2.721	1.777	2.117	2.783
1000	1.695	2.019	2.654	1.709	2.036	2.676	1.736	2.068	2.718
∞	1.645	1.960	2.576	1.645	1.960	2.576	1.645	1.960	2.576

From D. C. Montgomery (1985), *Introduction to Statistical Quality Control,* used by permission of John Wiley & Sons, Inc.

Appendix D2

k Values for One-Sided Normal Tolerance Limits

	90% Confidence that Percentage of Population is below (above) Limits Is			95% Confidence that Percentage of Population is below (above) Limits Is			99% Confidence that Percentage of Population is below (above) Limits Is		
n	90%	95%	99%	90%	95%	99%	90%	95%	99%
3	4.258	5.310	7.340	6.158	7.655	10.552			
4	3.187	3.957	5.437	4.163	5.145	7.042			
5	2.742	3.400	4.666	3.407	4.202	5.741			
6	2.494	3.091	4.242	3.006	3.707	5.062	4.408	5.409	7.334
7	2.333	2.894	3.972	2.755	3.399	4.641	3.856	4.730	6.411
8	2.219	2.755	3.783	2.582	3.188	4.353	3.496	4.287	5.811
9	2.133	2.649	3.641	2.454	3.031	4.143	3.242	3.971	5.389
10	2.065	2.568	3.532	2.355	2.911	3.981	3.048	3.739	5.075
11	2.012	2.503	3.444	2.275	2.815	3.852	2.897	3.557	4.828
12	1.966	2.448	3.371	2.210	2.736	3.747	2.773	3.410	4.633
13	1.928	2.403	3.310	2.155	2.670	3.659	2.677	3.290	4.472
14	1.895	2.363	3.257	2.108	2.614	3.585	2.592	3.189	4.336
15	1.866	2.329	3.212	2.068	2.566	3.520	2.521	3.102	4.224
16	1.842	2.299	3.172	2.032	2.523	3.463	2.458	3.028	4.124
17	1.820	2.272	3.136	2.001	2.486	3.415	2.405	2.962	4.038
18	1.800	2.249	3.106	1.974	2.453	3.370	2.357	2.906	3.961
19	1.781	2.228	3.078	1.949	2.423	3.331	2.315	2.855	3.893
20	1.765	2.208	3.052	1.926	2.396	3.295	2.275	2.807	3.832
21	1.750	2.190	3.028	1.905	2.371	3.262	2.241	2.768	3.776
22	1.736	2.174	3.007	1.887	2.350	3.233	2.208	2.729	3.727
23	1.724	2.159	2.987	1.869	2.329	3.206	2.179	2.693	3.680
24	1.712	2.145	2.969	1.853	2.309	3.181	2.154	2.663	3.638
25	1.702	2.132	2.952	1.838	2.292	3.158	2.129	2.632	3.601
30	1.657	2.080	2.884	1.778	2.220	3.064	2.029	2.516	3.446
35	1.623	2.041	2.833	1.732	2.166	2.994	1.957	2.431	3.334
40	1.598	2.010	2.793	1.697	2.126	2.941	1.902	2.365	3.250
45	1.577	1.986	2.762	1.669	2.092	2.897	1.857	2.313	3.181
50	1.560	1.965	2.735	1.646	2.065	2.863	1.821	2.296	3.124

From D. C. Montgomery (1985), *Introduction to Statistical Quality Control,* used by permission of John Wiley & Sons, Inc.

Appendix E1

Percentage Points of the F Distribution: Upper 5% Points

Denominator Degrees of Freedom (ν_2)	Numerator Degrees of Freedom (ν_1) 1	2	3	4	5	6	7	8	9	10	12	15	20	24	30	40	60	120	∞
1	161.4	199.5	215.7	224.6	230.2	234.0	236.8	238.9	240.5	241.9	243.9	245.9	248.0	249.1	250.1	251.1	252.2	253.3	254.3
2	18.51	19.00	19.16	19.25	19.30	19.33	19.35	19.37	19.38	19.41	19.40	19.43	19.45	19.45	19.46	19.47	19.48	19.49	19.50
3	10.13	9.55	9.28	9.12	9.01	8.94	8.89	8.85	8.81	8.79	8.74	8.70	8.66	8.64	8.62	8.59	8.57	8.55	8.53
4	7.71	6.94	6.59	6.39	6.26	6.16	6.09	6.04	6.00	5.96	5.91	5.86	5.80	5.77	5.75	5.72	5.69	5.66	5.63
5	6.61	5.79	5.41	5.19	5.05	4.95	4.88	4.82	4.77	4.74	4.68	4.62	4.56	4.53	4.50	4.46	4.43	4.40	4.36
6	5.99	5.14	4.76	4.53	4.39	4.28	4.21	4.15	4.10	4.06	4.00	3.94	3.87	3.84	3.81	3.77	3.74	3.70	3.67
7	5.59	4.74	4.35	4.12	3.97	3.87	3.79	3.73	3.68	3.64	3.57	3.51	3.44	3.41	3.38	3.34	3.30	3.27	3.23
8	5.32	4.46	4.07	3.84	3.69	3.58	3.50	3.44	3.39	3.35	3.28	3.22	3.15	3.12	3.08	3.04	3.01	2.97	2.93
9	5.12	4.26	3.86	3.63	3.48	3.37	3.29	3.23	3.18	3.14	3.07	3.01	2.94	2.90	2.86	2.83	2.79	2.75	2.71
10	4.96	4.10	3.71	3.48	3.33	3.22	3.14	3.07	3.02	2.98	2.91	2.85	2.77	2.74	2.70	2.66	2.62	2.58	2.54
11	4.84	3.98	3.59	3.36	3.20	3.09	3.01	2.95	2.90	2.85	2.79	2.72	2.65	2.61	2.57	2.53	2.49	2.45	2.40
12	4.75	3.89	3.49	3.26	3.11	3.00	2.91	2.85	2.80	2.75	2.69	2.62	2.54	2.51	2.47	2.43	2.38	2.34	2.30
13	4.67	3.81	3.41	3.18	3.03	2.92	2.83	2.77	2.71	2.67	2.60	2.53	2.46	2.42	2.38	2.34	2.30	2.25	2.21
14	4.60	3.74	3.34	3.11	2.96	2.85	2.76	2.70	2.65	2.60	2.53	2.46	2.39	2.35	2.31	2.27	2.22	2.18	2.13
15	4.54	3.68	3.29	3.06	2.90	2.79	2.71	2.64	2.59	2.54	2.48	2.40	2.33	2.29	2.25	2.20	2.16	2.11	2.07
16	4.49	3.63	3.24	3.01	2.85	2.74	2.66	2.59	2.54	2.49	2.42	2.35	2.28	2.24	2.19	2.15	2.11	2.06	2.01
17	4.45	3.59	3.20	2.96	2.81	2.70	2.61	2.55	2.49	2.45	2.38	2.31	2.23	2.19	2.15	2.10	2.06	2.01	1.96
18	4.41	3.55	3.16	2.93	2.77	2.66	2.58	2.51	2.46	2.41	2.34	2.27	2.19	2.15	2.11	2.06	2.02	1.97	1.92
19	4.38	3.52	3.13	2.90	2.74	2.63	2.54	2.48	2.42	2.38	2.31	2.23	2.16	2.11	2.07	2.03	1.98	1.93	1.88
20	4.35	3.49	3.10	2.87	2.71	2.60	2.51	2.45	2.39	2.35	2.28	2.20	2.12	2.08	2.04	1.99	1.95	1.90	1.84
21	4.32	3.47	3.07	2.84	2.68	2.57	2.49	2.42	2.37	2.32	2.25	2.18	2.10	2.05	2.01	1.96	1.92	1.87	1.81
22	4.30	3.44	3.05	2.82	2.66	2.55	2.46	2.40	2.34	2.30	2.23	2.15	2.07	2.03	1.98	1.94	1.89	1.84	1.78
23	4.28	3.42	3.03	2.80	2.64	2.53	2.44	2.37	2.32	2.27	2.20	2.13	2.05	2.01	1.96	1.91	1.86	1.81	1.76
24	4.26	3.40	3.01	2.78	2.62	2.51	2.42	2.36	2.30	2.25	2.18	2.11	2.03	1.98	1.94	1.89	1.84	1.79	1.73
25	4.24	3.39	2.99	2.76	2.60	2.49	2.40	2.34	2.28	2.24	2.16	2.09	2.01	1.96	1.92	1.87	1.82	1.77	1.71
26	4.23	3.37	2.98	2.74	2.59	2.47	2.39	2.32	2.27	2.22	2.15	2.07	1.99	1.95	1.90	1.85	1.80	1.75	1.69
27	4.21	3.35	2.96	2.73	2.57	2.46	2.37	2.31	2.25	2.20	2.13	2.06	1.97	1.93	1.88	1.84	1.79	1.73	1.67
28	4.20	3.34	2.95	2.71	2.56	2.45	2.36	2.29	2.24	2.19	2.12	2.04	1.96	1.91	1.87	1.82	1.77	1.71	1.65
29	4.18	3.33	2.93	2.70	2.55	2.43	2.35	2.28	2.22	2.18	2.10	2.03	1.94	1.90	1.85	1.81	1.75	1.70	1.64
30	4.17	3.32	2.92	2.69	2.53	2.42	2.33	2.27	2.21	2.16	2.09	2.01	1.93	1.89	1.84	1.79	1.74	1.68	1.62
40	4.08	3.23	2.84	2.61	2.45	2.34	2.25	2.18	2.12	2.08	2.00	1.92	1.84	1.79	1.74	1.69	1.64	1.58	1.51
60	4.00	3.15	2.76	2.53	2.37	2.25	2.17	2.10	2.04	1.99	1.92	1.84	1.75	1.70	1.65	1.59	1.53	1.47	1.39
120	3.92	3.07	2.68	2.45	2.29	2.17	2.09	2.02	1.96	1.91	1.83	1.75	1.66	1.61	1.55	1.50	1.43	1.35	1.25
∞	3.84	3.00	2.60	2.37	2.21	2.10	2.01	1.94	1.88	1.83	1.75	1.67	1.57	1.52	1.46	1.39	1.32	1.22	1.00

From M. Merrington and C. M. Thompson (1943), Tables of Percentage Points of the Inverted Beta (F) Distribution, *Biometrika,* used by permission of Oxford University Press.

Appendix E2

Percentage Points of the F Distribution: Upper 2.5% Points

Denominator Degrees of Freedom (ν_2)	Numerator Degrees of Freedom (ν_1)																		
	1	2	3	4	5	6	7	8	9	10	12	15	20	24	30	40	60	120	∞
1	647.8	799.5	864.2	899.6	921.8	937.1	948.2	956.7	963.3	968.6	976.7	984.9	993.1	997.2	1001.0	1006.0	1010.0	1014.0	1018.0
2	38.51	39.00	39.17	39.25	39.30	39.33	39.36	39.37	39.39	39.40	39.41	39.43	39.45	39.46	39.46	39.47	39.48	39.49	39.50
3	17.44	16.04	15.44	15.10	14.88	14.73	14.62	14.54	14.47	14.42	14.34	14.25	14.17	14.12	14.08	14.04	13.99	13.95	13.90
4	12.22	10.65	9.98	9.60	9.36	9.20	9.07	8.98	8.90	8.84	8.75	8.66	8.56	8.51	8.46	8.41	8.36	8.31	8.26
5	10.01	8.43	7.76	7.39	7.15	6.98	6.85	6.76	6.68	6.62	6.52	6.43	6.33	6.28	6.23	6.18	6.12	6.07	6.02
6	8.81	7.26	6.60	6.23	5.99	5.82	5.70	5.60	5.52	5.46	5.37	5.27	5.17	5.12	5.07	5.01	4.96	4.90	4.85
7	8.07	6.54	5.89	5.52	5.29	5.12	4.99	4.90	4.82	4.76	4.67	4.57	4.47	4.42	4.36	4.31	4.25	4.20	4.14
8	7.57	6.06	5.42	5.05	4.82	4.65	4.53	4.43	4.36	4.30	4.20	4.10	4.00	3.95	3.89	3.84	3.78	3.73	3.67
9	7.21	5.71	5.08	4.72	4.48	4.32	4.20	4.10	4.03	3.96	3.87	3.77	3.67	3.61	3.56	3.51	3.45	3.39	3.33
10	6.94	5.46	4.83	4.47	4.24	4.07	3.95	3.85	3.78	3.72	3.62	3.52	3.42	3.37	3.31	3.26	3.20	3.14	3.08
11	6.72	5.26	4.63	4.28	4.04	3.88	3.76	3.66	3.59	3.53	3.43	3.33	3.23	3.17	3.12	3.06	3.00	2.94	2.88
12	6.55	5.10	4.47	4.12	3.89	3.73	3.61	3.51	3.44	3.37	3.28	3.18	3.07	3.02	2.96	2.91	2.85	2.79	2.72
13	6.41	4.97	4.35	4.00	3.77	3.60	3.48	3.39	3.31	3.25	3.15	3.05	2.95	2.89	2.84	2.78	2.72	2.66	2.60
14	6.30	4.86	4.24	3.89	3.66	3.50	3.38	3.29	3.21	3.15	3.05	2.95	2.84	2.79	2.73	2.67	2.61	2.55	2.49
15	6.20	4.77	4.15	3.80	3.58	3.41	3.29	3.20	3.12	3.06	2.96	2.86	2.76	2.70	2.64	2.59	2.52	2.46	2.40
16	6.12	4.69	4.08	3.73	3.50	3.34	3.22	3.12	3.05	2.99	2.89	2.79	2.68	2.63	2.57	2.51	2.45	2.38	2.32
17	6.04	4.62	4.01	3.66	3.44	3.28	3.16	3.06	2.98	2.92	2.82	2.72	2.62	2.56	2.50	2.44	2.38	2.32	2.25
18	5.98	4.56	3.95	3.61	3.38	3.22	3.10	3.01	2.93	2.87	2.77	2.67	2.56	2.50	2.44	2.38	2.32	2.26	2.19
19	5.92	4.51	3.90	3.56	3.33	3.17	3.05	2.96	2.88	2.82	2.72	2.62	2.51	2.45	2.39	2.33	2.27	2.20	2.13
20	5.87	4.46	3.86	3.51	3.29	3.13	3.01	2.91	2.84	2.77	2.68	2.57	2.46	2.41	2.35	2.29	2.22	2.16	2.09
21	5.83	4.42	3.82	3.48	3.25	3.09	2.97	2.87	2.80	2.73	2.64	2.53	2.42	2.37	2.31	2.25	2.18	2.11	2.04
22	5.79	4.38	3.78	3.44	3.22	3.05	2.93	2.84	2.76	2.70	2.60	2.50	2.39	2.33	2.27	2.21	2.14	2.08	2.00
23	5.75	4.35	3.75	3.41	3.18	3.02	2.90	2.81	2.73	2.67	2.57	2.47	2.36	2.30	2.24	2.18	2.11	2.04	1.97
24	5.72	4.32	3.72	3.38	3.15	2.99	2.87	2.78	2.70	2.64	2.54	2.44	2.33	2.27	2.21	2.15	2.08	2.01	1.94
25	5.69	4.29	3.69	3.35	3.13	2.97	2.85	2.75	2.68	2.61	2.51	2.41	2.30	2.24	2.18	2.12	2.05	1.98	1.91
26	5.66	4.27	3.67	3.33	3.10	2.94	2.82	2.73	2.65	2.59	2.49	2.39	2.28	2.22	2.16	2.09	2.03	1.95	1.88
27	5.63	4.24	3.65	3.31	3.08	2.92	2.80	2.71	2.63	2.57	2.47	2.36	2.25	2.19	2.13	2.07	2.00	1.93	1.85
28	5.61	4.22	3.63	3.29	3.06	2.90	2.78	2.69	2.61	2.55	2.45	2.34	2.23	2.17	2.11	2.05	1.98	1.91	1.83
29	5.59	4.20	3.61	3.27	3.04	2.88	2.76	2.67	2.59	2.53	2.43	2.32	2.21	2.15	2.09	2.03	1.96	1.89	1.81
30	5.57	4.18	3.59	3.25	3.03	2.87	2.75	2.65	2.57	2.51	2.41	2.31	2.20	2.14	2.07	2.01	1.94	1.87	1.79
40	5.42	4.05	3.46	3.13	2.90	2.74	2.62	2.53	2.45	2.39	2.29	2.18	2.07	2.01	1.94	1.88	1.80	1.72	1.64
60	5.29	3.93	3.34	3.01	2.79	2.63	2.51	2.41	2.33	2.27	2.17	2.06	1.94	1.88	1.82	1.74	1.67	1.58	1.48
120	5.15	3.80	3.23	2.89	2.67	2.52	2.39	2.30	2.22	2.16	2.05	1.94	1.82	1.76	1.69	1.61	1.53	1.43	1.31
∞	5.02	3.69	3.12	2.79	2.57	2.41	2.29	2.19	2.11	2.05	1.94	1.83	1.71	1.64	1.57	1.48	1.39	1.27	1.00

From M. Merrington and C. M. Thompson (1943), Tables of Percentage Points of the Inverted Beta (F) Distribution, *Biometrika*, used by permission of Oxford University Press.

Appendix F

Critical Values of Hartley's Maximum *F* Ratio Test for Homogeneity of Variances

		Number of Groups (p)										
ν	α	2	3	4	5	6	7	8	9	10	11	12
2	.05	39.00	87.50	142.00	202.00	266.00	333.00	403.00	475.00	550.00	626.00	704.00
	.01	199.00	448.00	729.00	1036.00	1362.00	1705.00	2063.00	2432.00	2813.00	3204.00	3605.00
3	.05	15.40	27.80	39.20	50.70	62.00	72.90	83.50	93.90	104.00	114.00	124.00
	.01	47.50	85.00	120.00	151.00	184.00	216.00	249.00	281.00	310.00	337.00	361.00
4	.05	9.60	15.50	20.60	25.20	29.50	33.60	37.50	41.40	44.60	48.00	51.40
	.01	23.20	37.00	49.00	59.00	69.00	79.00	89.00	97.00	106.00	113.00	120.00
5	.05	7.15	10.80	13.70	16.30	18.70	20.80	22.90	24.70	26.50	28.20	29.90
	.01	14.90	22.00	28.00	33.00	38.00	42.00	46.00	50.00	54.00	57.00	60.00
6	.05	5.82	8.38	10.40	12.10	13.70	15.00	16.30	17.50	18.60	19.70	20.70
	.01	11.10	15.50	19.10	22.00	25.00	27.00	30.00	32.00	34.00	36.00	37.00
7	.05	4.99	6.94	8.44	9.70	10.80	11.80	12.70	13.50	14.30	15.10	15.80
	.01	8.89	12.10	14.50	16.50	18.40	20.00	22.00	23.00	24.00	26.00	27.00
8	.05	4.43	6.00	7.18	8.12	9.03	9.78	10.50	11.10	11.70	12.20	12.70
	.01	7.50	9.90	11.70	13.20	14.50	15.80	16.90	17.90	18.90	19.80	21.00
9	.05	4.03	5.34	6.31	7.11	7.80	8.41	8.95	9.45	9.91	10.30	10.70
	.01	6.54	8.50	9.90	11.10	12.10	13.10	13.90	14.70	15.30	16.00	16.60
10	.05	3.72	4.85	5.67	6.34	6.92	7.42	7.87	8.28	8.66	9.01	9.34
	.01	5.85	7.40	8.60	9.60	10.40	11.10	11.80	12.40	12.90	13.40	13.90
12	.05	3.28	4.16	4.79	5.30	5.72	6.09	6.42	6.72	7.00	7.25	7.48
	.01	4.91	6.10	6.90	7.60	8.20	8.70	9.10	9.50	9.90	10.20	10.60
15	.05	2.86	3.54	4.01	4.37	4.68	4.95	5.19	5.40	5.59	5.77	5.93
	.01	4.07	4.90	5.50	6.00	6.40	6.70	7.10	7.30	7.50	7.80	8.00
20	.05	2.46	2.95	3.29	3.54	3.76	3.94	4.10	4.24	4.37	4.49	4.59
	.01	3.32	3.80	4.30	4.60	4.90	5.10	5.30	5.50	5.60	5.80	5.90
30	.05	2.07	2.40	2.61	2.78	2.91	3.02	3.12	3.21	3.29	3.36	3.39
	.01	2.63	3.00	3.30	3.40	3.60	3.70	3.80	3.90	4.00	4.10	4.20
60	.05	1.67	1.85	1.96	2.04	2.11	2.17	2.22	2.26	2.30	2.33	2.36
	.01	1.96	2.20	2.30	2.40	2.40	2.50	2.50	2.60	2.60	2.70	2.70
∞	.05	1.00	1.00	1.00	1.00	1.00	1.00	1.00	1.00	1.00	1.00	1.00
	.01	1.00	1.00	1.00	1.00	1.00	1.00	1.00	1.00	1.00	1.00	1.00

From H. A. David (1952), Upper 5% and 1% Points of the Maximum F-Ratio, *Biometrika,* used by permission of Oxford University Press.

Appendix G

Table of Control Chart Constants

				$\overline{X}$ and R Charts			$\overline{X}$ and S Charts		
n	d_2	d_3	C_4	A_2	D_3	D_4	A_3	B_3	B_4
2	1.128	0.8525	0.7979	1.880	—	3.267	2.659	—	3.267
3	1.693	0.8884	0.8862	1.023	—	2.574	1.954	—	2.568
4	2.059	0.8798	0.9213	0.729	—	2.282	1.628	—	2.266
5	2.326	0.8798	0.9400	0.577	—	2.114	1.427	—	2.089
6	2.534	0.8480	0.9515	0.483	—	2.004	1.287	0.030	1.970
7	2.704	0.8332	0.9594	0.419	0.076	1.924	1.182	0.118	1.882
8	2.847	0.8198	0.9650	0.373	0.136	1.864	1.099	0.185	1.815
9	2.970	0.8078	0.9693	0.337	0.184	1.816	1.032	0.239	1.761
10	3.078	0.7971	0.9727	0.308	0.223	1.777	0.975	0.284	1.716
11	3.173	0.7873	0.9754	0.285	0.256	1.744	0.927	0.321	1.679
12	3.258	0.7785	0.9776	0.266	0.283	1.717	0.886	0.354	1.646
13	3.336	0.7704	0.9794	0.249	0.307	1.693	0.850	0.382	1.618
14	3.407	0.7630	0.9810	0.235	0.328	1.672	0.817	0.406	1.594
15	3.472	0.7562	0.9823	0.223	0.347	1.653	0.789	0.428	1.572
16	3.532	0.7499	0.9835	0.212	0.363	1.637	0.763	0.448	1.552
17	3.588	0.7441	0.9845	0.203	0.378	1.662	0.739	0.466	1.534
18	3.640	0.7386	0.9854	0.194	0.391	1.607	0.718	0.482	1.518
19	3.689	0.7335	0.9862	0.187	0.403	1.597	0.698	0.497	1.503
20	3.735	0.7287	0.9869	0.180	0.415	1.585	0.680	0.510	1.490
21	3.778	0.7272	0.9876	0.173	0.425	1.575	0.663	0.523	1.477
22	3.819	0.7199	0.9882	0.167	0.434	1.566	0.647	0.534	1.466
23	3.858	0.1759	0.9887	0.162	0.443	1.557	0.633	0.545	1.455
24	3.895	0.7121	0.9892	0.157	0.451	1.548	0.619	0.555	1.445
25	3.931	0.7084	0.9896	0.153	0.459	1.541	0.606	0.565	1.435

Glossary of Symbols

A, B, C	Factors
A_2	Factor in constructing $\overline{X}$ chart $(3/d_2\sqrt{n})$
A_3	Factor in constructing $\overline{X}$ chart $(3/c_4\sqrt{n})$
a, b, c	Number of levels of factors
α	In testing hypotheses, risk of rejecting the null hypotheses when true: also, level of significance
B_3	Factor in constructing S chart $(1 - 3\sqrt{1 - c_4^2}/c_4)$
B_4	Factor in constructing S chart $(1 + 3\sqrt{1 - c_4^2}/c_4)$
b	Intercept in linear regression
β	In testing hypotheses, risk of accepting the null hypotheses when false, power of a test $= 1 - \beta$
C_4	Factor in constructing S chart $[4(n - 1)/(4n - 3)]$
C_p	Capability index $[(USL - LSL)/6\sigma_{\text{short}}]$
C_{pk}	Capability index [smaller of $(USL - \overline{\overline{x}})/3\sigma_{\text{short}}$ or $(\overline{\overline{x}} - LSL)/3\sigma_{\text{short}}$]
CV	Coefficient of variation $(=RSD = \sigma/\mu)$
c	Number of defects per product
$\overline{c}$	Average number of defects per product
χ^2	Chi-square distribution
$\chi^2_{\alpha,n-1}$	The α point of a chi-square distribution with $(n - 1)$ degrees of freedom
D_3	Factor used to construct R chart $(1 - 3d_3/d_2)$
D_4	Factor used to construct R chart $(1 + 3d_3/d_2)$
d	Used in calculating sample size (Δ/σ)

d	Difference
$\overline{d}$	Average difference
d_2	Factor to estimate standard deviation ($\overline{R}/\sigma$)
d_3	Factor to estimate standard deviation of range (σ_R/σ)
Δ	In a t-test, the smallest difference in mean to be detected
ΔX	Short-run $\Delta X - mR$ chart ($X - T$)
$\Delta\overline{X}$	Short-run $\Delta\overline{X} - R$ chart ($\overline{X} - T$)
δ	Maximum probability of out-of-specification product for a modified limit chart
δ	In estimating σ, half-width of confidence interval
E_i	In testing normality, estimated frequency in cell i
e	The constant 2.718
e	Error
F	Ratio $s_1^2\sigma_2^2/s_2^2\sigma_1^2$ or s_1^2/s_2^2 under the hypothesis of equal variance, because the statistic has an F distribution
$f(X)$	Probability density function for random factor X
H_0	Null hypothesis
H_1	Alternate hypothesis
i	Used as a subscript to differentiate a factor
k	Constant
k	Factor to compute tolerance limits
L	Loss
LCL	Lower control limit
LSL	Lower specification limit
λ	Parameter of a Poisson distribution
M_t	Moving average at time t
MS	In analysis of variance, mean square
m	Slope of a regression line
mR	Moving range, $\lvert x_i - x_{i-1} \rvert$; also moving range chart
$m\overline{R}$	Average moving range
μ	Population mean
N	Population size
n	Sample size
ν	Degrees of freedom
O_i	In testing normality, observed frequency in cell i
P	Probability
P_p	Performance index $[(USL - LSL)/6\sigma_{\text{total}}]$
P_{pk}	Performance index [smaller of $(USL - \overline{\overline{x}})/3\sigma_{\text{total}}$ or $(\overline{\overline{x}} - LSL)/3\sigma_{\text{total}}$]
p	Fraction defective
$\overline{p}$	Average fraction defective
q	$(1 - p)$
R	Range; also range chart
R	Sample correlation coefficient

$\overline{R}$	Average range
RE	Relative error
RME	Relative measurement error
RSD	Relative standard deviation ($=CV = \sigma/\mu$)
ρ	Population correlation coefficient
S	Random factor representing standard deviation
S^2	Random factor representing variance
SS	Sum of squares
s	Sample standard deviation
s^2	Sample variance
s_{xy}	Sample covariance
σ	Population standard deviation
$\sigma_{\overline{X}}$	Population standard deviation of $\overline{X}$
σ^2	Population variance
$\sigma_{\overline{X}}^2$	Population variance of $\overline{X}$
T	Target
t	Time
t	t-distribution; also t value $[(\overline{x} - \mu)/(s/\sqrt{n})]$
$t_{\alpha,n-1}$	The α point of a t-distribution with $(n - 1)$ degrees of freedom
UCL	Upper control limit
USL	Upper specification limit
u	In u chart, number of defects per product
$\overline{u}$	Average number of defects per product
V	Variance
W	$(USL - LSL)/2$ or $\lvert\overline{\overline{x}} - \text{specification limit}\rvert$ for one-sided specification
W	Short-run $Z - W$ chart
$\overline{W}$	Short-run $\overline{Z} - \overline{W}$ chart
w	Span for a moving average control chart
w_i	Weights in weighted regression
X	Random factor representing individual observation
$\overline{X}$	Random factor representing average
x	Individual observation, observed value of X
$\overline{x}$	Sample average
$\overline{\overline{x}}$	Grand average of all data
$\hat{x}$	Estimated value of x
Y	Random factor like X
y	Observed value of Y
$\hat{y}$	Estimated value of y
Z	Z distribution, i.e., a normal distribution with $\mu = 0$ and $\sigma = 1$; also, Z value $[(x - \mu)/\sigma]$
Z	Short-run $Z - W$ chart
$\overline{Z}$	Short-run $\overline{Z} - \overline{W}$ chart
Z_α	The α point of the Z distribution

References

A.I.A.G. (1991). *Fundamental Statistical Process Control.* A.I.A.G., Detroit, MI.

AT&T Technologies (1985). *Statistical Quality Control Handbook.* AT&T Technologies, Indianapolis, IN.

Box, G. E. P., Hunter, W. G., and Hunter, J. S. (1978). *Statistics for Experimenters.* Wiley, New York.

Chrysler Corporation, Ford Motor Company, and General Motors Corporation (1995). *Measurement Systems Analysis.* Detroit, MI.

Davies, O. L. (1961). *Statistical Methods in Research and Production.* Oliver and Boyd, London.

Draper, N. R. and Smith, H. (1998). *Applied Regression Analysis.* Wiley, New York.

Duncan, A. J. (1986). *Quality Control and Industrial Statistics.* Irwin, Homewood, IL.

Ishikawa, K. (1984). *Guide to Quality Control.* Asian Productivity Organization, Tokyo, Japan. (Available from UNIPUB, New York.)

Johnson, N. L. and Kotz, S. (1969). *Discrete Distributions.* Houghton Mifflin, Boston, MA.

Johnson, N. L. and Kotz, S. (1970). *Continuous Univariate Distributions,* Vols. 1 and 2. Houghton Mifflin, Boston, MA.

Montgomery, D. C. (1985). *Introduction to Statistical Quality Control.* Wiley, New York.

Pande, P. S., Neuman, R. P., and Cavanagh, R. R. (2000). *The Six Sigma Way.* McGraw-Hill, New York.

Sahai, H. and Ageel, M. I. (2000). *The Analysis of Variance.* Birkhauser, Boston, MA.

Shewhart, W. A. (1931). *Economic Control of Quality of Manufactured Product.* Van Nostrand Reinhold, Princeton, NJ. (Republished in 1980 by ASQC Quality Press.)

Taguchi, G. (1987). *System of Experimental Design,* vols. 1 and 2. UNIPUB, Kraus International Publications, White Plains, NY.

Wheeler, D. J. (1991). *Short Run SPC.* SPC Press, Knoxville, TN.

Wheeler, D. J. and Chambers, D. S. (1992). *Understanding Statistical Process Control.* SPC Press, Knoxville, TN.

Wheeler, D. J. and Lyday, R. W. (1989). *Evaluating the Measurement Process.* SPC Press, Knoxville, TN.

Index

ISBN 0-471-20342-4